Advancements in Climate and Smart Environment Technology

Jamal Mabrouki
Faculty of Science, Mohammed V University in Rabat, Morocco

Mourade Azrour
Faculty of Sciences and Techniques, Moulay Ismail University of Meknes, Errachidia, Morocco

IGI Global
PUBLISHER of TIMELY KNOWLEDGE

A volume in the Advances in Environmental Engineering and Green Technologies (AEEGT) Book Series

Published in the United States of America by
IGI Global
Engineering Science Reference (an imprint of IGI Global)
701 E. Chocolate Avenue
Hershey PA, USA 17033
Tel: 717-533-8845
Fax: 717-533-8661
E-mail: cust@igi-global.com
Web site: http://www.igi-global.com

Copyright © 2024 by IGI Global. All rights reserved. No part of this publication may be reproduced, stored or distributed in any form or by any means, electronic or mechanical, including photocopying, without written permission from the publisher. Product or company names used in this set are for identification purposes only. Inclusion of the names of the products or companies does not indicate a claim of ownership by IGI Global of the trademark or registered trademark.

Library of Congress Cataloging-in-Publication Data

CIP DATA PROCESSING

Advancements in Climate and Smart Environment Technology
Jamal Mabrouki, Mourade Azrour
2024 Engineering Science Reference

ISBN: 9798369338070(hc) | ISBN: 9798369348734(sc) | eISBN: 9798369338087

This book is published in the IGI Global book series Advances in Environmental Engineering and Green Technologies (AEEGT) (ISSN: 2326-9162; eISSN: 2326-9170)

British Cataloguing in Publication Data
A Cataloguing in Publication record for this book is available from the British Library.

All work contributed to this book is new, previously-unpublished material. The views expressed in this book are those of the authors, but not necessarily of the publisher.

For electronic access to this publication, please contact: eresources@igi-global.com.

Advances in Environmental Engineering and Green Technologies (AEEGT) Book Series

Sang-Bing (Jason) Tsai
Zhongshan Institute, University of Electronic Science and Technology of China, China & Wuyi University, China
Ming-Lang Tseng
Lunghwa University of Science and Technology, Taiwan
Yuchi Wang
University of Electronic Science and Technology of China Zhongshan Institute, China

ISSN:2326-9162
EISSN:2326-9170

Mission

Growing awareness and an increased focus on environmental issues such as climate change, energy use, and loss of non-renewable resources have brought about a greater need for research that provides potential solutions to these problems. Research in environmental science and engineering continues to play a vital role in uncovering new opportunities for a "green" future.

The **Advances in Environmental Engineering and Green Technologies (AEEGT)** book series is a mouthpiece for research in all aspects of environmental science, earth science, and green initiatives. This series supports the ongoing research in this field through publishing books that discuss topics within environmental engineering or that deal with the interdisciplinary field of green technologies.

Coverage

- Pollution Management
- Air Quality
- Green Transportation
- Renewable Energy
- Alternative Power Sources
- Waste Management
- Industrial Waste Management and Minimization
- Biofilters and Biofiltration
- Sustainable Communities
- Water Supply and Treatment

IGI Global is currently accepting manuscripts for publication within this series. To submit a proposal for a volume in this series, please contact our Acquisition Editors at Acquisitions@igi-global.com or visit: http://www.igi-global.com/publish/.

The Advances in Environmental Engineering and Green Technologies (AEEGT) Book Series (ISSN 2326-9162) is published by IGI Global, 701 E. Chocolate Avenue, Hershey, PA 17033-1240, USA, www.igi-global.com. This series is composed of titles available for purchase individually; each title is edited to be contextually exclusive from any other title within the series. For pricing and ordering information please visit http://www.igi-global.com/book-series/advances-environmental-engineering-green-technologies/73679. Postmaster: Send all address changes to above address. Copyright © 2024 IGI Global. All rights, including translation in other languages reserved by the publisher. No part of this series may be reproduced or used in any form or by any means – graphics, electronic, or mechanical, including photocopying, recording, taping, or information and retrieval systems – without written permission from the publisher, except for non commercial, educational use, including classroom teaching purposes. The views expressed in this series are those of the authors, but not necessarily of IGI Global.

Titles in this Series

For a list of additional titles in this series, please visit: http://www.igi-global.com/book-series/advances-environmental-engineering-green-technologies/73679

Convergence Strategies for Green Computing and Sustainable Development
Vishal Jain (Sharda University, India) Murali Raman (Asia Pacific University of Technology, Malaysia) Akshat Agrawal (Amity University, Guragon, India) Meenu Hans (K.R. Mangalam University, India) and Swati Gupta (Center of Excellence, K.R. Mangalam University, India)
Engineering Science Reference • © 2024 • 332pp • H/C (ISBN: 9798369303382) • US $270.00

Biosorption Processes for Heavy Metal Removal
Pinki Saini (University of Allahabad, India)
Engineering Science Reference • © 2024 • 371pp • H/C (ISBN: 9798369316184) • US $275.00

Agriculture and Aquaculture Applications of Biosensors and Bioelectronics
Alex Khang (Global Research Institute of Technology and Engineering, USA)
Engineering Science Reference • © 2024 • 576pp • H/C (ISBN: 9798369320693) • US $265.00

Innovations in Engineering and Food Science
Shilpa Mehta (Auckland University of Technology, New Zealand) Fakhar Islam (NUR International University, Pakistan) and Ali Imran (Government College University, Pakistan)
Engineering Science Reference • © 2024 • 424pp • H/C (ISBN: 9798369308196) • US $265.00

Modeling and Monitoring Extreme Hydrometeorological Events
Carmen Maftei (Transilvania University of Brasov, Romania) Radu Muntean (Transilvania University of Brașov, Romania) and Ashok Vaseashta (International Clean Water Institute, USA)
Engineering Science Reference • © 2024 • 338pp • H/C (ISBN: 9781668487716) • US $250.00

Novel AI Applications for Advancing Earth Sciences
Sudesh Yadav (Department of Higher Education, Government PG College, Ateli, India) Satya Prakash Yadav (GL Bajaj Institute of Technology and Management, India) Pethuru Raj (Edge AI Division, Reliance Jio Platforms Ltd., Bangalore, India) Prabhakar Tiwari (Madan Mohan Malaviya University of Technology, India) and Victor Hugo C. de Albuquerque (Department of Teleinformatics Engineering (DETI), Federal University of Ceará, Brazil)
Engineering Science Reference • © 2024 • 405pp • H/C (ISBN: 9798369318508) • US $265.00

IGI Global
PUBLISHER of TIMELY KNOWLEDGE

701 East Chocolate Avenue, Hershey, PA 17033, USA
Tel: 717-533-8845 x100 • Fax: 717-533-8661
E-Mail: cust@igi-global.com • www.igi-global.com

Table of Contents

Detailed Table of Contents

Neglected and underutilized (NUS) crops are increasingly recognized for their agricultural sustainability, playing a crucial role in solving environmental problems and improving food Security in MENA's region. This study conducts an in-depth assessment of the environmental impacts of NUS cultivation, focusing particularly on carbon emissions and water use in this region. By integrating life cycle assessment (LCA) and empirical field data, this research compares the carbon capture capacities, greenhouse gas emissions and water efficiency of NUS with crops traditional. The results suggest that NUS, thanks to their ability to thrive in less fertile soils of MENA's region and minimal resource demands, generally has a lower carbon footprint. Many NUS varieties also exhibit drought resistance, leading to more water-efficient agricultural practices. In addition, the study examines how NUS can boost biodiversity, improve soil quality, and strengthen the resilience of agricultural systems against the effects of climate change. This chapter highlights the environmental benefits of NUS cultivation, with the aim of guiding policy makers, farmers and relevant stakeholders towards sustainable agricultural practices and environmental conservation. The research highlights the urgency for continued research and development in this area, particularly to refine growing methods that maximize environmental benefits.

Ozone pollution poses a significant environmental and health challenge in Marrakech, Morocco, intensified by urban and industrial growth. This study compares two statistical models, SARIMA and Facebook Prophet, for accurate O3-level forecasting in Marrakech. Through time series analysis, we assess their performance using metrics like MAE, MAPE, and RMSE. The findings highlight the Facebook Prophet model's superior accuracy, attributed to its better handling of seasonality and adaptability to unusual events. This research aids environmental and public health efforts in Marrakech by offering insights into ozone trends, supporting informed policymaking to combat ozone pollution.

The rising influence of greenhouse gas (GHG) emissions from burning fossil fuels on climate change has led to a demand for clean energy. The only consequence of H2's combustion is water vapor, making it a clean energy source. As a clean energy source, hydrogen contributes significantly to energy and environmental sustainability and helps mitigate the world's energy problem and climate change. Microalgae have recently been found to hold tremendous promise for producing biodiesel and biohydrogen. Under certain conditions, sunlight is used by green microalgae to split water particles, resulting in the production of molecular hydrogen and oxygenation. With an emphasis on green microalgae, this review outlined possible algal strains for hydrogen production. In addition, a comprehensive review procedure aimed at presenting the importance of using microalgae to produce green hydrogen. More significantly, this review included an overview of present constraints and potential future directions for the sustainable synthesis of biohydrogen from microalgae.

This chapter delves into the pivotal role of AI and smart technologies in tackling water scarcity through sustainable water resource management. It commences with an overview of water scarcity's background, underscoring its impact on the environment and society and emphasizing the significance of sustainable water resource management catalyzed by AI and smart technologies. The challenges posed by water scarcity, including contributing factors and effects on ecosystems, agriculture, and human health, are thoroughly analyzed. Furthermore, it discusses the principles of sustainable water resource management, integrating environmental, social, and economic factors, alongside successful initiatives. The chapter also highlights opportunities for future research, fostering innovation, stakeholder collaboration, and implementation strategies. The conclusion emphasizes the ongoing necessity for innovative approaches to combat water scarcity, showcasing the enduring environmental, societal, and economic benefits gleaned from advanced technologies in sustainable water management.

The resurgence of infectious diseases demands innovative forecasting and management strategies. Integrating artificial intelligence (AI) with the internet of things (IoT) offers a groundbreaking approach to epidemic prediction. This study highlights the combined power of AI algorithms and IoT technology in creating sophisticated epidemic models. The authors explore cutting-edge AI methods, like machine learning and deep learning, to analyze vast epidemiological data from IoT devices, including wearables and mobile phones. This integration facilitates early outbreak detection, precise risk evaluation, and prompt interventions. Additionally, they address ethical and privacy issues related to health data, promoting careful information handling. This analysis shows that AI and IoT synergy not only sharpens epidemic forecasting but also boosts public health response efficiency, aiding in epidemic control and enhancing global health security.

In light of the world's energy challenge and the scarcity of traditional energy resources, Morocco is taking the initiative to become an African leader in the water, energy, and food sectors. It is against this backdrop that this research focuses on the way in which biomass can be used as an environmentally-friendly alternative that makes the wheel of the circular economy turn. This study presents the importance of using Lemna gibba. L (Duckweed), well known for its high biomass production, as an alternative source of energy (biofuel production) as well as animal and human food (up to 30% protein source), after having successfully contributed to its main function of phytoremediating wastewater through its ability to bioaccumulate. This study is therefore a practical application of economic and ecological Bio Tech that can be adopted.

challenges faced in adopting these practices and providing insights into future directions and emerging technologies in this field.

The chapter explores aquaponics and hydroponics, two innovative and environmentally friendly farming techniques. It provides a comprehensive overview of the concepts, their history, and scientific principles. The chapter also explores the structure and elements of these systems, emphasizing their ecological and financial benefits. It also discusses plant growth dynamics in hydroponic systems, focusing on nutrient management and growth variables. The chapter also covers fish species selection, health, and nutrition in aquaponics, integrated pest management, and technological advancements in system optimization. It provides an impartial evaluation of the sustainability and environmental consequences of these systems, showcasing case studies and practical applications.

Geospatial techniques are being increasingly used in agriculture to optimize production, manage resources, and mitigate climate change impacts. These techniques include GPS, GIS, and remote sensing, which map and monitor crop growth and yield variability, allowing for more efficient application of inputs like fertilizers, water, and pesticides. They also create models that simulate crop growth under different climate scenarios, allowing farmers to plan for future conditions and adapt their practices accordingly. Geospatial data is used to create detailed maps of soil properties, helping farmers make better decisions about crop selection, planting, and nutrient management. These techniques also help monitor and manage water resources, particularly in areas prone to drought or flooding, by mapping groundwater reserves, analyzing precipitation patterns, and predicting crop water demand. Geospatial data can evaluate climate hazards, guide crop selection and management, and analyze land use patterns, aiding farmers and policymakers in decision-making.

The world has always been divided based on access to technology. Since the dawn of humanity, the idea of evolving and utilizing the best of resources gave birth to the advent of technological evolution. Amidst all of this, our human lives have been transformed by technology in the form of personal belongings, like a phone, which is now smart enough to do almost any function that you can imagine. It states that technology has definitely made our lives easier and engaging for the most part; however, one must not forget that there are challenges of the technological prowess as well. Technology is now taking the next step already where data is the fuel and based on that, technology is stepping up to play the role in the world of government policies and governance initiative. This chapter wants to highlight the importance of technology and governance of urban resources especially with the onset of the more intense discussions on climate change, global warming, etc.

In this chapter, we explore the integration of artificial intelligence (AI) in enhancing environmental monitoring and managing the COVID-19 pandemic in Morocco. Utilizing AI-based event detection techniques, the authors extend their investigation from analyzing COVID-19 trends to applying predictive algorithms for environmental surveillance. This approach not only demonstrates AI's capacity in public health realms but also underscores its potential in environmental stewardship. Through detailed analysis, the authors illustrate how machine learning and data analytics significantly improve response strategies against pandemic threats and environmental degradation. Their findings reveal AI's critical role in developing resilient and sustainable monitoring systems, offering valuable insights for policymakers, researchers, and practitioners. This comprehensive study highlights the synergy between technological innovation and multifaceted crisis management, positioning AI as a cornerstone in advancing smart technologies for the betterment of public health and environmental sustainability in Morocco.

Preface

Welcome to *Advancements in Climate and Smart Environment Technology*, a comprehensive reference book that explores the intersection of cutting-edge technology and environmental sustainability. In a world where the challenges of climate change and environmental degradation are becoming increasingly urgent, the need for innovative solutions has never been more apparent.

This book is the result of collaborative efforts from experts in various fields, including artificial intelligence, environmental engineering, healthcare, and sustainable development. Their insights and research contributions have been instrumental in shaping the content of this book, which aims to provide a holistic view of the latest advancements and strategies for addressing environmental and healthcare issues.

The chapters in this book cover a wide range of topics, from advanced technology for smart environments and healthcare to sustainable management of natural resources and the impact of climate change on socio-economic systems. Key themes include the role of artificial intelligence and smart technologies in creating sustainable solutions, the importance of nature-based approaches, and the challenges and opportunities in sectors such as agriculture, energy, and water management.

Throughout these pages, readers will find valuable insights, case studies, and practical applications that demonstrate the potential of technology to drive positive change. Whether you are a researcher, practitioner, policymaker, or student, we hope that this book will serve as a valuable resource for understanding and contributing to the ongoing efforts towards a more sustainable and resilient future.

Thank you to all the contributors who have shared their expertise and knowledge, and we hope that this book inspires further innovation and collaboration in the vital field of climate and smart environment technology.

Chapter 1
Carbon and Water Efficiency Study on Growing Neglected and Underutilized Crops and Its Environmental Benefits in the MENA Region

Soukaina Agoud
Mohammed V University in Rabat, Morocco

Jamal Mabrouki
https://orcid.org/0000-0002-3841-7755
Mohammed V University in Rabat, Morocco

ABSTRACT

Neglected and underutilized (NUS) crops are increasingly recognized for their agricultural sustainability, playing a crucial role in solving environmental problems and improving food Security in MENA's region. This study conducts an in-depth assessment of the environmental impacts of NUS cultivation, focusing particularly on carbon emissions and water use in this region. By integrating life cycle assessment (LCA) and empirical field data, this research compares the carbon capture capacities, greenhouse gas emissions and water efficiency of NUS with crops traditional. The results suggest that NUS, thanks to their ability to thrive in less fertile soils of MENA's region and minimal resource demands, generally has a lower carbon footprint. Many NUS varieties also exhibit drought resistance, leading to more water-efficient agricultural practices. In addition, the study examines how NUS can boost biodiversity, improve soil quality, and strengthen the resilience of agricultural systems against the effects of climate change. This chapter highlights the environmental benefits of NUS cultivation, with the aim of guiding policy makers, farmers and relevant stakeholders towards sustainable agricultural practices and environmental conservation. The research highlights the urgency for continued research and development in this area, particularly to refine growing methods that maximize environmental benefits.

DOI: 10.4018/979-8-3693-3807-0.ch001

Copyright © 2024, IGI Global. Copying or distributing in print or electronic forms without written permission of IGI Global is prohibited.

INTRODUCTION

Neglected and underutilized, they are a focal point of agricultural research and policy discussions, due to their potential to address pressing issues related to agricultural sustainability, environmental conservation and food security in the MENA region.

In the face of growing environmental challenges in this region, such as climate change and resource scarcity, it has become increasingly imperative to explore innovative agricultural practices (Ebert, 2022). This chapter embarks on a comprehensive review of the environmental impacts associated with growing NUS crops, with a particular focus on carbon emissions and water use. Through the integration of contemporary life cycle assessment methodologies and empirical field data, this study aims to elucidate carbon sequestration capacities, greenhouse gas emissions and water efficiency NUS cultures, making a critical comparison with conventional cultures (Frelat, 2016) The results of this research not only highlight the remarkable potential of NUS crops to minimize their carbon footprint, through their adaptability to less fertile soils and resource-efficient growth, but also highlight their intrinsic ability to withstand harsh conditions. Thus promoting more sustainable water management and management practices (Parmesan, 2003).

Furthermore, this chapter explores the broader ecological benefits of NUS crops, including their contributions to biodiversity conservation, soil improvement and fortification of agricultural systems against the adverse effects of climate change (Tadesse, 2021).

MATERIALS AND METHODS

To evaluate the environmental impacts of NUS, specifically focusing on carbon emissions and water use efficiency, we adopted a holistic research approach.

A systematic literature review was conducted to identify a diverse range of NUS crops with documented cultivation practices, giving priority to crops recognized for their environmental sustainability (Ayer, 2019). Selection criteria included their underutilized status, adaptability to varying agroecological conditions, and documented carbon sequestration potential. Empirical field data were meticulously gathered from strategically selected study sites representing diverse geographical regions and climate zones (Bernholt, 2021). Data collection encompassed critical parameters, including crop yield, carbon capture capacity, and water use efficiency. Stringent protocols were followed to ensure the accuracy and reliability of field data (Eakin, 2021).

Figure 1. Integration of the underutilized prickly pear in the carbon cycle dynamics amidst the climatic challenges of the MENA region

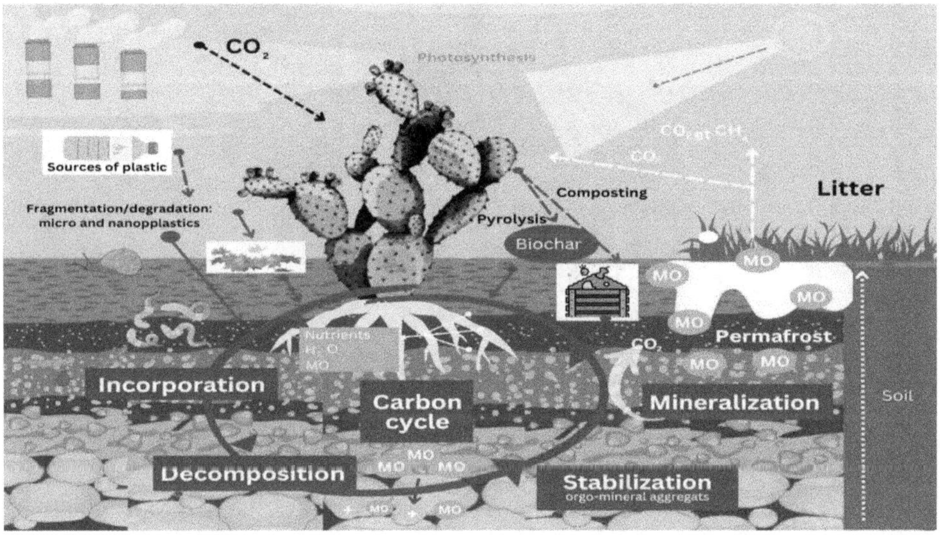

LCA is as a robust methodology for assessing environmental impacts, was employed to evaluate NUS crop cultivation (De Ruiter, 2018). The LCA process entailed a comprehensive assessment of inputs like seeds, fertilizers and pesticides, cultivation practices like irrigation, tillage, and harvesting, and outputs like carbon emissions and water consumption.

Conventional crops with analogous growth requirements and geographical suitability were selected as reference points for comparative analysis. Rigorous LCA analyses were conducted to assess carbon emissions and water use efficiency in NUS crop cultivation compared to conventional crop cultivation (Van Der Voet, 2019). Biodiversity indices were calculated to measure the impact of NUS crop cultivation on local ecosystems, with a focus on bird and insect populations (Foley, 2021).

RESULTS AND DISCUSSION

The MENA region is especially susceptible to the adverse effects of climate change, including increased temperatures, desertification, and water scarcity. These changes threaten agricultural productivity, water supplies, and overall human security in the region. Although the MENA region only contributes to 7% of global carbon emissions, the impact of climate change on this area is disproportionately high due to its geographical and climatic conditions (Sghaier & Labbene, 2016).

Figure 2. MENA region carbon emissions in percentage

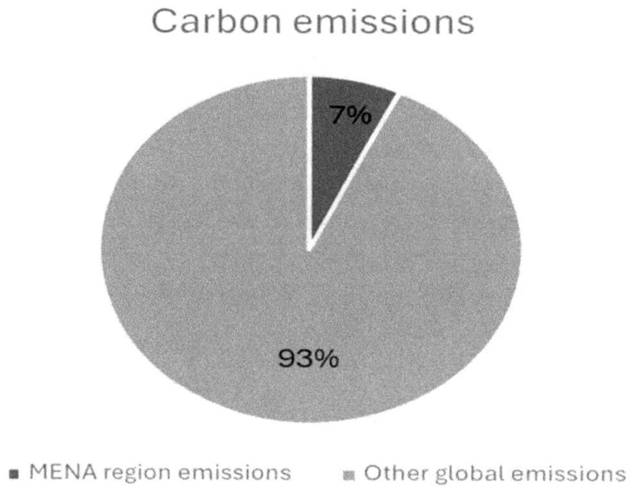

Our study found compelling evidence that NUS crops exhibit considerable potential for carbon emissions reduction when compared to conventional crops. The adaptability of NUS crops to less fertile soils and their minimal resource requirements contribute to a lower carbon footprint (Groot, 2021).

This large percentage of gas emissions in the region leads us to question more and more about the distribution of these gas emissions in this region as well as the absorption power of each. The following figure will present the largest and most significant players in the MENA region in terms of carbon emissions.

Figure 3. Total gross emissions of greenhouse gases
(Sghaier & Labbene, 2016)

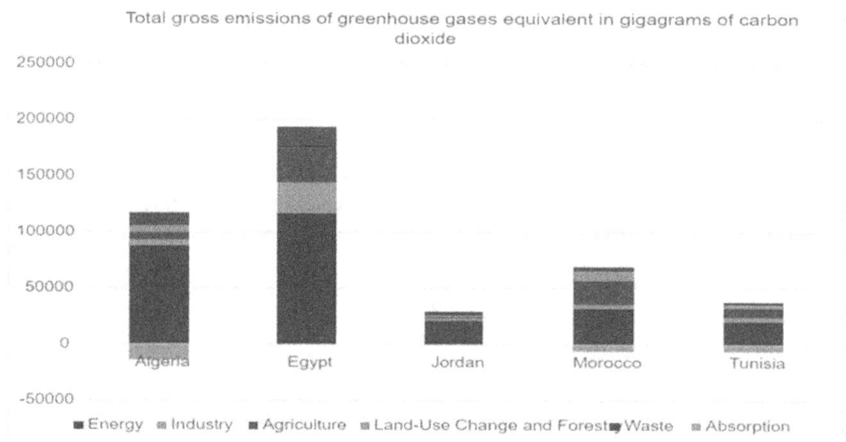

Solving this problem calls upon the culture of the NUS, as shown in the figure 2 the carbon capture capacity of certain NUS crops can play a crucial role in sequestering atmospheric carbon dioxide, mitigating the effects of climate change. These findings align with global efforts to enhance agricultural sustainability and reduce greenhouse gas emissions.

Our study found compelling evidence that NUS crops exhibit considerable potential for carbon emissions reduction when compared to conventional crops. The adaptability of NUS crops to less fertile soils and their minimal resource requirements contribute to a lower carbon footprint (Groot, 2021).

This research also highlighted the superior water use efficiency of many NUS crops, particularly in regions facing water scarcity challenges (Kutcher, 2020). Their ability to thrive with reduced irrigation and minimal water inputs underscores their potential in addressing the increasing pressure on freshwater resources. This aligns with the Sustainable Development Goals (SDGs) by promoting responsible water management in agriculture (FAO, 2021). The ecological assessments conducted in our study revealed that NUS crop cultivation can positively impact biodiversity, fostering the presence of diverse bird and insect populations (Matson, 2021). This aligns with the growing recognition of the role of NUS crops in promoting biodiversity and ecosystem services (Mbow, 2019). It underscores their potential in contributing to more resilient and sustainable agricultural landscapes.

Figure 4. Millimeters of water annually for irrigation in MENA's region
(Ministry of Agriculture, Fisheries, Rural Development, Water, and Forests, 2019; Food and Agriculture Organization of the United Nations, 2022; OECD/FAO, 2020; Botanical online: Botanical Online; Food and Agriculture Organization of the United Nations, 2021).

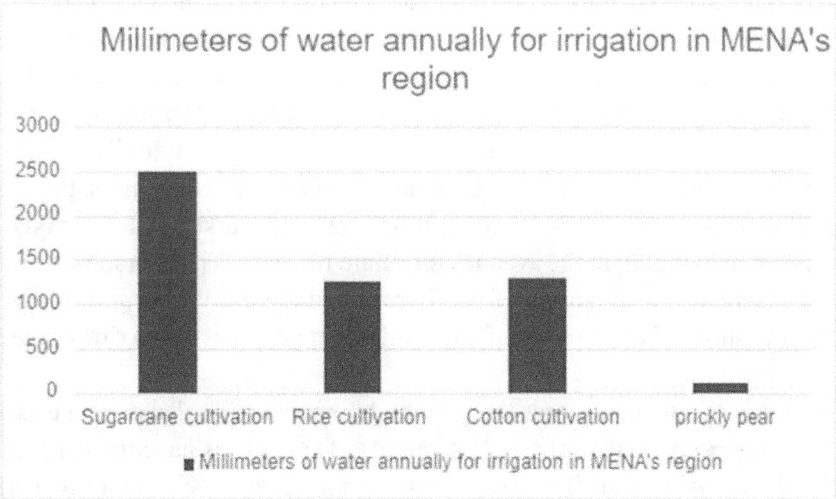

The statistics of sugarcane, rice, and cotton are based on general agricultural knowledge and practices identified by FAO. The cultivation of sugarcane, cotton, and rice in the MENA region poses several environmental challenges due to their high water consumption and the region's existing water scarcity. This extensive use of water leads to soil leaching and increased salinity, adversely affecting water quality and other agricultural activities. Monoculture practices associated with these crops reduce biodiversity, impacting ecosystem health and crop resilience. Additionally, these crops sensitivity to climate change introduces unpredictability in yields, affecting farmers livelihoods. On the other hand prickly pear's low water requirement and resilience to harsh climates exemplify how NUS can contribute to food security and sustainable development.

Cultivating NUS like prickly pear diversifies crops, reduces reliance on traditional water-intensive crops, and better utilizes limited water resources. Moreover, these crops can aid in soil restoration and combat desertification, common issues in the MENA region.

The findings have significant policy and practical implications. Policymakers and agricultural stakeholders should consider the promotion of NUS crops as a key strategy in achieving sustainable agriculture and environmental conservation. Incorporating NUS crops into crop diversification efforts can enhance food security and resilience in the face of climate change (Lipper, 2020). Strategies to support NUS crop adoption should be developed, including incentives for farmers, improved access to NUS crop varieties, and knowledge sharing.

CONCLUSION

NUS crops, offer substantial potential for enhancing food security and bolstering agricultural resilience in the face of a changing climate. Their adaptability to diverse agroecological zones and minimal resource requirements make them valuable assets in ensuring food availability and accessibility . Our research has emphasized the superior water use efficiency of cactus as a NUS crops, aligning with the need for sustainable resource management in agriculture . Their reduced reliance on irrigation and minimal water footprint contribute to more responsible water usage. Also we demonstrate that NUS crops have a significant role to play in carbon emissions reduction and climate change mitigation. Their ability to sequester carbon in both aboveground biomass and soil highlights their potential in the global effort to combat climate change. That's proving the positive impact of NUS crop cultivation on biodiversity conservation, aligning with our findings. NUS crops can provide refuge and sustenance for diverse wildlife, promoting ecosystem health. The implications for policy and development initiatives are profound. The need for targeted policy support is essential to promote the adoption of NUS crops. Policymakers should prioritize the development of incentives, breeding programs, and market access for NUS crops. As we conclude, it is imperative to highlight the avenues for future research. That's emphasizes the importance of comprehensive genetic characterization of NUS crops and the development of climate-resilient varieties. Further studies should focus on optimizing cultivation practices to maximize the environmental benefits of NUS crops.

In conclusion, our research underscores the profound potential of neglected and underutilized crops in addressing the complex challenges of the 21st century. These crops have the capacity to contribute significantly to food security, sustainable resource management, carbon emissions reduction, and biodiversity conservation. As we move forward, it is paramount that policymakers, researchers, and agricultural stakeholders collaborate to harness the full potential of NUS crops and integrate them into global efforts toward a more sustainable and resilient agricultural future.

REFERENCES

Ayer, N. W. (2019). Life cycle assessment of a novel low-input agricultural system for nutrient management. *Journal of Cleaner Production*, *212*, 401–410.

Bernholt, S. (2021). Neglected and Underutilized Crop Species: A Multidimensional Analysis of Data Availability. *Sustainability, 13*(14), 7787.

de Ruiter, H. (2018). The Potential of Multi-Criteria Decision Analysis for Evaluating Sustainable Agriculture at the Farm Level. *Sustainability 10*(11).

Eakin, H. (2021). The Global South's Contribution to Global Crop Diversity and the Role of International Agricultural Research. *Environmental Science & Policy*, *118*, 236–243.

Ebert, A. W. (2022). Neglected and underutilized crop species: An overview on prospects for improving nutrition and food security. *Sustainability, 14*.

FAO. (2021). *Sustainable Crop Production.* FAO Agricultural Development Economics Briefs.

Foley, J. A. (2021). Our share of the planetary pie. *Proceedings of the National Academy of Sciences of the United States of America*, *118*, 26. PMID:17646656

Food and Agriculture Organization of the United Nations. (2022). *The 2022 Statistical Yearbook of the FAO*. FAO. www.fao.org

Food and Agriculture Organization of the United Nations (FAO). (2021). *FAO*. Ecology, Cultivation and Uses of the Prickly Pear.

Frelat, R., Lopez-Ridaura, S., Giller, K. E., Herrero, M., Douxchamps, S., Djurfeldt, A. A., Erenstein, O., Henderson, B., Kassie, M., Paul, B. K., Rigolot, C., Ritzema, R. S., Rodriguez, D., van Asten, P. J. A., & van Wijk, M. T. (2016). Drivers of household food availability in sub-Saharan Africa based on big data from small farms. *Proceedings of the National Academy of Sciences of the United States of America*, *113*(2), 458–463. doi:10.1073/pnas.1518384112 PMID:26712016

Groot, R. S. (2021). Neglected and Underutilized Crop Species: The Key to Sustainable Food and Nutrition Security in a Changing World? *Sustainability, 19*, 10777.

Kutcher, H. R. (2020). The Role of Alternative Crops in the Long-term Resilience of Prairie Agricultural Systems: Implications for Food Security. *Frontiers in Sustainable Food Systems*, *4*, 146.

Lipper, L. (2020). Climate-smart agriculture for food security. *Nature Climate Change*, *10*(12), 1074–1076. PMID:33262808

Matson, P. A. (2021). Agriculture's contribution to biodiversity conservation. *Science*, *371*(6531), 239–240.

Mbow, C. (2019). Agroforestry solutions to address food security and climate change challenges in Africa. *Current Opinion in Environmental Sustainability*, *41*, 61–67.

Ministry of Agriculture, Fisheries, Rural Development, Water and Forests, Kingdom of Morocco. (2019). *Sugar Sector*. Ministère de l'agriculture.

OECD/FAO. (2020). *OECD-FAO Agricultural Outlook 2020-2029, Sugar*. OECD iLibrary.

Botanical Online. (n.d.). *Cultivation of the Prickly Pear*. Botanical Onlin..

Parmesan, C., & Yohe, G. (2003). A globally coherent fingerprint of climate change impacts across natural systems. *Nature*, *421*(6918), 37–42. doi:10.1038/nature01286 PMID:12511946

Sghaier, N., & Labbene, M. (2016). *PROJET MENA-DELP*. Middle East & North Africa - Desert Ecosystems and Livelihoods Program. March 2016.

Tadesse, T. W. (2021). Adaptation to climate change in agriculture: Analysis of potential determinants for investment in Ethiopia. *World Development*, *140*, 105365.

van der Voet, H. (2019). Methodology for the systematic analysis of trade-offs in sustainability assessment: make fewer assumptions. *Sustainability, 11*, 499.

Chapter 2

Forecasting Ozone Levels in Morocco: A Comparative Study of SARIMA and FB Prophet Models

Abdellatif Bekkar
https://orcid.org/0000-0002-8373-0038
Faculty of Science and Techniques, University Hassan II-Casablanca, Mohammedia, Morocco

Badr Hssina
Faculty of Science and Techniques, University Hassan II-Casablanca, Mohammedia, Morocco

Samira Douzi
FMPR, Mohammed V University in Rabat, Morocco

Khadija Douzi
Faculty of Science and Techniques, University Hassan II-Casablanca, Mohammedia, Morocco

ABSTRACT

Ozone pollution poses a significant environmental and health challenge in Marrakech, Morocco, intensified by urban and industrial growth. This study compares two statistical models, SARIMA and Facebook Prophet, for accurate O3-level forecasting in Marrakech. Through time series analysis, we assess their performance using metrics like MAE, MAPE, and RMSE. The findings highlight the Facebook Prophet model's superior accuracy, attributed to its better handling of seasonality and adaptability to unusual events. This research aids environmental and public health efforts in Marrakech by offering insights into ozone trends, supporting informed policymaking to combat ozone pollution.

DOI: 10.4018/979-8-3693-3807-0.ch002

Copyright © 2024, IGI Global. Copying or distributing in print or electronic forms without written permission of IGI Global is prohibited.

INTRODUCTION

The quest for precise ozone (O3) concentration forecasting epitomizes the confluence of environmental stewardship and public health advocacy, underscored by the compound's significant impacts. Traditional statistical methodologies, despite their foundational role in the realm of O3 prediction, have frequently encountered predictive accuracy limitations, propelling the exploration of advanced techniques (Comrie, 1997; Chattopadhyay & Bandyopadhyay, 2007).

The infusion of Artificial Intelligence (AI) and Machine Learning (ML) into ozone forecasting heralds a paradigm shift, with methodologies such as Support Vector Machines (SVM) and Decision Trees (DT) offering enhanced predictive precision. These advancements underscore the potential of computational intelligence in navigating the complexities inherent in ozone level prediction (Gardner & Dorling, 1999; Robeson & Steyn, 1990). Among these advancements, Bekkar et al. (2022) introduced a spatiotemporal prediction model based on IoT sensors, leveraging the Facebook Prophet model to forecast particulate matter concentrations in Taiwan with significant improvements in accuracy, showcasing the model's efficacy in environmental applications.

Despite the promise of advanced models, the accurate prediction of extreme ozone events remains a challenge, often limited by the scope of observational data (Zhang et al., 2012). Ensemble algorithms emerged as a response to these challenges, amalgamating various predictive models to improve overall forecast accuracy (Singh, Gupta, & Rai, 2013; Debry & Mallet, 2014).

The proliferation of Machine Learning (ML) and Deep Learning (DL) models, particularly those incorporating hybrid approaches like CNNs combined with LSTM networks, highlights the critical role of integrating spatial and temporal data in air quality prediction. These models, by leveraging comprehensive data sets, pave the way for more precise ozone forecasting methods (Jumin et al., 2020; Yılmaz, 2021). In a novel approach, Bekkar et al. (2023) employed a transformer-based model for multi-horizon forecasting of ozone in Marrakech, Morocco, illustrating the transformative potential of neural networks in air quality management with significant improvements in predictive metrics.

In Morocco, the momentum in air quality research is palpable, with studies addressing urban, rural, and industrial pollution levels. The emergence of localized studies and the implementation of advanced forecasting models are essential in navigating the complex air quality management landscape in the region (Razouki et al., 2018; Badri et al., 2016 ; Bekkar et al., 2023).

This literature review synthesizes the evolution and current state of ozone forecasting methodologies, transitioning from traditional statistical analyses to sophisticated SARIMA and Facebook Prophet models, and now, to the cutting-edge transformer-based models. By situating this study within the Moroccan context and beyond, it endeavors to contribute to the enhancement of air quality management practices in regions confronting pollution challenges. The comparative analysis of SARIMA and Facebook Prophet models, along with the introduction of transformer-based models for ozone forecasting, aims to offer novel insights into the efficacy of these models in forecasting ozone levels, marking a significant contribution to the ongoing discourse in environmental science and public health.

LITERATURE REVIEW

The quest for precise ozone (O3) concentration forecasting epitomizes the confluence of environmental stewardship and public health advocacy, underscored by the compound's significant impacts. Traditional statistical methodologies, despite their foundational role in the realm of O3 prediction, have frequently encountered predictive accuracy limitations, propelling the exploration of advanced techniques (Comrie, 1997; Chattopadhyay & Bandyopadhyay, 2007).

The infusion of Artificial Intelligence (AI) and Machine Learning (ML) into ozone forecasting heralds a paradigm shift, with methodologies such as Support Vector Machines (SVM) and Decision Trees (DT) offering enhanced predictive precision. These advancements underscore the potential of computational intelligence in navigating the complexities inherent in ozone level prediction (Gardner & Dorling, 1999; Robeson & Steyn, 1990). Despite their promise, the accurate prediction of extreme ozone events remains a challenge, often limited by the scope of observational data (Zhang et al., 2012).

Ensemble algorithms emerged as a response to these challenges, amalgamating various predictive models to improve overall forecast accuracy. Techniques such as boosting, bagging, and stacking have demonstrated efficacy in air pollution prediction, though not without inheriting the limitations of their constituent algorithms, especially in forecasting ozone concentration exceedances (Singh, Gupta, & Rai, 2013; Debry & Mallet, 2014).

In the specific context of ozone forecasting, the exploration of diverse models reveals a spectrum of capabilities and limitations. The Seasonal Autoregressive Integrated Moving Average (SARIMA) model, for instance, is celebrated for its adeptness at capturing seasonal variations in time series data, offering a solid theoretical foundation for its application in environmental forecasting (Hoi, K. I., Yuen, K. V., & Mok, K. M., 2014). Concurrently, the Facebook Prophet model, known for its flexibility and ability to accommodate seasonal fluctuations, presents a novel approach to time series forecasting, demonstrating its utility in environmental applications (Taylor & Letham, 2018; Zhu & Shyy, 2017; Chouit, E. M., Rachdi, M., Bellafkih, M., & Raouyane, B. 2022).

The proliferation of Machine Learning (ML) and Deep Learning (DL) models, particularly those incorporating hybrid approaches like CNNs combined with LSTM networks, highlights the critical role of integrating spatial and temporal data in air quality prediction. These models, by leveraging comprehensive data sets, pave the way for more precise ozone forecasting methods (Jumin et al., 2020; Yılmaz, 2021; Bekkar et al., 2022).

In Morocco, the momentum in air quality research is palpable, with studies addressing urban, rural, and industrial pollution levels. The emergence of localized studies and the implementation of advanced forecasting models are essential in navigating the complex air quality management landscape in the region (Razouki et al., 2018; Badri et al., 2016).

This literature review synthesizes the evolution and current state of ozone forecasting methodologies, transitioning from traditional statistical analyses to sophisticated SARIMA and Facebook Prophet models. By situating this study within the Moroccan context, it endeavors to contribute to the enhancement of air quality management practices in regions confronting pollution challenges. The comparative analysis of SARIMA and Facebook Prophet models, against the backdrop of Morocco's unique environmental and climatic conditions, aims to offer novel insights into the efficacy of these models in forecasting ozone levels, marking a significant contribution to the ongoing discourse in environmental science and public health.

MATERIALS, METHODS, AND METHODOLOGY

Materials

The foundation of this study is the analysis of ozone levels in Marrakech, Morocco, through a secondary dataset meticulously compiled by Sadiq, Abderrahmane; ELFAZZIKI, A.; OUARZAZI, J. (2016). This dataset, made accessible via figshare, offers a comprehensive overview of air quality measurements from three monitoring stations across Marrakech, capturing a critical period from June 1, 2009, to November 28, 2010.

Data Utilization for Ozone Levels in Marrakech, Morocco

Rather than conducting primary data collection, this study leverages the extensive dataset curated by Sadiq et al. (2016). This approach enables a robust analysis of historical air quality data, including ozone levels, across different parts of Marrakech. The dataset includes hourly measurements of key pollutants (PM10, O3, NO2, CO, SO2) and meteorological variables (humidity, temperature, wind speed, solar radiation), providing a rich basis for understanding the dynamics of ozone concentration in relation to various environmental factors.

Overview of Monitoring Stations

The dataset encompasses data from three strategically placed air quality monitoring stations within Marrakech: Jamae Lafnae (ST-1), Mhamid (ST-2), and Daoudiat (ST-3). These stations, equipped with automated physico-chemical analyzers, are situated in locations that offer a broad perspective on the urban air quality, from central city areas to residential districts.

Figure 1. Location of air quality monitoring stations in Marrakech

Data Summary

Our study is anchored in a comprehensive multivariate dataset derived from three strategically situated air quality monitoring stations within Marrakech: Jamaa El Fna (JMF), Mhamid, and Dawdiat. This dataset, systematically recorded on an hourly basis from June 1, 2009, to November 28, 2010, provides a granular view of air quality fluctuations over time, facilitating a nuanced analysis of pollution trends within an urban environment.

The dataset encompasses readings of five critical air pollutants: carbon monoxide (CO), nitrogen dioxide (NO2), ozone (O3), particulate matter with a diameter less than ten microns (PM10), and sulfur dioxide (SO2). These pollutant measurements are complemented by four essential meteorological variables: humidity ratio (HR), ambient temperature in Celsius (TC), wind speed in meters per second (WS), and solar radiation intensity in watts per square meter (SR). Collectively, these nine variables constitute the dataset's core components, offering a multidimensional perspective on the environmental factors that influence air quality in Marrakech.

Table 1. Characteristics of air quality monitoring stations

STATION	ENVIRONMENT	LATITUDE	LONGITUDE	OPERATION START
MHAMID	Urban	31.5962	8.0436	01/06/2009
JEF	Urban	31.6202	7.9888	01/06/2009
DAOUDIAT	Urban	31.6532	7.9956	01/03/2010

A detailed characterization of the monitoring stations, including their specific locations and the commencement dates of their operation, is delineated in ***Table 1***. This table serves as a critical reference point for contextualizing the data collection framework and ensuring the geographical and temporal relevance of the dataset to the study's objectives.

Figure 2. Graphically represents the temporal sequence

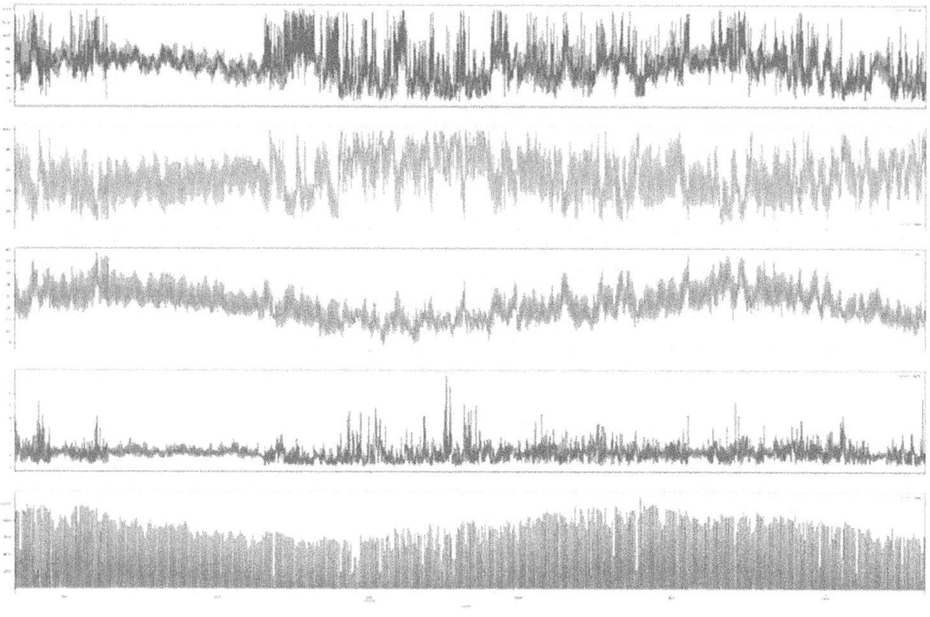

Figure 2 graphically represents the temporal sequence of all input and target variables, excluding the temporal component. This visual aid underscores the dataset's complexity and the intricate relationships between the measured pollutants and meteorological conditions over time.

The primary aim of this research is to employ advanced statistical models to predict future ozone (O3) levels based on historical data and concurrent meteorological conditions. By decoding the patterns inherent in the dataset, the study seeks to contribute to the broader body of knowledge on air quality prediction and management, particularly within the rapidly urbanizing context of Marrakech. This endeavor highlights the potential of empirical data analysis in informing targeted interventions and policy measures to mitigate urban air pollution.

Methods

Autoregressive Integrated Moving Average (Arima) Model

The ARIMA model, pioneered by Box et al., stands as a cornerstone in the field of univariate time series forecasting. This model posits that future values in a series can be predicted as a linear amalgamation of past observations and errors. The ARIMA model is mathematically encapsulated by the equation:

$$y_t = \phi_1 y_{t-1} + \phi_2 y_{t-2} + \ldots + \phi_p y_{t-p} + ,_1\epsilon_{t-1} + ,_2\epsilon_{t-2} + \ldots + ,_q\epsilon_{t-q} + \epsilon_t \quad (1)$$

Here, y_t represents the actual value at time *t,εt* is the error term, and ϕ *and* θ *ar*e the coefficients of the autoregressive and moving average parts, respectively. The model uses the differencing operator (1–B)d t^o achieve stationarity:

$$(1-B)^d y_t = y_t - dy_{t-1} \tag{2}$$

The backward shift operator B is defined to facilitate this transformation, where $B^p y_t = y_{t-p}$ and $B^q \epsilon t_= \epsilon t-_{q.}$ The ARIMA model, thus, integrates autoregressive (AR) and moving average (MA) components with differencing to model and predict time series data effectively.

Seasonal Autoregressive Integrated Moving Average (Sarima) Model

Expanding on the ARIMA framework, the SARIMA model incorporates seasonal fluctuations into the predictive model, as presented by Box et al. This model captures the seasonality in data by adding seasonal differencing and seasonal AR and MA components to the non-seasonal ARIMA model, thereby enhancing its forecasting capability for data with periodic patterns.

The SARIMA model is expressed as:

$$\left(1-\sum_{i=1}^{p}\phi_i B^i\right)\left(1-\sum_{i=1}^{P} {}_i B^{iS}\right)^D y_t = \left(1+\sum_{i=1}^{q} {}_i B^i\right)\left(1+\sum_{i=1}^{Q} {}_i B^{iS}\right)\epsilon_t \tag{3}$$

In this equation, D and d denote the orders of seasonal and non-seasonal differencing, respectively, and S is the seasonal period length. The parameters p, q, P, and Q represent the orders of the non-seasonal and seasonal autoregression and moving average processes.

The estimation of SARIMA model parameters involves initially making the series stationary through differencing. The orders of the autoregressive and moving average components p, q, P, and Q are then determined based on the autocorrelation function (ACF) and the partial autocorrelation function (PACF) of the differenced series. This meticulous parameter estimation process ensures the SARIMA model's effectiveness in forecasting time series data that exhibit both non-stationary and seasonal characteristics.

1. **Achieving Stationarity:** The first step in SARIMA modeling involves ensuring that the time series data are stationary. This typically requires differencing the data, which can be non-seasonal differencing (d) to address trends and seasonal differencing (D) to handle seasonality. The order of differencing is determined by examining plots of the data and using statistical tests for stationarity (e.g., Augmented Dickey-Fuller test).
2. **Determining AR and MA Orders:** Once the series is made stationary, the next step is to identify the orders of the AR and MA components. This process involves two sets of parameters: non-seasonal (p, q) and seasonal (P, Q).
    - **Autocorrelation Function (ACF) and Partial Autocorrelation Function (PACF) Analysis**: These tools are instrumental in determining the appropriate orders for the AR and MA components. For non-seasonal components, the PACF can help identify the order of the AR part (p) by observing where the PACF plot cuts off. Similarly, the ACF plot can suggest the order of the MA part (q) by its cutoff point.
    - **Seasonal Components:** The same analysis is applied to the seasonal components of the data. Here, the seasonal period (S) plays a crucial role. The PACF and ACF plots should be examined at multiples of the seasonal period to determine the seasonal AR (P) and MA (Q) orders.

3. **Model Specification**: With the orders of differencing (d, D) and the AR and MA components (p, q, P, Q) identified, the SARIMA model can be specified. This is often an iterative process, where several models with different combinations of parameters are fitted to the data.
4. Model Fitting and Diagnostic Checking:
   - **Model Fitting:** Using statistical software, the SARIMA model is fitted to the data. This involves estimating the coefficients of the AR and MA components that minimize the error of the model.
   - **Diagnostic Checking:** After fitting the model, diagnostics are performed to assess its adequacy. This includes analyzing the residuals to ensure they resemble white noise. Tools such as the Ljung-Box test can be used to test for autocorrelation in the residuals. Plots of the residuals should show no apparent patterns or seasonality.
   - **Information Criteria:** Model selection is often guided by information criteria such as the Akaike Information Criterion (AIC) or the Bayesian Information Criterion (BIC). These criteria balance model fit with model complexity, where lower values indicate a better model. The model with the lowest AIC or BIC among the candidates is typically chosen.
5. **Validation:** The final step involves validating the selected SARIMA model. This can include forecasting on a hold-out sample of the data and comparing the forecasts to the actual values to assess the model's predictive accuracy.

Facebook Prophet

The Prophet model, developed by Facebook in 2017 (Taylor & Letham, 2018), is a sophisticated open-source algorithm designed for forecasting time series data. Its utility spans various applications, from predicting daily confirmed COVID-19 cases to modeling business and environmental time series. Available for use in both R and Python, Prophet simplifies time series forecasting by accommodating non-linear trends, seasonality, and holiday effects with minimal preprocessing required from the user.

The core of the Prophet model is encapsulated in the formula:

$$y(t) = g(t) + s(t) + h(t) + \epsilon_t \tag{4}$$

Here, $y(t)$ represents the forecasted value at time t, capturing the dynamic interplay of the model's components:

- $g(t)$ is the trend function, modeling non-periodic changes and adapting to various growth behaviors, including saturating growth and linear trends.
- $s(t)$ accounts for seasonality, using Fourier series to model periodic changes, such as weekly patterns in the data.
- $h(t)$ adjusts for holidays and significant events, recognizing their potential to disrupt regular patterns.
- ϵt_denotes idiosyncratic changes not captured by the model, encompassing the random variation inherent in time series data.

Trend Component *g*(*t*)

The trend in the Prophet model is typically modeled as a piecewise linear or logistic growth curve, allowing for flexibility in capturing various growth behaviors:

$$g(t) = \frac{C}{1+e^{-k(t-m)}} \tag{5}$$

This logistic growth formulation includes parameters for carrying capacity *C*, growth rate *K*, and an offset *m*, enabling the model to adapt to saturation limits and points of inflection in the time series.

Seasonality Component *s*(*t*)

Seasonality is captured through a series of Fourier terms, allowing the model to fit seasonal effects with varying frequencies:

$$s(t) = \sum_{n=1}^{N} \left(a_n \cos\left(\frac{2\pi nt}{P}\right) + b_n \sin\left(\frac{2\pi nt}{P}\right) \right) \tag{6}$$

P denotes the period (e.g., 365.25 for yearly seasonality), and a_n, b_n are coefficients that the model fits to the data.

Holidays and Events Component *h*(*t*)

The model accounts for holidays and events with irregular timings that could influence the time series:

$$h(t) = \sum_i I(t \in D_i)^{\circ}{}_i \tag{7}$$

I(*tϵDi*$_)$ is an indicator function that identifies the occurrence of event i at time t, and k*i* represents the effect of the event on the predicted value.

Evaluation of Prediction Performance

In this study, we employed three statistical metrics to evaluate the prediction performances of the ARIMA, and Prophet models: Mean Absolute Error (MAE), Mean Absolute Percentage Error (MAPE), and Root Mean Square Error (RMSE). These metrics are crucial for assessing the accuracy of the models, with lower values indicating better prediction performance. The formulas for these evaluation indices are expressed as follows:

Mean Absolute Error (MAE) measures the average magnitude of the errors in a set of forecasts, without considering their direction:

$$MAE = \frac{1}{n}\sum_{i=1}^{n}\left|y_i - \widehat{y_i}\right| \tag{8}$$

Mean Absolute Percentage Error (MAPE) quantifies the prediction accuracy as a percentage, offering an intuitive measure of the model's performance:

$$MAPE = \frac{1}{n}\sum_{i=1}^{n}\left|\frac{y_i - \widehat{y_i}}{y_i}\right| \times 100 \tag{9}$$

Root Mean Square Error (RMSE) calculates the square root of the average of squared differences between the predicted and observed values, providing insight into the error distribution:

$$RMSE = \sqrt{\frac{1}{n}\sum_{i=1}^{n}\left(y_i - \widehat{y_i}\right)^2} \tag{10}$$

where:

- y_i is the observed value,
- $\widehat{y_i}$ represents the predicted value,
- and n is the total number of observations in the sequence.

By leveraging MAE, MAPE, and RMSE as benchmarks, we can rigorously assess the predictive accuracy of the ARIMA, and Prophet models, thereby determining their efficacy in forecasting the time series data of interest.

RESULTS

Data Preprocessing

Outlier Detection and Removal

In our dataset preparation, significant attention was given to outliers, which could distort the analysis. An extreme instance, such as a PM10 measurement of 4187 µg/m^3, clearly deviates from plausible environmental conditions. To systematically filter out such outliers, we applied the Interquartile Range (IQR) method. This technique identifies outliers as data points that fall more than 1.5 times the IQR distance from the quartiles. Detected outliers were subsequently marked as NaN values. Through box and whisker plots, this step assured the removal of inaccuracies or anomalies, preserving the dataset's integrity for further analytical processing.

Missing Data Imputation

For the imputation of missing values, we adopted the k-Nearest Neighbors (k-NN) algorithm, a method chosen for its efficacy in maintaining the intrinsic structure of the data. The k-NN algorithm imputes missing values by identifying the 'k' nearest neighbors to a data point with a missing value and calculates the imputation value based on the mean or median of these neighbors. This approach is particularly advantageous for time series data, where temporal proximity often implies similarity in value, making k-NN a suitable choice for our dataset. The application of k-NN ensures a comprehensive dataset, crucial for the accurate forecasting of ozone pollution levels and subsequent analysis.

Sarima Model

Stationarity Testing

The Augmented Dickey-Fuller (ADF) test was employed to determine the stationarity of the time series data for ozone (O3) concentrations. The results are as follows:

- ADF test statistic: -3.090888
- p-value: 0.027235
- Number of lags used: 3
- Number of observations: 542
- Critical value (1%): -3.442473
- Critical value (5%): -2.866887
- Critical value (10%): -2.569618

Given that the ADF test statistic (-3.090888) is less than the critical value at 5% (-2.866887), we have strong evidence against the null hypothesis. Thus, we reject the null hypothesis, concluding that the data has no unit root and is stationary. This stationarity is a prerequisite for the effective use of SARIMA models, as it implies that the time series does not have time-dependent structures such as trends or seasonality that could bias the forecast.

ACF and PACF Analysis

To fine-tune the SARIMA model's parameters, meticulous analysis of the Autocorrelation Function (ACF) and Partial Autocorrelation Function (PACF) plots was conducted. These functions are instrumental in determining the model's autoregressive (AR) and moving average (MA) orders. ACF helps identify the extent of the correlation between the time series and its lagged versions, while PACF isolates the correlation of the time series with its lags after removing the effects of shorter lags. By examining these plots, we can discern the point at which correlations become insignificant, which is indicative of the optimal lag values for the AR and MA components.

Figure 3. Auto correlation and partial correlation graphs

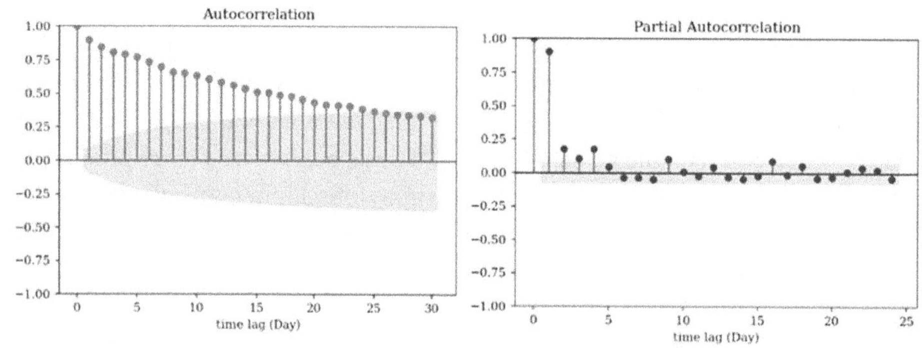

The graphs showcasing ACF and PACF provided an empirical basis for selecting the SARIMA model's parameters p (autoregressive order), d (degree of differencing), and q (moving average order). After inferring these key parameters, our workflow transitioned into the data partitioning phase, where the dataset was divided into training and testing sets—a critical step that forms a solid foundation for subsequent model evaluation.

Subsequently, model construction commenced, utilizing the SARIMAX library, a versatile tool that extends the SARIMA model to include seasonality through the integration of seasonal terms. This extension is particularly important given the presence of seasonal patterns in our data. The SARIMAX library's capabilities streamline the model fitting process, enabling a more nuanced adjustment of the model to the specific characteristics.

Parameter Selection

The SARIMA model parameters chosen were (p, d, q) x (P, D, Q, s) = (1, 1, 1) x (0, 1, 1, 15), which includes:

p = 1: Suggests one autoregressive term based on the PACF plot.

d = 1: Indicates that the data was made stationary by differencing once.

q = 1: Implies one moving average term as per the ACF plot.

P = 0, D = 1, Q = 1: These are the seasonal elements of the model, with one seasonal differencing (D) to ensure stationarity and one seasonal moving average (Q) component.

s = 15: This denotes the seasonal period, which suggests that the data has a seasonal pattern that repeats every 15 time periods.

Sarimax Results

The SARIMA(1, 1, 1)x(0, 1, 1, 15) model was fit to the O3 concentration data, yielding the following results:

- Log Likelihood: -1318.485
- Akaike's Information Criterion (AIC): 2644.970
- Bayesian Information Criterion (BIC): 2660.966

- Hannan-Quinn Information Criterion (HQIC): 2651.303

Figure 4. Statistical analysis of SARIMA

Dep. Variable:	O3	**No. Observations:**	436
Model:	SARIMAX(1, 1, 1)x(0, 1, 1, 15)	**Log Likelihood**	-1318.485
Date:	Tue, 27 Feb 2024	**AIC**	2644.970
Time:	01:26:29	**BIC**	2660.966
Sample:	06-01-2009	**HQIC**	2651.303
	- 08-10-2010		
Covariance Type:	opg		

	coef	**std err**	**z**	**P>|z|**	**[0.025**	**0.975]**
ar.L1	0.3709	0.091	4.066	0.000	0.192	0.550
ma.L1	-0.6795	0.071	-9.541	0.000	-0.819	-0.540
ma.S.L15	-1.0000	421.549	-0.002	0.998	-827.221	825.221
sigma2	36.8505	1.55e+04	0.002	0.998	-3.04e+04	3.05e+04

Ljung-Box (L1) (Q):	0.02	**Jarque-Bera (JB):**	192.22
Prob(Q):	0.89	**Prob(JB):**	0.00
Heteroskedasticity (H):	1.92	**Skew:**	0.64
Prob(H) (two-sided):	0.00	**Kurtosis:**	6.13

The coefficients for the AR and MA components were significant, with the AR coefficient at 0.3709 and the MA coefficient at -0.6795. Interestingly, the seasonal MA component at lag 15 (ma.S.L15) had a coefficient of -1.0000 with a very high standard error, which could suggest potential issues with the model such as overfitting or identification problems.

The diagnostics like the Ljung-Box test and the Jarque-Bera test showed that the residuals were not significantly autocorrelated (Prob(Q) = 0.89) and were not normally distributed (Prob(JB) = 0.00), with skewness and kurtosis indicating deviation from normality. The heteroskedasticity test (Prob(H) = 0.00) with a value of 1.92 indicates the presence of heteroskedasticity.

These diagnostics suggest that while the model parameters are statistically significant, there may be some issues with the model's assumptions, and further refinement or validation of the model may be necessary.

Table 2. Perfermence metics of SARIMA

Mean Absolute Error	**5.584**
Root Mean Square Error	7.179
Mean Absolute Percentage Error	24.700

These metrics serve as indicators of the model's forecasting accuracy. The MAE value of 5.584 suggests that the model's predictions are, on average, within approximately 5.584 units of the actual values. The RMSE of 7.179, being higher than the MAE, points to the presence of some larger errors, which could be outliers affecting the forecast. Lastly, the MAPE of 24.70% implies that the model's predictions deviate from the actual values by an average of nearly a quarter, which highlights potential variability in the model's performance across the dataset.

FB Prophet Model

In this study, the dataset was thoughtfully prepared to conform to the input specifications of the Prophet model, emphasizing its design for univariate time series forecasting. This preparation involved organizing the dataset into two columns: 'ds' for the date and 'y' for the observed values. The model's capability to include additional regressors enhances its predictive power despite its primary design for univariate analysis. The 'ds' column was constructed by merging the year, month, and day fields into a singular date format, while the 'y' column consisted of observed values with missing entries imputed to ensure data completeness.

The study covers the period from June 1, 2009, to November 28, 2010, totaling 546 observations. These were strategically split into training and testing sets, with 80% of the observations (approximately 437) allocated for training and the remaining 20% (approximately 109) designated for testing. This distribution facilitated the model's development on the training set and its subsequent validation and evaluation on the testing set.

Hyperparameter tuning was conducted to optimize the settings within the Prophet model, focusing on key parameters such as the changepoint prior scale for trend flexibility, the holidays prior scale for holiday effects, and the number of changepoints for precise trend adjustments. The selection of a 'multiplicative' seasonality mode was pivotal in accurately modeling the dataset's seasonal patterns. Configurations were made to account for the observed weekly, daily, and yearly seasonality, with a 95% confidence interval established for the forecast. This meticulous setup enabled the nuanced forecasting of ozone levels, leveraging the specific dataset spanning from June 1, 2009, to November 28, 2010. The approach was designed to optimize model performance, providing a comprehensive understanding of the seasonal and trend dynamics within the ozone level data.

Prophet can generate both weekly and yearly seasonality out of daily data. The ***Figure 5*** shows the trend and seasonality.

Figure 5. Plot trend, weekly seasonality, and yearly seasonality

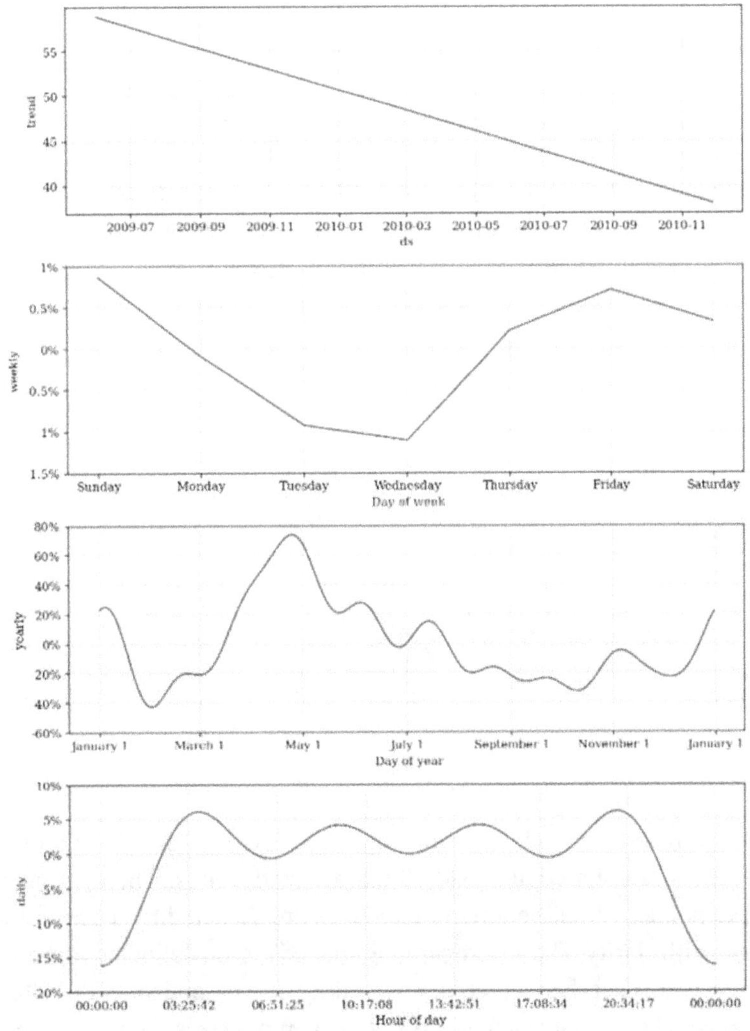

The trend depicted in the data demonstrates a clear decline in ozone (O3) levels from July 2009 to November 2010. This suggests a reduction in O3 concentrations over the period, which could be attributed to various factors such as environmental policies, changes in emissions, or atmospheric conditions.

The weekly seasonality highlights minor fluctuations in O3 levels throughout the week, with variations ranging from a decrease of roughly 1.5% to an increase of slightly more than 1%. This pattern may be due to weekly industrial cycles, traffic patterns, or other anthropogenic activities that influence ozone production and degradation.

When examining the annual cycle, there is a discernible pattern with O3 levels increasing to a peak around the middle of the year, particularly in July, which may coincide with higher temperatures and increased sunlight contributing to more ozone formation. Subsequently, there is a noticeable decrease toward the end of the year, aligning with the winter months when conditions for ozone formation are less favorable.

The diurnal pattern of O3 levels shows significant variation throughout a 24-hour period, peaking around midday, which is consistent with the photochemical production of ozone being highest when sunlight is most intense. The levels drop during the night when the absence of sunlight limits the formation of ozone.

The observed data underscores the importance of considering both additive and multiplicative models in forecasting O3 levels. The additive model would interpret seasonality as a fixed increase or decrease in O3 levels, while the multiplicative model suggests that these seasonal effects are proportional to the overall trend at any given time. Therefore, the impact of seasonality on ozone concentrations can vary depending on the underlying trend and the time of year or day.

Figure 6. Forecast of ozone levels using the Prophet model, displaying historical data points, predicted values, and confidence intervals

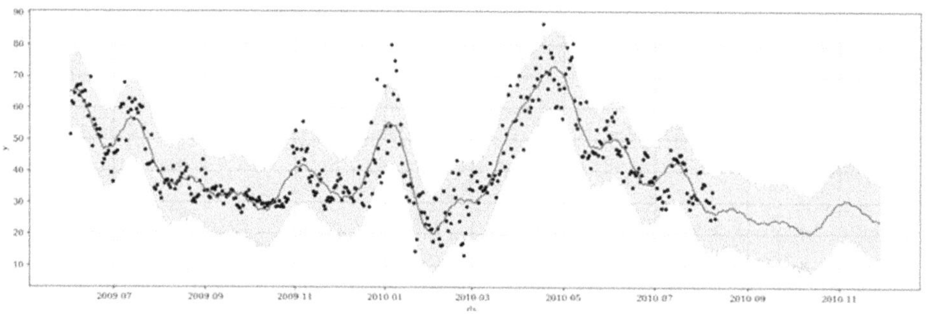

The comprehensive forecasting model provided by Prophet is visualized in Figure 6, where the time series data of ozone levels are plotted alongside the forecasted trends. The historical observations are represented as discrete points, while the forecasted values are depicted by a continuous line, capturing the underlying patterns with clarity. The accompanying shaded region delineates the confidence intervals, offering a visual representation of the predictive uncertainty as it extends into the future. This figure not only illustrates the expected seasonal fluctuations but also emphasizes the model's capacity to adapt to the intricacies of environmental data. The inclusion of this visual aid underscores the nuanced interplay between observed values and predicted outcomes, facilitating a deeper understanding of the model's analytical prowess

Table 3. Performance metrics of FB prophet

Mean Absolute Error	**5.066**
Root Mean Square Error	6.194
Mean Absolute Percentage Error	20.354

The forecast accuracy of the ozone levels using the Prophet model is measured by the Mean Absolute Error (MAE) at 5.066, indicating average prediction deviations. The Root Mean Square Error (RMSE)

is 6.194, pointing to occasional larger errors, while the Mean Absolute Percentage Error (MAPE) of 20.354% reflects the average percentage difference from actual values. These metrics collectively assess the model's precision and reliability in forecasting.

DISCUSSION

In the quest to forecast ozone (O3) levels, the choice of the predictive model plays a pivotal role in the accuracy and reliability of the forecasts. This study pits the Seasonal Autoregressive Integrated Moving Average (SARIMA) against the more contemporary Facebook Prophet model in a head-to-head comparison to evaluate their performance in O3 level prediction.

The evaluation metrics reveal a distinct edge for the Facebook Prophet model over SARIMA in our dataset. The Prophet model demonstrates a lower Mean Absolute Error (MAE) of 5.066 compared to SARIMA's MAE of 5.584, indicating that Prophet's predictions are, on average, closer to the true values. This is particularly noteworthy in the context of environmental studies, where precision is paramount for planning and response strategies.

Moreover, the Root Mean Square Error (RMSE), another critical measure of prediction accuracy, is markedly lower for Prophet at 6.194, as opposed to SARIMA's 7.179. The RMSE is sensitive to large errors, suggesting that Prophet is better suited to handle outliers and extreme values, which are common in environmental data due to sudden changes in weather patterns and other anomalies.

The Mean Absolute Percentage Error (MAPE) further solidifies Prophet's supremacy with a score of 20.354%, against SARIMA's 24.700%. In terms of relative error, the Prophet model's forecasts are closer to the actual values, making it a more reliable tool for stakeholders who rely on percentage differences, such as policy makers and environmental agencies.

Figure 7. Comparative forecast analysis of ozone levels using SARIMA and facebook prophet models

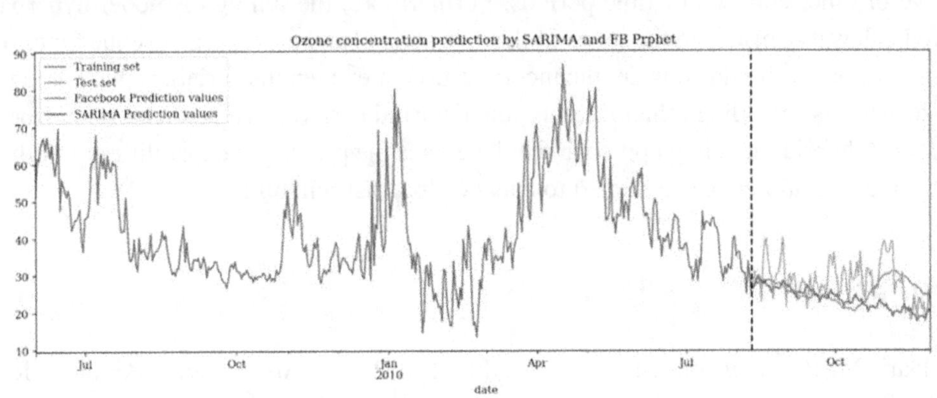

Visual inspection of the forecast plot Figure 7 aligns with the quantitative assessment, where the Prophet model's forecasts (green line) exhibit a commendable conformity to the actual test set (orange line). Conversely, the SARIMA model (red line), albeit capturing the general trend, tends to stray from the actual values more frequently. The graphical representation not only provides an intuitive comparison but also augments the numerical analysis, offering a comprehensive view of the models' performance.

In conclusion, while SARIMA holds its merit as a traditional model with established applications, the Facebook Prophet model emerges as the more adept forecasting tool in this study. Prophet's ability to seamlessly handle seasonality and trends with less manual intervention positions it as an advantageous alternative for O3 level prediction. Future research could explore the integration of exogenous variables, which might further enhance the forecasting prowess of the Prophet model, or potentially improve SARIMA's performance, warranting a continued exploration of these methodologies.

CONCLUSION

In this chapter, we engaged in an empirical evaluation of the Seasonal Autoregressive Integrated Moving Average (SARIMA) and the Facebook Prophet models, with the specific objective of forecasting tropospheric ozone (O3) concentrations. The comparative analysis was grounded on several performance metrics, which included Mean Absolute Error (MAE), Root Mean Square Error (RMSE), and Mean Absolute Percentage Error (MAPE).

The assessment yielded that the Facebook Prophet model manifested a more accurate forecasting ability with a MAE of 5.066, RMSE of 6.194, and MAPE of 20.354%. In comparison, the SARIMA model recorded a MAE of 5.584, RMSE of 7.179, and MAPE of 24.700%. The discrepancy in the performance of these models can be principally attributed to the Prophet model's robustness in handling the intrinsic seasonality and trend patterns associated with atmospheric ozone levels. This robustness is further reinforced by the model's resilience to irregularities in the data, such as missing values and anomalies, which are prevalent in environmental datasets.

It is, however, pertinent to recognize the limitations intrinsic to our analysis. The dataset encompassed observations from a singular geographic locale over a specific temporal span, which may not be entirely representative of other contexts or time periods. Furthermore, the tuning of model hyperparameters, albeit methodical, was confined to the scope of the data at hand and presents an avenue for optimization.

Prospective research should consider the incorporation of exogenous variables that could potentially influence ozone levels, including vehicular emissions data, industrial activity indicators, and meteorological parameters. A hybrid modeling approach that leverages the predictive capabilities of both SARIMA and Prophet models could also be explored to enhance forecast reliability.

REFERENCES

Badri, W., Dari, M. B., Agunaou, M., & Bounakhla, M. (2016). Atmospheric Pollution Research in Morocco: An Overview. *Journal of Materials & Environmental Sciences*, *7*(6), 2178–2185.

Bekkar, A., Hssina, B., Douzi, S., & Douzi, K. (2022). Air Quality Forecasting using decision trees algorithms. In *2022 2nd International Conference on Innovative Research in Applied Science, Engineering and Technology (IRASET)* (pp. 1-4). IEEE. 10.1109/IRASET52964.2022.9737814

Bekkar, A., Hssina, B., Douzi, S., & Douzi, K. (2022). Spatiotemporal Prediction of Concentrations Based on IoT Sensors. In Z. Boulouard, M. Ouaissa, M. Ouaissa, & S. El Himer (Eds.), *AI and IoT for Sustainable Development in Emerging Countries. Lecture Notes on Data Engineering and Communications Technologies* (Vol. 105). Springer. doi:10.1007/978-3-030-90618-4_10

Bekkar, A., Hssina, B., Douzi, S., & Douzi, K. (2023). *Transformer-Based Model for Multi-Horizon Forecasting Ozone in Marrakech city, Morocco.* 2023 14th International Conference on Intelligent Systems: Theories and Applications (SITA), Casablanca, Morocco. 10.1109/SITA60746.2023.10373756

Box, G. E. P., Jenkins, G. M., & Reinsel, G. C. (1994). *Time Series Analysis: Forecasting and Control* (3rd ed.). Prentice-Hall.

Chattopadhyay, S., & Bandyopadhyay, G. (2007). Artificial neural network with backpropagation learning to predict mean monthly total ozone in Arosa, Switzerland. *International Journal of Remote Sensing, 28*(20), 4471–4482. doi:10.1080/01431160701250440

Chouit, F. M., Rachdi, M., Bellafkih, M., & Raouyane, B. (2022). Forecasting of the epidemiological situation: Case of COVID-19 in Morocco. *Journal of the Nigerian Society of Physical Sciences, 4*(4), 843. doi:10.46481/jnsps.2022.843

Comrie, A. C. (1997). Comparing neural networks and regression models for ozone forecasting. *Journal of the Air & Waste Management Association, 47*(6), 653–663. doi:10.1080/10473289.1997.10463925

Debry, E., & Mallet, V. (2014). Ensemble forecasting with machine learning algorithms for ozone, nitrogen dioxide and PM10 on the Prev'Air platform. *Atmospheric Environment, 91*, 71–84. doi:10.1016/j.atmosenv.2014.03.049

Facebook, Inc. (2017). *Prophet: Automatic forecasting procedure [Computer software].* GitHub repository. https://github.com/facebook/prophet

Gardner, M. W., & Dorling, S. R. (1999). Neural network modelling and prediction of hourly NOx and NO2 concentrations in urban air in London. *Atmospheric Environment, 33*(5), 709–719. doi:10.1016/S1352-2310(98)00230-1

Hoi, K. I., Yuen, K. V., & Mok, K. M. (2014). Prediction of daily maximum ozone concentrations by seasonally integrated time series models. *Atmospheric Environment, 89*, 168–175.

Jumin, E., Zaini, N., Ahmed, A. N., Abdullah, S., Ismail, M., Sherif, M., Sefelnasr, A., & El-Shafie, A. (2020). Machine learning versus linear regression modelling approach for accurate ozone concentrations prediction. *Engineering Applications of Computational Fluid Mechanics, 14*(1), 713–725. doi:10.1080/19942060.2020.1758792

Razouki, S., Kousksou, T., Allaoui, T., Rhafiki, T. E. M., & Jamil, A. (2018). Air quality indicators in Morocco - Application to urban, rural, and industrial areas. *Energy Procedia, 139*, 696–701.

Robeson, S. M., & Steyn, D. G. (1990). Evaluation and comparison of statistical forecast models for daily maximum ozone concentrations. *Atmospheric Environment. Part B, Urban Atmosphere, 24*(2), 303–312. doi:10.1016/0957-1272(90)90036-T

Sadiq, A. (2016). Air quality monitoring stations data - Marrakech - Morocco. figshare. Dataset. doi:10.6084/m9.figshare.3113515.v1

Singh, K. P., Gupta, S., & Rai, P. (2013). Identifying pollution sources and predicting urban air quality using ensemble learning methods. *Atmospheric Environment*, *80*, 426–437. doi:10.1016/j.atmosenv.2013.08.023

Taylor, S. J., & Letham, B. (2017). *Prophet: Forecasting at scale [Computer software manual]*. Github. https://facebook.github.io/prophet/

Taylor, S. J., & Letham, B. (2018). Forecasting at scale. PeerJ Preprints, 5:e3190v2. https://peerj.com/preprints/3190/

Yılmaz, A. (2021). Ozone level prediction with machine learning algorithms. *Journal of Aeronautics and Space Technologies*, *14*(2), 177–183.

Zhang, Y., Bocquet, M., Mallet, V., Seigneur, C., & Baklanov, A. (2012). Real-time air quality forecasting, part I: History, techniques, and current status. *Atmospheric Environment*, *60*, 632–655. doi:10.1016/j.atmosenv.2012.06.031

Zhu, Y., & Shyy, G. (2017). *Sub-Daily Rainfall Forecasting by Using Facebook Prophet*. In *Proceedings of the 2017 International Conference on Computational Science and Computational Intelligence (CSCI)*, Las Vegas, USA.

Chapter 3
From Microalgal Biomass to Biohydrogen:
A Novel Approach to Environmentally Sustainable Technology

Khadija El-Moustaqim
https://orcid.org/0009-0007-9193-6638
Faculty of Science, Kenitra, Morocco

Jamal Mabrouki
https://orcid.org/0000-0002-3841-7755
Mohammed V University, Morocco

Driss Hmouni
Faculty of Science, Kenitra, Morocco

ABSTRACT

The rising influence of greenhouse gas (GHG) emissions from burning fossil fuels on climate change has led to a demand for clean energy. The only consequence of H2's combustion is water vapor, making it a clean energy source. As a clean energy source, hydrogen contributes significantly to energy and environmental sustainability and helps mitigate the world's energy problem and climate change. Microalgae have recently been found to hold tremendous promise for producing biodiesel and biohydrogen. Under certain conditions, sunlight is used by green microalgae to split water particles, resulting in the production of molecular hydrogen and oxygenation. With an emphasis on green microalgae, this review outlined possible algal strains for hydrogen production. In addition, a comprehensive review procedure aimed at presenting the importance of using microalgae to produce green hydrogen. More significantly, this review included an overview of present constraints and potential future directions for the sustainable synthesis of biohydrogen from microalgae.

DOI: 10.4018/979-8-3693-3807-0.ch003

Copyright © 2024, IGI Global. Copying or distributing in print or electronic forms without written permission of IGI Global is prohibited.

INTRODUCTION

In terms of resources and environmental impact, the idea of "green chemistry" addresses concerns about future development. With the increase in the amount of oil used, especially for transportation, the primary fossil energy source has experienced a decline in resource availability(Das & Veziroglu, 2008; Hemschemeier et al., 2009; Umoh et al., 2024). In response to the growing demand for energy, scientific communities and major research bodies must develop a variety of energy production schemes that minimize environmental impact (El-Moustaqim, El Bakraoui, et al., 2023; El-Moustaqim, Mabrouki, et al., 2023). Due to its high energy value, its environmentally friendly nature and potential to improve social, economic, and environmental conditions, hydrogen may prove to be the fuel of the next generation (El-Moustaqim et al., 2024). Future sustainable energy systems may result from the development of a hydrogen economy that unifies the production and use of hydrogen as a fuel source. Future sustainable energy systems may result from the development of a hydrogen economy that unifies producing and using hydrogen as a fuel source (Jiao et al., 2024; Omer et al., 2024). We can develop hydrogen-based production methods that use less energy and are, most importantly, more ecologically friendly to reduce CO_2 emissions overall (Goria et al., 2024). These innovative green technologies have the potential to turn a profit since they can use cheap biomass as a raw material (Sankaranath et al., 2023). The main cause of increased pollution, global warming, and climate change is the emission of greenhouse gases from the use of non-renewable fuels (Sankaranath et al., 2023). Therefore, to reduce these emissions, alternate energy sources are needed (S. Li et al., 2021). It has been acknowledged that hydrogen is a sustainable energy source for the future and that it is the most promising alternative to meet the necessary energy demands (Arimbrathodi et al., 2023). Hydrogen can be produced from biomass, fossil fuels or both by a variety of thermochemical and biological processes (Umoh et al., 2024). In order to mitigate the environmental problems caused by the fuel crisis and the emission of greenhouse gases, H_2 energy may be helpful. Developing the H_2 economy is, however, significantly hampered by the high cost of H_2 generation. Steam reforming methane is the main process for creating H_2, and it consumes a significant quantity of fossil fuel (Ekins & Hughes, 2009; Encinar et al., 1996; Kumar et al., 2016). The synthesis of H_2 from power using electrolysis technology may have an overall efficacy of only 60%. Global energy security is threatened by the depletion of fossil fuels, which has prompted the search for alternatives such as biofuels (Chisti, 2008; Oumsalem & Zanoun, 2019; Zhao et al., 2022). Microalgae have been explored as one of the possible sources of biofuels due to their ability it can be used for the storage of biomolecules such as proteins and carbohydrates, which can be used for the production of biofuels or food (Ho et al., 2013; Pandit & Fulekar, 2019; Yu et al., 2020). As a result, significant efforts have been made to promote the transition of industrial processes from fossil fuels to renewable energy sources through technological breakthroughs (C. Li & Fang, 2007; Scharlemann & Laurance, 2008). Biofuels including bioethanol, hydrogen, and biodiesel can be made from microalgae (El-Moustaqim et al., 2024). Microalgae's fast growth rate, adaptability to a variety of environments, ability to balance the needs for food and fuel production, and ability to absorb and use ambient carbon dioxide makes them especially well-suited for producing hydrogen (Alleau et al., 2000; Marbán & Valdés-Solís, 2007). As a result, the production of biohydrogen from microalgae has garnered a lot of interest as a clean, sustainable fuel that can help achieve carbon neutrality and sustainability in the environment (Algapani et al., 2016; Granovskii et al., 2007).

In order to achieve this, the review paper focuses on recent findings regarding the production of hydrogen using microalgae, mechanisms of sustainable hydrogen production, variables influencing the production of hydrogen using microalgae, design of bioreactors and hydrogen production, and sophisticated techniques to increase the productivity of hydrogen production using microalgae. The objective of this review is to compile the latest developments and information regarding the production of biohydrogen from microalgae and encourage the use of biohydrogen instead of traditional hydrocarbon biofuels, so accelerating the attainment of the most environmentally friendly carbon neutrality target.

MICROALGAL H_2 PRODUCTION MECHANISM

Depending on sunlight, fermentation, and thermochemical methods including biomass conversion, algal H_2 generation can occur directly or indirectly as mentioned in **Fig.1**. Anaerobic digestion, microbial electrolysis cells, light and dark fermentation, and direct and indirect biophotolysis are the most widely used biological techniques for producing hydrogen from algae (Umoh et al., 2024). Biohydrogen is the word used to describe the hydrogen created as a result of biological processes (Sahota et al., 2024). Biohydrogen can also be is specified as hydrogen obtained by a range of biomass materials through biological, chemical, thermochemical, biophotolytic, and fermentative processes. This is because biomass is involved in the synthesis of biohydrogen. Microalgae have the ability to convert solar energy into chemical energy in a direct biophotolysis system (Arimbrathodi et al., 2023; Omer et al., 2024). This procedure has the advantage of being abundantly available and environmentally friendly in nature. For instance, sunlight is a highly efficient source of energy that produces less CO_2 than it does when used as a source. As has been the case over the previous fifty years, the O_2 sensitivity of the hydrolysis process, and complicated, unpredictable routes make it is a tedious process to put into practice. Indirect biophotolysis, which used two-stage mechanisms for the temporary separation of H_2 evolution and photosynthesis, was shown to produce sustained amounts of H_2 (Maneeruttanarungroj et al., 2010; Sillero et al., 2022). Consequently, decomposition of H_2 and O_2 evolution reactions into distinct steps, connected by the fixation and evolution of CO_2 is a component of indirect biophotolysis processes. Despite the initial notion that the O_2 produced was absorbed through respiration, this process was actually indirect biophotolysis (Faraji & Saidi, 2021; Sridhar et al., 2022).

Figure 1. Advantages and disadvantages of several types of biological mechanisms-based biohydrogen production

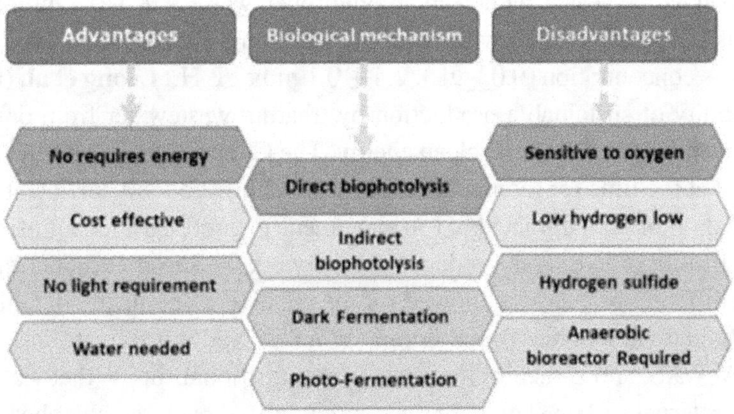

First-stage microalgae development is impacted by a number of factors, including light, temperature, pH, nutrients, carbon source, and bioreactor type. In the second stage, these components hold similar significance (Chandolias et al., 2016; Chum & Overend, 2001; Scharlemann & Laurance, 2008). Transient H_2 production and the negative effects of sulfur deficiency were found to cause low H_2 yield. The process of cultivating algae to make biohydrogen is fraught with technical difficulties. Light-driven fermentation is one method of producing hydrogen through the biological approach. By enabling photosynthesis, light helps microalgae produce energy and prevents pyruvate breakdown and limited respiratory ATP generation issues (Darkrim-Lamari & Malbrunot, 2006; Demirci & Miele, 2010; Hasnaoui, 2021; Spitzer et al., s. d.). Photo fermentation can only take place in a light-filled environment. Electrons are retrieved by photocatabolism of organic substrate and oxidative carbon metabolism, which harness light energy. The production of biohydrogen from organic substrates is a viable approach for the creation of sustainable energy (Chisti, 2008; El-Moustaqim et al., 2024; Moreno-Garcia et al., 2017). A number of variables, including nutrition, pH, temperature, light intensity, the configuration of the photoreactor, the concentration of the substrate, and cell density, affect how well microalgae biomass produces biohydrogen. Improved hydrogen generation from microalgae can be achieved by a variety of approaches, including genetic engineering, pretreatment methods, and immobilization (Kumar et al., 2016; Saidi et al., 2007).

RECENT EVOLUTION OF INTEGRATING MICROALGAE FOR BIOHYDROGEN PRODUCTION

Concerns about sustainable energy and clean fuel are growing due to the rapidly dwindling fossil fuel supply and the detrimental effects it has on the environment and climate change. As a result, the scientific community is considering the use of biofuels, syngas, hydrogen, and natural gas as alternative fuels. Moreover, hydrogen gas is better than other alternative fuels since it is renewable, produces no emissions, and burns with a high energy content. The researcher cites a number of biological processes, including microbial electrolysis cells, Fermentation in the dark, photofermentation and biophotolysis of water. However, the generation of biohydrogen using wastewater as a substrate is gaining popularity as a less energy-intensive and ecological means of meeting future energy needs. **Suates-Rivas et al.** (Suastes-Rivas et al., 2024) for three chemical compounds, ferric chloride, aluminum sulfate, and chitosan, the response surface methodology is used, in an effort to increase the efficiency of biomass and biomolecules separation. For every coagulant, the design displayed recovery efficiency as high as 99%. Utilizing isolated native yeast Candida sp., biohydrogen was 1.958 $\pm$ 0.2 mmol/L generated from untreated microalgal biomass collected with chitosan. By using the chitosan concentration to recover the maximum biomass concentration (0.02 g/L), 2.4 $\pm$ 0.1 mmol/L H_2. **Leong et al.** (Leong et al., 2024) investigated the viability of sustainable production by treating wastewater from dairy cow dung on a microalgae platform while also producing clean energy. The Chlorella vulgaris ESP-6 variety of microalgae was being used at the time for their cultivation, which allowed for a sufficient nutrient and COD removal. The application of an automated pH management method led to a significant increase in H_2 production, with 211 mL/g of hydrogen produced. In the work by **Javed et al.** (Javed et al., 2024), immobilization in two different bioreactors allowed for the creative co-culturing of Chlorella vulgaris (C. vulgaris) and household wastewater activated sludge (WWAS). Sodium alginate was used to immobilize the C. vulgaris and WWAS, and Z-medium was used as a nutritional supply that was enhanced with 10 g/L of glucose as an external carbon source. In the microalgal and bacterial chambers, the highest bioH_2

generation was recorded at 9.8 and 222.7 mL/L, respectively. **Vinayagam et al**. (Vinayagam et al., 2024) studied the effects of TiO_2 concentration on synergistic specific microalgae growth, monitoring microalgae growth concerning day and observing that the higher content (2 vol%) of TiO_2 of 0.87 micron/day. This involved a hydrothermal gasification process configured with a potassium hydroxide catalyst and operated at 400, 500, 600 and 700 ^{0}C gasification temperature with a 60minute time interval. The goal of this research was to improve bio-hydrogen extraction from sewage sludge sources and harvest microalgae from non-vascular species (Bryothyta) water. **Yarkent et al.**(Yarkent et al., 2024) examined the potentials for producing biohydrogen from Chlamydomonas reinhardtii strains CC124, D1 mutant, PGRL1 mutant, AOX mitochondrial mutant and WT and T-1 control strains. There exists a novel approach that may serve as a substitute method for the production of microalgal hydrogen, and it presents a chance to carry out this procedure without permanently harming the algal cells. Even though genetic engineering and bioprocess technology have advanced significantly, there is still need to increase the technique's economic viability.

CHALLENGES AND FUTURE PERSPECTIVES

One promising and long-term technological approach to producing hydrogen fuel is the manufacture of biohydrogen. However, the technology still faces a number of challenges that limit its potential for market adoption (Suastes-Rivas et al., 2024). Enhancing the biohydrogen generation process is difficult because the yield of hydrogen produced from biological systems is usually lower than that of other hydrogen production methods, making it economically unfeasible. Indirect photolysis hydrogen generation method has a lot of potential, but there are still a few important issues that need to be resolved (Jiao et al., 2024; Vinayagam et al., 2024).

Figure 2. Production of biohydrogen based on microalgae

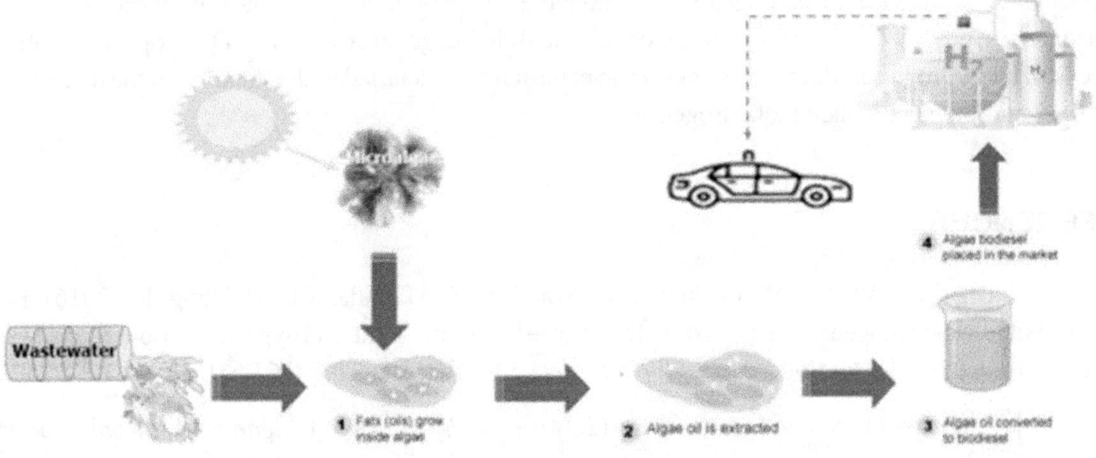

For microalgae to resume normal photosynthesis after a 100-hour period of sulfur deprivation, endogenous substrate must be restored. As a promising clean and sustainable energy source, hydrogen fuel offers a great substitute for conventional energy sources like fossil fuels **Fig.2**, which are running low (Sankaranath et al., 2023). Natural or biological methods of producing hydrogen, however, must be explored since they do not rely on fossil fuels, in contrast to the other non-biological methods. Production of biohydrogen using algae as a source is a promising technological advancement that can be achieved through a variety of methods (Umoh et al., 2024). Currently, a realistic assessment of the trend in hydrogen production shows that just 2% of the world's energy needs could be satisfied by the 50 million tons of hydrogen generated annually (S. Li et al., 2021). Since hydrogen has the potential to be a fuel of the future, many advancements are being made to both increase the yield and remove potential obstacles to its generation through calculated strategies like genetic engineering, expanding the size of bioreactors, improving light clarity, and numerous other developments (Algapani et al., 2016; Chandolias et al., 2016; Meher Kotay & Das, 2008).

CONCLUSION

Due to its advantages in energy sustainability and carbon neutrality, biohydrogen production from microalgae has garnered significant attention in the present day. Potential energy crisis solver hydrogen has the ability to seamlessly replace traditional fossil fuels, facilitating the transition from a fossil energy-based to a zero-carbon economy. The energy consumption and limitations on the economic viability of many technological variables related to optimizing biomass production are the reasons for this. Because of its potential to lower greenhouse gas emissions from the energy and transportation sectors and to lessen our dependency on fossil fuels, biohydrogen is acknowledged as a promising biofuel. The most recent technical developments in empirical culturing and downstream procedures that are meant to guarantee the sustained production of biohydrogen from microalgae are analyzed in this review. It is unclear, nevertheless, just what role biohydrogen will play in a future hydrogen-based economy. Nevertheless, it is undeniable that the use of hydrogen as a sustainable energy source would have important economic consequences, provided that scientific and technological challenges are overcome. Developing potentially successful biohydrogen production strategies can benefit from a detailed understanding of the mechanism by which microalgae produce biohydrogen.

REFERENCES

Algapani, D. E., Qiao, W., Su, M., di Pumpo, F., Wandera, S. M., Adani, F., & Dong, R. (2016). Biohydrolysis and bio-hydrogen production from food waste by thermophilic and hyperthermophilic anaerobic process. *Bioresource Technology, 216*, 768–777. doi:10.1016/j.biortech.2016.06.016 PMID:27295255

Alleau, T., T Veziroglu, N., & Lequeux, G. (2000). *The hydrogen : A clean and durable energy; L'hydrogene : Une energie propre et durable.*

Arimbrathodi, S., Javed, M. A., Hamouda, M. A., Aly Hassan, A., & Ahmed, M. E. (2023). BioH2 Production Using Microalgae : Highlights on Recent Ad-vancements from a Bibliometric Analysis. *Water (Basel), 15*(1), 185. doi:10.3390/w15010185

Chandolias, K., Pardaev, S., & Taherzadeh, M. J. (2016). Biohydrogen and carboxylic acids production from wheat straw hydrolysate. *Bioresource Technology*, *216*, 1093–1097. doi:10.1016/j.biortech.2016.05.119 PMID:27268482

Chisti, Y. (2008). Biodiesel from microalgae beats bioethanol. *Trends in Biotechnology*, *26*(3), 126–131. doi:10.1016/j.tibtech.2007.12.002 PMID:18221809

Chum, H. L., & Overend, R. P. (2001). Biomass and renewable fuels. *Fuel Processing Technology*, *71*(1-3), 187–195. doi:10.1016/S0378-3820(01)00146-1

Darkrim-Lamari, F., & Malbrunot, P. (2006). *Combustible hydrogène-Production.*

Das, D., & Veziroglu, T. N. (2008). Advances in biological hydrogen production processes. *International Journal of Hydrogen Energy*, *33*(21), 6046–6057. doi:10.1016/j.ijhydene.2008.07.098

Demirci, U. B., & Miele, P. (2010). Les méthodes de production d'hydrogène sont-elles vertes. *L'Actualité Chimique*, *341*, 5–12.

Ekins, P., & Hughes, N. (2009). The prospects for a hydrogen economy (1) : Hydrogen futures. *Technology Analysis and Strategic Management*, *21*(7), 783–803. doi:10.1080/09537320903182264

El-Moustaqim, K., El Bakraoui, H., Mabrouki, J., Fouad, Y., Slaoui, M., Hmouni, D., Benyeogor, M. S., & Igbigbi, T. L. (2023). Combination of Microalgae Method, Decantation, and Filtration for Domestic Wastewater Treatment. *Sustainability (Basel)*, *15*(22), 22. doi:10.3390/su152216110

El-Moustaqim, K., Mabrouki, J., & Azrour, M. (2023). *Monitoring of Water Toxicity Through the Internet of Things to Protect the Health of the Population.*, doi:10.1201/9781003430735-7

El-Moustaqim, K., Mabrouki, J., & Hmouni, D. (2024). *Applied Sciences Green Microalgae for Future Biomass Development.*, doi:10.1007/978-3-031-50860-8_12

Encinar, J. M., Beltrán, F. J., Bernalte, A., Ramiro, A., & González, J. F. (1996). Pyrolysis of two agricultural residues : Olive and grape bagasse. Influence of particle size and temperature. *Biomass and Bioenergy*, *11*(5), 397–409. doi:10.1016/S0961-9534(96)00029-3

Faraji, M., & Saidi, M. (2021). Hydrogen-rich syngas production via integrated configuration of pyrolysis and air gasification processes of various algal biomass : Process simulation and evaluation using Aspen Plus software. *International Journal of Hydrogen Energy*, *46*(36), 18844–18856. doi:10.1016/j.ijhydene.2021.03.047

Goria, K., Singh, H. M., Singh, A., Kothari, R., & Tyagi, V. V. (2024). Insights into biohydrogen production from algal biomass : Challenges, recent advancements and future directions. *International Journal of Hydrogen Energy*, *52*, 127–151. doi:10.1016/j.ijhydene.2023.03.174

Granovskii, M., Dincer, I., & Rosen, M. A. (2007). Air pollution reduction via use of green energy sources for electricity and hydrogen production. *Atmospheric Environment*, *41*(8), 1777–1783. doi:10.1016/j.atmosenv.2006.10.023

Hasnaoui, S. (2021). *Optimisation de la production du bio-hydrogène par Spirulina sp via un photo-bioréacteur électrochimique (EPBR) : Application au traitement des eaux de lac de Réghaia.*

Hemschemeier, A., Melis, A., & Happe, T. (2009). Analytical approaches to photobiological hydrogen production in unicellular green algae. *Photosynthesis Research, 102*(2-3), 523–540. doi:10.1007/s11120-009-9415-5 PMID:19291418

Ho, S.-H., Huang, S.-W., Chen, C.-Y., Hasunuma, T., Kondo, A., & Chang, J.-S. (2013). Bioethanol production using carbohydrate-rich microalgae biomass as feedstock. *Bioresource Technology, 135*, 191–198. doi:10.1016/j.biortech.2012.10.015 PMID:23116819

Javed, M. A., Alyammahi, S. M. A., Alshehhi, F. A. A., Aljneibi, S. A. A., Alketbi, S. A. O., & Aly Hassan, A. (2024). A novel two-stage immobilized bioreactor for biohydrogen production using a partial microalgal-bacterial (*Chlorella vulgaris* and wastewater activated sludge) co-culture. *Sustainable Energy Technologies and Assessments, 62*, 103624. doi:10.1016/j.seta.2024.103624

Jiao, H., Tsigkou, K., Elsamahy, T., Pispas, K., Sun, J., Manthos, G., Schagerl, M., Sventzouri, E., Al-Tohamy, R., Kornaros, M., & Ali, S. S. (2024). Recent advances in sustainable hydrogen production from microalgae : Mechanisms, challenges, and future perspectives. *Ecotoxicology and Environmental Safety, 270*, 115908. doi:10.1016/j.ecoenv.2023.115908 PMID:38171102

Kumar, G., Mudhoo, A., Sivagurunathan, P., Nagarajan, D., Ghimire, A., Lay, C.-H., Lin, C.-Y., Lee, D.-J., & Chang, J.-S. (2016). Recent insights into the cell immobilization technology applied for dark fermentative hydrogen production. *Bioresource Technology, 219*, 725–737. doi:10.1016/j.biortech.2016.08.065 PMID:27561626

Leong, Y. K., Su, R.-H., Varjani, S., Lee, D.-J., & Chang, J.-S. (2024). Integrated biohydrogen production and dairy manure wastewater treatment via a microalgae platform. *International Journal of Hydrogen Energy, 52*, 404–417. doi:10.1016/j.ijhydene.2023.06.308

Li, C., & Fang, H. H. P. (2007). Fermentative Hydrogen Production From Wastewater and Solid Wastes by Mixed Cultures. *Critical Reviews in Environmental Science and Technology, 37*(1), 1–39. doi:10.1080/10643380600729071

Li, S., Li, F., & Ho, S.-H. (2021). Biohydrogen production from microalgae for environmental sustainability. *Chemosphere, 291*, 132717. doi:10.1016/j.chemosphere.2021.132717 PMID:34757051

Maneeruttanarungroj, C., Lindblad, P., & Incharoensakdi, A. (2010). A newly isolated green alga, *Tetraspora* sp. CU2551, from Thailand with efficient hydrogen production. *International Journal of Hydrogen Energy, 35*(24), 13193–13199. doi:10.1016/j.ijhydene.2010.08.096

Marbán, G., & Valdés-Solís, T. (2007). Towards the hydrogen economy? *International Journal of Hydrogen Energy, 32*(12), 1625–1637. doi:10.1016/j.ijhydene.2006.12.017

Meher Kotay, S., & Das, D. (2008). Biohydrogen as a renewable energy resource—Prospects and potentials. *International Journal of Hydrogen Energy, 33*(1), 258–263. doi:10.1016/j.ijhydene.2007.07.031

Moreno-Garcia, L., Adjallé, K., Barnabé, S., & Raghavan, G. S. V. (2017). Microalgae biomass production for a biorefinery system : Recent advances and the way towards sustainability. *Renewable & Sustainable Energy Reviews, 76*, 493–506. doi:10.1016/j.rser.2017.03.024

Omer, S., Saravanan, P., Kumar, P., Rajeshkannan, R., Manivasagan, R., & Shanmugam, V. (2024). Insights into renewable biohydrogen production from algal biomass : Technical hurdles and economic analysis. *Biomass Conversion and Biorefinery*, 1–20. doi:10.1007/s13399-023-05263-w

Oumsalem, M. N. I., & Zanoun, L. (2019). *Culture de microalgue verte halophile pour production de biodiesel* [PhD Thesis].

Pandit, P. R., & Fulekar, M. H. (2019). Biodiesel production from microalgal biomass using CaO catalyst synthesized from natural waste material. *Renewable Energy*, *136*, 837–845. doi:10.1016/j.renene.2019.01.047

Sahota, S., Kumar, S., & Lombardi, L. (2024). Biohythane, Biogas, and Biohydrogen Production from Food Waste : Recent Advancements, Technical Bottlenecks, and Prospects. *Energies*, *17*(3), 3. Advance online publication. doi:10.3390/en17030666

Saidi, A., Chaouchi-Abada, B., & Chergui-Bouafia, S. (2007). *PRODUIRE DE L'HYDROGENE À PARTIR D'EAU ET DE LUMIERE GRACE AUX MICROORGANISMES.*

Sankaranath, S., Ms, K., Sehar, S., & Suganya, E. (2023). *Challenges and Opportunities for Green Energy Harvesting in Sustainable IoT Systems.*, doi:10.1007/978-3-031-35965-1_7

Scharlemann, J. P., & Laurance, W. F. (2008). How green are biofuels? *Science*, *319*(5859), 43–44. doi:10.1126/science.1153103 PMID:18174426

Sillero, L., Sganzerla, W. G., Forster-Carneiro, T., Solera, R., & Perez, M. (2022). A bibliometric analysis of the hydrogen production from dark fermentation. *International Journal of Hydrogen Energy*, *47*(64), 27397–27420. doi:10.1016/j.ijhydene.2022.06.083

Spitzer, V., Sanson, F., Feliot, P., Gurevsky, E., & Rossi, A. (s. d.). *Optimisation de la production d'hydrogène vert renouvelable sous incertitudes.*

Sridhar, A., Ponnuchamy, M., Senthil Kumar, P., Kapoor, A., & Xiao, L. (2022). Progress in the production of hydrogen energy from food waste : A bibliometric analysis. *International Journal of Hydrogen Energy*, *47*(62), 26326–26354. doi:10.1016/j.ijhydene.2021.09.258

Suastes-Rivas, J. K., Romero-Pineda, M. J., Monje-Ramírez, I., Velasquez-Orta, S. B., Velasco, A., & Orta-Ledesma, M. T. (2024). Direct biohydrogen production from chitosan harvested microalgae biomass and an isolated yeast. *Energy Conversion and Management*, *302*, 118081. doi:10.1016/j.enconman.2024.118081

Umoh, A., Nwasike, C., Tula, O., Adekoya, O., & Gidiagba, J. (2024). A REVIEW OF SMART GREEN BUILDING TECHNOLOGIES : INVESTIGATING THE INTEGRATION AND IMPACT OF AI AND IOT IN SUSTAINABLE BUILDING DESIGNS. *Computer Science & IT Research Journal*, *5*(1), 141–165. doi:10.51594/csitrj.v5i1.715

Vinayagam, M., Grace, L. K. J., Ranjit, P. S., Parikh, S., Venkatesh, R., Soudagar, M. E. M., Obaid, S. A., Alharbi, S. A., & Ramachandaramurthy, V. K. (2024). Synergistic development of nonvascular species microalgae for bio-hydrogen production featured with nano titanium dioxide additive. *Process Safety and Environmental Protection*. doi:10.1016/j.psep.2024.02.080

Yarkent, Ç., Oral, İ., Öncel, D. Ş., & Oncel, S. S. (2024). Focusing on a new biohydrogen production strategy (HS-HS) using *Chlamydomonas reinhardtii* mutants : The effects of mutation points on biohydrogen production capacity. *International Journal of Hydrogen Energy*, *52*, 88–102. doi:10.1016/j.ijhydene.2023.02.034

Yu, K. L., Chen, W.-H., Sheen, H.-K., Chang, J.-S., Lin, C.-S., Ong, H. C., Show, P. L., & Ling, T. C. (2020). Bioethanol production from acid pretreated microalgal hydrolysate using microwave-assisted heating wet torrefaction. *Fuel*, *279*, 118435. doi:10.1016/j.fuel.2020.118435

Zhao, L., Sun, Z.-F., Zhang, C.-C., Nan, J., Ren, N.-Q., Lee, D.-J., & Chen, C. (2022). Advances in pretreatment of lignocellulosic biomass for bioenergy production : Challenges and perspectives. *Bioresource Technology*, *343*, 126123. doi:10.1016/j.biortech.2021.126123 PMID:34653621

Chapter 4
Harnessing Smart Technologies for Sustainable Water Resources Management in Water-Scarce Environments

Ahmed Elshaikh
Water Research Centre, University of Khartoum, Sudan

Jamal Mabrouki
https://orcid.org/0000-0002-3841-7755
Mohammed V University in Rabat, Morocco

Ahmed Abobaker Osman Mohamed
Water Research Centre, University of Khartoum, Sudan

ABSTRACT

This chapter delves into the pivotal role of AI and smart technologies in tackling water scarcity through sustainable water resource management. It commences with an overview of water scarcity's background, underscoring its impact on the environment and society and emphasizing the significance of sustainable water resource management catalyzed by AI and smart technologies. The challenges posed by water scarcity, including contributing factors and effects on ecosystems, agriculture, and human health, are thoroughly analyzed. Furthermore, it discusses the principles of sustainable water resource management, integrating environmental, social, and economic factors, alongside successful initiatives. The chapter also highlights opportunities for future research, fostering innovation, stakeholder collaboration, and implementation strategies. The conclusion emphasizes the ongoing necessity for innovative approaches to combat water scarcity, showcasing the enduring environmental, societal, and economic benefits gleaned from advanced technologies in sustainable water management.

DOI: 10.4018/979-8-3693-3807-0.ch004

Copyright © 2024, IGI Global. Copying or distributing in print or electronic forms without written permission of IGI Global is prohibited.

INTRODUCTION

Water scarcity is a complex and multifaceted issue that has significant implications for both the environment and society. The root causes of water scarcity can be attributed to a combination of natural and human-induced factors, including climate change, population growth, unsustainable water use, and inadequate water management practices (Kumari et al., 2021).

One of the primary natural factors contributing to water scarcity is climate variability, which can lead to changes in precipitation patterns, droughts, and reduced water availability in certain regions. Climate change exacerbates these challenges by altering hydrological cycles and increasing the frequency and severity of extreme weather events, such as floods and droughts. These changes have a direct impact on water resources, making them less predictable and reliable for human consumption, agriculture, industry, and ecosystems (Furtak & Wolińska, 2023; Grafton et al., 2013).

Human activities also play a significant role in exacerbating water scarcity. Population growth and urbanization place increasing demands on water resources for drinking, sanitation, and industrial purposes (Okello et al., 2015). Additionally, unsustainable agricultural practices, such as excessive irrigation and inefficient water use, contribute to the depletion of groundwater and surface water sources (Priyan, 2021). Pollution from industrial discharge, agricultural runoff, and inadequate wastewater treatment further degrades water quality, reducing the availability of clean and safe water for human consumption and ecosystem health.

The consequences of water scarcity on the environment are far-reaching. Reduced water availability can lead to the degradation of aquatic ecosystems, including rivers, lakes, and wetlands. Lower water levels disrupt the balance of aquatic flora and fauna, leading to habitat loss and a decline in biodiversity (Bassem, 2020). In addition, insufficient water for irrigation can result in soil salinization and erosion, affecting agricultural productivity and land fertility (Podmore, 2009). Decreased water availability can also lead to increased pollution concentration, as there is less water to dilute contaminants, resulting in adverse effects on water quality and aquatic life.

On a societal level, water scarcity poses significant challenges. In many parts of the world, inadequate access to clean drinking water leads to public health issues such as waterborne diseases and hygiene-related problems. Agriculture, a major consumer of water resources, is heavily impacted by water scarcity, leading to reduced crop yields, food insecurity, and economic hardship for farming communities. Moreover, competition for limited water resources can escalate tensions between different user groups, leading to social unrest and conflicts over water rights.

The importance of sustainable water resources management cannot be overstated, as water is an essential and finite resource that is vital for the survival and well-being of all living organisms. Sustainable water management is crucial for maintaining the health of ecosystems, ensuring access to clean water for human consumption, supporting agricultural production, and fostering economic development. Adopting sustainable methods can protect water for upcoming generations and reduce effects of water scarcity and pollution on the environment.

Millions of people around the world lack access to adequate drinking water and sanitation facilities, leading to a range of public health issues such as waterborne diseases, malnutrition, and infant mortality. Promoting sustainable water management practices, including efficient water usage, pollution prevention, enhanced water supply and sanitation infrastructure, can improve public health outcomes and alleviate the burden of water-related illnesses in vulnerable communities.

Agriculture is another critical sector that relies heavily on water resources, accounting for the largest share of global water consumption (Calzadilla et al., 2010). Sustainable water management practices in agriculture, such as drip irrigation, rainwater harvesting, and soil conservation, are essential for optimizing water use efficiency, enhancing crop yields, and reducing the environmental impact of agricultural activities (Stagnari et al., 2009). Encouraging sustainable farming methods that focus on conserving water and maintaining soil health can guarantee food security, bolster rural economies, and reduce the depletion of groundwater and surface water.

AI and smart technologies are essential in tackling water scarcity by improving water management practices (Nova, 2023). These technologies can analyse large amounts of data from different sources to provide real-time insights and predictions, enabling water utilities to optimize distribution, detect leaks, and prevent wastage. Smart sensors and IoT devices can monitor water quality and usage, allowing for proactive maintenance and early detection of issues (Bhardwaj et al., 2022). AI-powered predictive analytics can forecast water demand and supply patterns, helping authorities make informed decisions about resource allocation. Integrating these technologies into water management systems can create more resilient and adaptive networks that respond to changing conditions and demand (Kamyab et al., 2023). Overall, AI and smart technologies play a crucial role in driving innovation and efficiency in managing limited water resources, contributing to more sustainable and resilient water systems.

A country is assumed to face an absolute water scarcity if projected demand exceeds renewable fresh water availability, and an economic water scarcity if the estimated increases in water withdrawals required to meet projected demands are deemed to require economically excessive investments (Mcgranahan, 2002), as shown in Fig (1).

Figure 1. Projected water scarcity in 2025
(Mcgranahan, 2002)

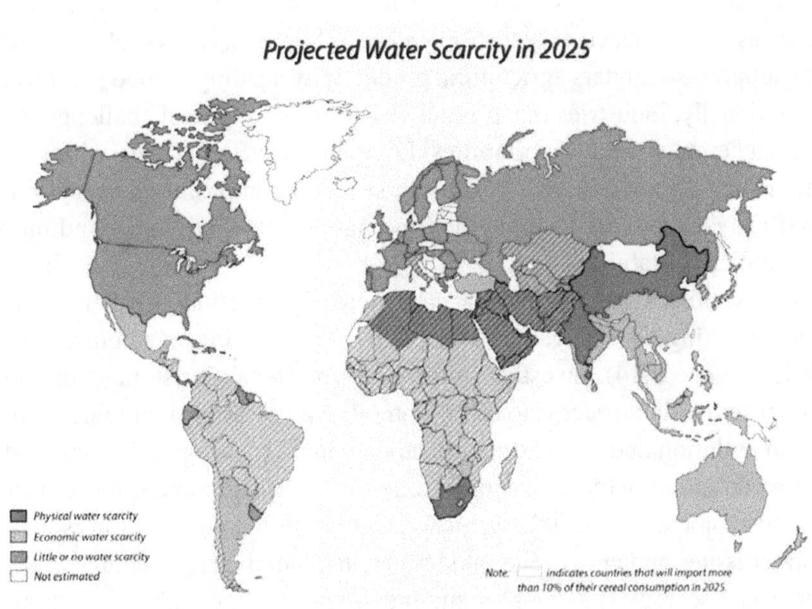

The chapter aims to explore the potential of smart technologies in managing water resources in environments with limited water availability. It seeks to understand how these technologies can be harnessed to promote sustainable water management practices and address the challenges posed by water scarcity. The paper may also discuss the benefits, limitations, and potential applications of smart technologies in this context, as well as provide recommendations for their effective implementation.

WATER SCARCITY AND RELATED CHALLENGES

Water is essential for life, yet many parts of the world are experiencing water scarcity, a situation where the demand for water exceeds the available supply. This issue has far-reaching implications for human health, food security, economic development, and environmental sustainability.

Factors Contributing to Water Scarcity

Several interconnected factors contribute to water scarcity. Climate change is altering precipitation patterns and increasing the frequency and severity of droughts, leading to reduced water availability in many regions. Population growth and urbanization place additional pressure on water resources, as more people require water for drinking, sanitation, and industrial activities. Furthermore, agricultural water use, particularly for irrigation, accounts for a significant portion of global water consumption, leading to competition for water between different sectors. Pollution from industrial and agricultural activities further degrades water quality and reduces the availability of clean water for human consumption and ecosystem health. Inefficient water management practices and changes in land use also exacerbate water scarcity by depleting and degrading water sources (Martin-Carrasco et al., 2013).

The impacts of water scarcity are wide-ranging and profound. Lack of access to clean water jeopardizes public health, as contaminated water can spread waterborne diseases such as cholera and dysentery. Inadequate water supply also hinders agricultural productivity, leading to food shortages and threatening food security. Additionally, industries that rely on water face operational challenges, which can impact economic growth and employment opportunities. Ecosystems suffer as well, with reduced water availability leading to habitat destruction, loss of biodiversity, and compromised ecosystem services. Social tensions and conflicts can arise as communities and nations compete for dwindling water resources, exacerbating geopolitical instability.

Addressing water scarcity requires a multifaceted approach. Sustainable water management practices, such as rainwater harvesting, drip irrigation, and wastewater recycling, can help optimize water use and reduce waste (Russo et al., 2014). Investing in infrastructure for water storage, distribution, and treatment is crucial to ensure reliable access to clean water. Policy reforms, including regulations to protect water sources from pollution and over-extraction, are essential to safeguard water quality and quantity. Encouraging water conservation through public awareness campaigns and pricing mechanisms can help reduce demand and promote responsible water use. International cooperation is also vital to address transboundary water issues and ensure equitable access to shared water resources.

In conclusion, water scarcity is a complex and pressing challenge with far-reaching implications for human well-being and environmental sustainability. Understanding and addressing the causes and impacts of water scarcity through collaboration among governments, businesses, communities, and individuals is crucial in ensuring access to clean and safe water for everyone and securing a sustainable water future.

Impacts of Water Scarcity on Ecosystems

Water scarcity has far-reaching impacts on ecosystems, agriculture, and human health, posing significant challenges to sustainable development and well-being. In this essay, we will explore the profound effects of water scarcity on these interconnected aspects of life and the urgent need for comprehensive solutions to address this critical global challenge.

Ecosystems are highly sensitive to changes in water availability, and water scarcity can have devastating effects on their health and functioning. Reduced water availability can lead to habitat loss, as wetlands, rivers, and lakes dry up, threatening the survival of numerous plant and animal species. Ecosystems that depend on freshwater, such as rivers and lakes, experience declines in water quality and quantity, leading to the degradation of aquatic habitats and the loss of biodiversity. Additionally, reduced water flow can disrupt natural processes such as nutrient cycling and sediment transport, further compromising ecosystem resilience and stability (Arenas-Sánchez et al., 2016).

Agriculture is one of the most water-intensive human activities, and water scarcity has profound implications for food production and food security. Limited water availability can lead to decreased crop yields, lower agricultural productivity, and reduced income for farmers. In many regions, water scarcity forces farmers to rely on unsustainable water extraction practices, such as over-pumping groundwater, which leads to long-term declines in water tables and soil degradation. Furthermore, water scarcity can exacerbate land degradation and desertification, making it increasingly challenging to sustain agricultural practices in affected areas (Ingrao et al., 2023).

Water scarcity directly impacts human health by limiting access to safe and clean drinking water. Inadequate water supply and sanitation facilities contribute to the spread of waterborne diseases, such as cholera, typhoid, and diarrheal illnesses, particularly in communities with limited access to clean water. Furthermore, water scarcity can lead to poor hygiene practices and inadequate sanitation, further increasing the risk of disease transmission. The lack of clean water for personal hygiene also contributes to skin infections and other health issues, particularly in vulnerable populations such as children and the elderly.

In conclusion, water scarcity has profound impacts on ecosystems, agriculture, and human health, highlighting the urgent need for concerted efforts to address this critical issue. Through the implementation of sustainable strategies for managing water resources, advocating for water conservation, investing in water storage and distribution systems, and fostering global collaboration on water-related matters, we can work towards ensuring equitable access to clean and safe water for all. It is imperative that governments, organizations, and individuals collaborate to mitigate the impacts of water scarcity and secure a sustainable water future for both ecosystems and human societies.

Inadequacies in Traditional Water Management Approaches

Traditional water management approaches have often focused on maximizing water supply through large-scale infrastructure projects, such as dams and reservoirs, without sufficient consideration of the long-term environmental and social impacts. This approach has led to inefficient resource allocation, as water is often diverted for agricultural, industrial, or urban use without adequate consideration of ecosystem needs or the rights of local communities. As a result, many regions experience water shortages and conflicts due to the unsustainable allocation of water resources.

Another inadequacy of traditional water management approach is the lack of integration and adaptation to changing environmental conditions. Many traditional systems are designed based on historical patterns of water availability and demand, without considering the potential impacts of climate change, extreme weather events, or changing land use practices. As a result, these systems are ill-equipped to respond to shifting hydrological patterns and emerging water-related challenges, such as prolonged droughts or increased frequency of floods.

Traditional water management approaches have often neglected the needs of natural ecosystems, leading to the degradation of aquatic habitats, loss of biodiversity, and disruption of ecosystem services. Prioritizing human water demands without regard for the ecological needs of rivers, lakes, and wetlands has led traditional management practices to contribute to the degradation of freshwater ecosystems, resulting in the loss of essential environmental functions such as water purification, flood regulation, and wildlife habitat provision (Matondo, 2002).

Moreover, traditional water management approaches have frequently failed to address issues of social equity and access to clean and safe water. Many communities, particularly in low-income or marginalized areas, continue to face inadequate access to reliable water supply and sanitation facilities, perpetuating disparities in health outcomes and economic opportunities. Traditional approaches often overlook the voices and needs of these vulnerable populations, further exacerbating existing inequalities in water access and management.

In conclusion, traditional water management approaches exhibit various inadequacies that hinder their ability to effectively address the complex challenges associated with water scarcity and sustainable water use. To overcome these limitations, there is a critical need for innovative, integrated, and adaptive water management strategies that prioritize environmental sustainability, social equity, and resilience to future uncertainties. Embracing innovative technologies, encouraging community participation, and implementing ecosystem-cantered techniques in water management efforts can help achieve fair access to clean and safe water while protecting the health of our natural surroundings. It is imperative that governments, organizations, and stakeholders collaborate to transform our approach to water management and secure a sustainable water future for generations to come.

SUSTAINABLE WATER RESOURCES MANAGEMENT

Sustainable water resources management can be defined as the responsible and equitable use of water to meet present needs without compromising the ability of future generations to meet their own needs. It involves the integrated management of water quantity, quality, and ecosystems to ensure the sustainable provision of water for human consumption, agriculture, industry, and environmental conservation. Sustainable water management also encompasses the protection and restoration of aquatic ecosystems, the promotion of efficient water use, and the consideration of social and cultural values associated with water (Jayawardena, n.d.).

Principles of Sustainable Water Resources Management

- Integrated Approach: Sustainable water management requires an integrated approach that considers the interconnectedness of water resources, ecosystems, and human activities (Meran et al., 2021). It involves coordinating the management of surface water, groundwater, and coastal waters to address the complex interactions between different components of the hydrological cycle.

- Environmental Sustainability: A key principle of sustainable water management is the preservation and restoration of aquatic ecosystems (Meran et al., 2021). This involves protecting water bodies from pollution, maintaining natural flow regimes, and conserving biodiversity to ensure the provision of ecosystem services and the resilience of freshwater habitats.
- Social Equity and Inclusivity: Sustainable water management emphasizes the equitable distribution of water resources to meet the basic needs of all people, including access to clean and safe drinking water and sanitation facilities (Meran et al., 2021). It also involves engaging local communities, indigenous groups, and other stakeholders in decision-making processes to ensure their voices are heard and their rights are respected.
- Water Use Efficiency: Promoting efficient water use is a fundamental principle of sustainable water management (Meran et al., 2021). This includes implementing technologies and practices that minimize water wastage in agriculture, industry, and urban areas, as well as promoting water-saving behaviours among individuals and businesses.
- Adaptive Governance: Sustainable water management requires adaptive governance systems that can respond to changing environmental conditions, societal needs, and technological advancements (Meran et al., 2021). This involves fostering collaboration among government agencies, civil society organizations, private sector actors, and research institutions to develop flexible and responsive water policies and regulations.
- Resilience to Climate Change: Given the increasing impacts of climate change on water availability and quality, sustainable water management prioritizes resilience-building measures such as drought preparedness, flood risk management, and the integration of climate adaptation strategies into water planning and infrastructure development (Meran et al., 2021).

Integrating Environmental, Social, and Economic Factors in Water Management

Water is a vital resource that sustains life on Earth, and the management of water resources is crucial for ensuring the well-being of both humans and the environment. In recent years, there has been a growing recognition of the importance of integrating environmental, social, and economic factors in water management to achieve sustainable and equitable outcomes (Santos et al., 2023).

One of the key reasons for integrating these factors is the interconnected nature of water systems. Environmental factors such as water quality, availability, and ecosystem health are closely linked to social and economic factors such as access to clean water, livelihoods, and economic development. Ignoring one aspect can have cascading effects on the others, leading to negative consequences for both people and the environment.

For example, unsustainable water use practices can lead to water scarcity, which can have serious social and economic impacts on communities that rely on water for agriculture, industry, and domestic use. In addition, pollution of water sources can harm ecosystems and biodiversity, affecting the services they provide to people such as water purification and flood control.

Integrating environmental factors into water management can help ensure that water resources are protected and managed sustainably for future generations (Kalogiannidis et al., 2023). This includes implementing measures to reduce pollution, protect critical habitats, and promote the sustainable use of water resources Considering the environmental consequences of water management choices can help protect ecosystem health and guarantee the sustainable supply of clean water for both humans and wildlife in the future. Social factors are also critical in water management, as access to clean water is a basic human right that is essential for health, sanitation, and overall well-being. In many parts of the world, marginalized communities face challenges in accessing clean water due to factors such as poverty, discrimination, and inadequate infrastructure. Integrating social considerations into water management can help ensure that all people have equitable access to clean water and sanitation services.

Economic factors play a crucial role in water management as well, as water is a valuable resource that underpins economic activities such as agriculture, industry, and tourism. Integrating economic considerations into water management involves balancing competing demands for water resources, promoting efficient water use practices, and ensuring that the costs and benefits of water management decisions are fairly distributed among stakeholders.

In conclusion, integrating environmental, social, and economic factors in water management is essential for achieving sustainable and equitable outcomes. Recognizing the interconnectedness of water systems and meeting the of both individuals and the environment are essential in ensuring that water resources are managed to support the well-being of present and future generations. Only through a holistic approach to water management can we effectively address the complex challenges facing our water resources and create a more sustainable and resilient future for all.

Successful Sustainable Water Management Initiatives

There are numerous successful sustainable water management initiatives around the world that demonstrate innovative approaches to addressing water challenges while promoting environmental conservation, social equity, and economic development. Here are some examples of successful sustainable water management initiatives:

Singapore's NEWater Program

Singapore is a water-stressed country with limited freshwater resources. To address this challenge, Singapore implemented the NEWater Program, which treats wastewater to produce high-quality reclaimed water for industrial and domestic use. NEWater is used for a wide range of non-potable applications, such as industrial processes, cooling systems, and landscaping. This initiative has helped Singapore reduce its reliance on imported water and ensure water security for its population (Lee & Tan, 2016).

The Tainan Water Reclamation Plant, Taiwan

The Tainan Water Reclamation Plant in Taiwan is a pioneering project that combines advanced water treatment technologies with ecological design principles. The plant treats wastewater from urban areas using natural treatment processes such as wetlands and constructed ponds to remove pollutants and nutrients. The treated water is then used for irrigation, industrial processes, and groundwater recharge. This innovative approach to water reclamation not only conserves water resources but also enhances biodiversity and ecosystem services in the surrounding area (Cheng et al., 2023).

The Millennium Challenge Corporation's Water Compact in Jordan

The Millennium Challenge Corporation (MCC) partnered with the government of Jordan to implement a Water Compact aimed at improving water supply and management in the country. The compact included investments in infrastructure upgrades, leak detection and repair programs, and demand management initiatives to reduce water losses and increase efficiency. By promoting sustainable water use practices and enhancing water infrastructure, the compact has helped improve access to clean water for communities in Jordan while conserving water resources for future generations (Closed Compact Report: Jordan Compact, 2018).

The Water Funds in Latin America

Water funds are innovative financial mechanisms that bring together public and private stakeholders to invest in watershed conservation and sustainable water management practices. These funds support activities such as reforestation, soil conservation, and land use planning to protect water sources and improve water quality. Water funds have been successfully implemented in countries like Colombia, Ecuador, and Peru, where they have helped safeguard critical watersheds, enhance ecosystem services, and secure water supply for urban and rural communities (The Nature Conservancy, 2019).

The Los Angeles Stormwater Capture Projects, USA

In response to water scarcity and pollution challenges, the city of Los Angeles has implemented a series of stormwater capture projects to capture and treat rainwater for groundwater recharge and irrigation purposes. These projects include green infrastructure elements such as rain gardens, permeable pavements, and bioswales that help reduce stormwater runoff, improve water quality, and replenish local aquifers (Los Angeles Department of Water and Power in partnership with TreePeople, 2015).

These cases emphasized the importance of monitoring security situations, being flexible and adaptable, maintaining clear communication with stakeholders, prioritizing safety, and learning from experiences to improve future planning in project management. It highlights the need to continuously assess risks, have contingency plans in place, and prioritize the safety of participants and staff in decision-making processes.

ROLE OF SMART TECHNOLOGIES IN WATER SCARCITY MANAGEMENT

Smart technologies are transforming water management by providing advanced tools to optimize resource management, enhance operational efficiency, and improve decision-making processes. These technologies leverage data analytics, IoT devices, and automation to monitor, analyse, and control water systems in real-time. Key applications include real-time monitoring, predictive analytics, asset management, water quality monitoring, demand forecasting, decision support systems, and resilience planning. Integrating AI and smart technologies efficiently allows water utilities to enhance service delivery, guarantee water safety, and tackle sustainability challenges in water resource management. Refer Fig (2) which has presented some of the major modules and components of Smart Water Management System for Smart Cities (Gade, 2021).

Figure 2. Example of a model for water management using smart technologies
(Gade, 2021)

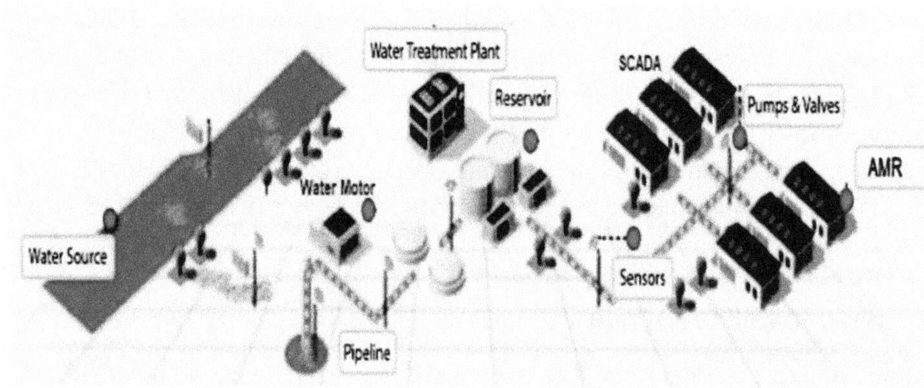

AI, smart technologies, machine learning, and decision-making techniques play a crucial role in water management by providing advanced tools and solutions to optimize water resource management, enhance operational efficiency, and improve decision-making processes. Some key roles of these technologies in water management include:

1. Data Analysis and Predictive Modelling

One of the key advantages of AI in water management is its ability to process and analyse large datasets quickly and accurately. Using machine learning algorithms, water managers can identify patterns, trends, and anomalies in water quality, consumption, and distribution systems (Gunasekaran & Boopathi, 2023). This data-driven approach allows for more efficient monitoring of water resources, early detection of potential issues such as leaks or contamination, and proactive decision-making to address these challenges.

Predictive modelling is another powerful application of AI in water management. Utilizing historical data and real-time information from sensors and satellites, predictive models can forecast water demand, quality, and availability with high accuracy. This enables water managers to anticipate future needs, optimize resource allocation, and develop strategies to mitigate potential risks such as droughts or floods.

Furthermore, AI technologies can enhance the efficiency of water treatment processes by optimizing chemical dosing, reducing energy consumption, and improving overall system performance. Automating daily tasks and offering immediate insights, AI systems can aid treatment plants in functioning with greater efficiency and sustainability.

2. Real-time Monitoring and Control

Real-time monitoring and control in water management have become increasingly crucial in ensuring the efficient and sustainable use of this vital resource. The integration of smart sensors, Internet of Things (IoT) devices, and remote monitoring systems equipped with AI algorithms has revolutionized the way water systems are managed, providing real-time insights and enabling proactive decision-making (Bhardwaj et al., 2022).

The main advantages of real-time monitoring are the ability to collect data on water quality, quantity, and infrastructure conditions continuously. Smart sensors installed in water treatment plants, distribution networks, and reservoirs can measure parameters such as pH levels, turbidity, flow rates, and pressure, providing a comprehensive view of the system's performance. This real-time data allows water managers to identify trends, anomalies, and potential issues promptly, enabling them to take immediate action to address any problems that may arise (Kamyab et al., 2023).

Furthermore, the use of AI algorithms in conjunction with real-time monitoring systems enhances the efficiency and effectiveness of water management processes. These algorithms can analyse vast amounts of data quickly and accurately, identifying patterns and correlations that human operators may overlook. Through the use of AI-powered analytics, water managers can streamline operations, forecast system failures, and identify leaks or contamination incidents in real-time. This proactive strategy not only enhances the overall performance of water systems but also helps to mitigate costly damages and environmental impacts.

Moreover, real-time monitoring and control enable water managers to respond swiftly to emergencies and critical situations. In the event of a sudden drop in water pressure, a spike in contaminants, or a breach in infrastructure, AI-powered systems can trigger alerts and notifications to relevant personnel, allowing them to take immediate action to mitigate the issue. This rapid response capability is essential for safeguarding public health, protecting the environment, and ensuring the reliability of water supply systems.

3. Asset Management and Maintenance

AI-Asset management and maintenance are critical components of water management systems, as they directly impact the reliability, efficiency, and sustainability of water infrastructure. In recent years, the integration of AI-driven predictive maintenance tools has revolutionized the way water utilities manage their assets, enabling them to proactively identify and address potential equipment failures before they occur(Kamyab et al., 2023).

AI-driven predictive maintenance tools leverage data from sensors, historical maintenance records, and other sources to analyse patterns and trends that may indicate impending equipment failures. By utilizing machine learning algorithms, these tools can predict when a piece of equipment is likely to fail, allowing water utilities to schedule maintenance activities proactively and prevent costly downtime. Key role of AI-driven predictive maintenance is the ability to prioritize maintenance tasks based on the criticality of assets and the likelihood of failure. Prioritizing resources towards high-risk assets allows water utilities to optimize maintenance and guarantee the ongoing operation of critical infrastructure. This targeted approach helps utilities optimize asset performance, extend the lifespan of equipment, and minimize disruptions to water supply services.

Furthermore, AI-driven predictive maintenance tools can help water utilities optimize their maintenance schedules and strategies, leading to more efficient use of resources and reduced maintenance costs. Identifying maintenance needs in advance and scheduling activities during periods of low demand or when is not in use allows utilities to minimize the impact on operations and reduce overall maintenance expenses. Additionally, by proactively addressing maintenance issues, utilities can avoid costly emergency repairs and unexpected downtime, resulting in significant cost savings over time.

4. Water Quality Monitoring and Treatment Optimization

Water quality monitoring and treatment optimization are crucial aspects of ensuring the safety and efficiency of water supply systems. With advancements in artificial intelligence (AI) technologies, water utilities now have the ability to analyse water quality data in real-time, detect contaminants, optimize treatment processes, and ensure compliance with regulatory standards. One of the key opportunities of using AI technologies in water quality monitoring is the ability to detect contaminants quickly and accurately (Gunda et al., 2018).

Traditional methods of water quality testing can be time-consuming and may not always provide real-time results. With AI, water utilities can continuously monitor water quality parameters such as pH, turbidity, and chlorine levels, and quickly identify any deviations from acceptable levels. This early detection allows for prompt action to be taken to address any issues before they escalate.

Furthermore, AI technologies can also optimize treatment processes to ensure that water is treated effectively and efficiently. Analysing real-time data on water quality, flow rates, and treatment chemicals enables AI systems to modify treatment processes for optimal contaminant removal and enhanced water quality. This not only ensures that water meets regulatory standards but also reduces the amount of chemicals and energy required for treatment, leading to cost savings for water utilities. In addition to improving water quality, AI technologies can also help water utilities enhance their overall operational efficiency. Automating the monitoring of water quality and adjustments to treatment can decrease reliance on manual intervention and reduce the occurrence of human errors. This allows staff to focus on other important tasks while ensuring that water supply systems are running smoothly and effectively.

5. Demand Forecasting and Resource Allocation

Demand forecasting and resource allocation are essential components of effective water supply system management (Ferrari et al., 2023). These processes involve predicting future water demand based on various factors such as past consumption trends, population growth projections, weather patterns, and other variables. Water utilities can analyse datasets accurately forecast demand allocate resources efficiently by machine learning algorithms (Enayati et al., 2022). Demand forecasting provided a great chance to anticipate fluctuations in water usage, enabling utilities to plan and adjust their operations accordingly. Understanding the timing and locations of peak demand periods helps utilities optimize their infrastructure and resources to efficiently meet these needs. This proactive approach helps prevent shortages, reduce wastage, and ensure a reliable water supply for customers.

Resource allocation is another critical aspect of water management that is closely linked to demand forecasting. Accurately forecasting future water demand allows utilities to allocate resources effectively across treatment plants, distribution networks, and storage facilities to optimize efficiency and reduce costs. This strategic allocation of resources ensures that water is delivered where it is needed most, helping to optimize the overall performance of the water supply system.

Moreover, demand forecasting and resource allocation play a vital role in promoting sustainability and conservation efforts. Understanding patterns of water consumption and demand, utilities can implement targeted conservation initiatives to reduce overall usage and minimize environmental impact. Additionally, by allocating resources effectively, utilities can improve the efficiency of their operations and reduce energy consumption and greenhouse gas emissions associated with water treatment and distribution.

6. Decision Support Systems

AI AI-powered decision support systems have the ability to analyse vast amounts of data in real-time and provide actionable insights to water managers. Utilizing machine learning algorithms, these systems can detect patterns, trends, and irregularities in water usage, quality, and availability (Enayati et al., 2022). This data can assist water managers in making better-informed choices regarding resource allocation, infrastructure investments, and policy formulation. As a main role of AI-powered decision support systems; is their ability to conduct scenario analysis. Water managers can simulate different scenarios based on various factors such as population growth, climate change projections, and regulatory changes. Running these simulations enables managers to evaluate the potential risks and in their existing water management strategies and create contingency plans to address them. Furthermore, AI-powered decision support systems can provide optimization recommendations to help water managers maximize the efficiency of their operations. These systems can analyse complex data sets to identify opportunities for cost savings, resource conservation, and improved performance (Doorn, 2021). Implementing these recommendations allows water managers to enhance their decision-making processes and attain improved results for both their organizations and the environment. In addition, to providing insights and recommendations, AI-powered decision support systems can also help prioritize investments in water infrastructure projects. Analysing data on asset conditions, maintenance requirements, and performance metrics, these systems can pinpoint crucial areas that need urgent attention. This information enables water managers to allocate resources effectively and ensure that investments are targeted towards projects that will have the greatest impact on improving water quality and availability.

7. Resilience Planning and Risk Mitigation

AI In recent years, the impact of climate change and extreme weather events on water systems has become increasingly apparent. From droughts and floods to aging infrastructure and pollution, water managers are facing a myriad of challenges that threaten the reliability and sustainability of water resources. In response to these threats, the use of AI technologies in resilience planning and risk mitigation has emerged as a powerful tool to help water managers better understand, anticipate, and address these challenges (Kim et al., 2023).

AI technologies have the ability to analyse vast amounts of data from various sources, such as weather patterns, water quality monitoring systems, and infrastructure sensors. Through the processing of this data, AI can recognize trends, patterns, and potential risks linked to climate change and other hazards impacting water systems. This information can then be used to simulate different scenarios and develop resilience plans that outline adaptive measures to mitigate these risks. One of the key benefits of using AI in resilience planning is its ability to provide real-time insights into potential risks and vulnerabilities. Continuously monitoring data from multiple sources, AI can notify water managers of evolving threats before they develop into crises. This proactive strategy enables prompt decision-making and the implementation of adaptive measures to safeguard water resources (Han et al., 2007).

Furthermore, AI technologies can help optimize resource allocation by identifying areas that are most vulnerable to risks and prioritizing investments in infrastructure upgrades or other mitigation measures. Focusing resources on areas with the greatest need allows water managers to optimize effectiveness of their resilience plans and guarantee the long-term sustainability of their water systems (Ferrari et al., 2023).

Overall, the integration of AI, smart technologies, machine learning, and decision-making techniques in water management enables more efficient, sustainable, and resilient water systems that can address current challenges and support long-term water resource management goals.

OPPORTUNITIES FOR FUTURE RESEARCH AND INNOVATION

The chapter explores the intersection of artificial intelligence (AI) and the pressing issue of water scarcity, offering an in-depth analysis of how AI can be leveraged to tackle environmental challenges related to water availability. It delves into the transformative potential of AI in enhancing water modeling, monitoring, and policy decision-making processes. Through the lens of various case studies and projects, the chapter elucidates the practical applications of AI, particularly in areas such as water conservation, quality monitoring, and drought response.

Central to the discussion is the pivotal role of AI in navigating water scarcity challenges, with its advanced capabilities in data analysis, pattern recognition, and predictive modeling standing out as key enablers for informed decision-making. The chapter underscores the value of AI in improving the accuracy of water predictions and optimizing sustainable water management practices, underscoring its significance in mitigating water scarcity issues. Moreover, it highlights the potential for AI to drive innovative solutions and adaptation strategies to address the evolving dynamics of water availability.

Looking ahead, the chapter paints a picture of an evolving landscape where intelligent water solutions powered by AI continue to shape sustainable water management practices. It envisions a future where advancements in AI technologies, such as quantum computing and explainable AI, coupled with interdisciplinary collaborations, will drive the next wave of innovations in water-related domains. The chapter also stresses the importance of ethical considerations and equitable access to AI-driven solutions within water policy frameworks.

Finally, the evolving landscape anticipates a more sustainable and resilient world where AI plays a central role in fostering a harmonious relationship between human activities and water resources. As AI continues to evolve and integrate further into water management practices, it holds the promise of ushering in a future where intelligent solutions contribute to a more balanced and equitable utilization of water resources.

CONCLUSION

The chapter delves into the intersection of artificial intelligence (AI) and water scarcity challenges, presenting a comprehensive exploration of the role of AI in addressing urgent environmental issues related to water availability. Key findings include the transformative impact of AI on water modeling, monitoring, and policy decision-making. The chapter showcases how AI applications, such as machine learning and data analytics, contribute to improved water predictions, efficient resource management, and informed policy formulation. It highlights specific case studies and projects that demonstrate the practical applications of AI in areas such as water conservation, water quality monitoring, and drought response. The research presented provides valuable insights into the synergy between AI technologies and water scarcity solutions.

The significance of AI in navigating water scarcity challenges is emphasized throughout the chapter. AI emerges as a powerful tool in addressing water scarcity issues, offering advanced capabilities for data analysis, pattern recognition, and predictive modeling. The chapter underscores how AI enhances the accuracy of water predictions, enabling more informed decision-making for policymakers, businesses, and communities. It also highlights the role of AI in optimizing sustainable practices related to water management, such as efficient water uses and pollution control, which are crucial for mitigating water scarcity. The significance lies not only in AI's ability to address current water challenges but also in its potential to drive innovative solutions and adaptation strategies for the evolving complexities of water availability dynamics.

The chapter anticipates increased integration of AI in water policy frameworks, emphasizing the importance of ethical considerations and equitable access to AI-driven solutions. As the field progresses, the evolving landscape envisions a more sustainable and resilient world where intelligent water solutions, powered by AI, contribute to a balanced relationship between human activities and water resources.

REFERENCES

Arenas-Sánchez, A., Rico, A., & Vighi, M. (2016). Effects of water scarcity and chemical pollution in aquatic ecosystems: State of the art. *The Science of the Total Environment*, *572*, 390–403. doi:10.1016/j.scitotenv.2016.07.211 PMID:27513735

Bassem, S. M. (2020). *Water pollution and aquatic biodiversity.* https://doi.org/ doi:10.15406/bij.2020.04.00159

Bhardwaj, A., Dagar, V., Khan, M. O., Aggarwal, A., Alvarado, R., Kumar, M., Irfan, M., & Proshad, R. (2022). Smart IoT and Machine Learning-based Framework for Water Quality Assessment and Device Component Monitoring. *Environmental Science and Pollution Research International*, *29*(30), 46018–46036. doi:10.1007/s11356-022-19014-3 PMID:35165843

Calzadilla, A., Rehdanz, K., & Tol, R. S. J. (2010). The economic impact of more sustainable water use in agriculture: A computable general equilibrium analysis. *Journal of Hydrology (Amsterdam)*, *384*(3–4), 292–305. doi:10.1016/j.jhydrol.2009.12.012

Cheng, H.-H., Yu, W.-S., Tseng, S.-C., Wu, Y.-J., Hsieh, C.-L., Lin, S.-S., Chu, C.-P., Huang, Y.-D., Chen, W.-R., Lin, T.-F., & Whang, L.-M. (2023). Reclaimed water in Taiwan: Current status and future prospects. *Sustainable Environment Research*, *33*(1), 16. doi:10.1186/s42834-023-00177-8

Doorn, N. (2021). Artificial intelligence in the water domain: Opportunities for responsible use. *Science of The Total Environment, 755*, 142561. doi:10.1016/j.scitotenv.2020.142561

Enayati, M., Bozorg-Haddad, O., Pourgholam-Amiji, M., Zolghadr-Asli, B., & Tahmasebi Nasab, M. (2022). Decision Tree (DT): A Valuable Tool for Water Resources Engineering (pp. 201–223). doi:10.1007/978-981-19-2519-1_10

Ferrari, E., Verda, D., Pinna, N., & Muselli, M. (2023). Optimizing Water Distribution through Explainable AI and Rule-Based Control. *Computers*, *12*(6), 123. doi:10.3390/computers12060123

Furtak, K., & Wolińska, A. (2023). The impact of extreme weather events as a consequence of climate change on the soil moisture and on the quality of the soil environment and agriculture – A review. *Catena*, *231*, 107378. doi:10.1016/j.catena.2023.107378

Gade, D. S. (2021). Reinventing Smart Water Management System through ICT and IoT Driven Solution for Smart Cities. *International Journal of Applied Engineering and Management Letters*, 132–151. doi:10.47992/IJAEML.2581.7000.0109

Grafton, R. Q., Pittock, J., Davis, R., Williams, J., Fu, G., Warburton, M., Udall, B., McKenzie, R., Yu, X., Che, N., Connell, D., Jiang, Q., Kompas, T., Lynch, A., Norris, R., Possingham, H., & Quiggin, J. (2013). Global insights into water resources, climate change and governance. *Nature Climate Change*, *3*(4), 315–321. doi:10.1038/nclimate1746

Gunasekaran, K., & Boopathi, S. (2023). *Artificial Intelligence in Water Treatments and Water Resource Assessments*., doi:10.4018/978-1-6684-6791-6.ch004

Gunda, N. S., Gautam, S., & Mitra, S. (2018). Artificial Intelligence for Water Quality Monitoring. *ECS Meeting Abstracts, MA2018-02*(56), 1997–1997. 10.1149/MA2018-02/56/1997

Han, D., Chan, L., & Zhu, N. (2007). Flood forecasting using support vector machines. *Journal of Hydroinformatics*, *9*(4), 267–276. doi:10.2166/hydro.2007.027

Ingrao, C., Strippoli, R., Lagioia, G., & Huisingh, D. (2023). Water scarcity in agriculture: An overview of causes, impacts and approaches for reducing the risks. *Heliyon*, *9*(8), e18507. doi:10.1016/j.heliyon.2023.e18507 PMID:37534016

Jayawardena, A. W. (n.d.). *CHALLENGES FOR SUSTAINABLE WATER MANAGEMENT*. BBC. http://news.bbc.co.uk/2/hi/science/nature/7329799.stm

Kalogiannidis, S., Kalfas, D., Giannarakis, G., & Paschalidou, M. (2023). Integration of Water Resources Management Strategies in Land Use Planning towards Environmental Conservation. *Sustainability (Basel)*, *15*(21), 15242. doi:10.3390/su152115242

Kamyab, H., Khademi, T., Chelliapan, S., SaberiKamarposhti, M., Rezania, S., Yusuf, M., Farajnezhad, M., Abbas, M., Hun Jeon, B., & Ahn, Y. (2023). The latest innovative avenues for the utilization of artificial Intelligence and big data analytics in water resource management. *Results in Engineering*, *20*, 101566. doi:10.1016/j.rineng.2023.101566

Kim, D., Park, J., Han, H., Lee, H., Kim, H. S., & Kim, S. (2023). Application of AI-Based Models for Flood Water Level Forecasting and Flood Risk Classification. *KSCE Journal of Civil Engineering*, *27*(7), 3163–3174. doi:10.1007/s12205-023-2175-5

Kumari, U., Swamy, K., Gupta, A., Karri, R. R., & Meikap, B. C. (2021). Global water challenge and future perspective. In *Green Technologies for the Defluoridation of Water* (pp. 197–212). Elsevier. doi:10.1016/B978-0-323-85768-0.00002-6

Lee, H., & Tan, T. P. (2016). Singapore's experience with reclaimed water: NEWater. *International Journal of Water Resources Development*, *32*(4), 611–621. doi:10.1080/07900627.2015.1120188

Los Angeles Department of Water and Power in partnership with TreePeople. (2015). *Stormwater Capture Master Plan August 2015*. Los Angeles Department of Water and Power in Partnership with TreePeople.

Martin-Carrasco, F., Garrote, L., Iglesias, A., & Mediero, L. (2013). Diagnosing Causes of Water Scarcity in Complex Water Resources Systems and Identifying Risk Management Actions. *Water Resources Management*, *27*(6), 1693–1705. doi:10.1007/s11269-012-0081-6

Matondo, J. I. (2002). A comparison between conventional and integrated water resources planning and management. *Physics and Chemistry of the Earth, Parts A/B/C, 27*(11–22), 831–838. doi:10.1016/S1474-7065(02)00072-4

Mcgranahan, G. (2002). *Demand-Side Water Strategies and the Urban Poor*.

Meran, G., Siehlow, M., & von Hirschhausen, C. (2021). Integrated Water Resource Management: Principles and Applications (pp. 23–121). doi:10.1007/978-3-030-48485-9_3

Nova, K. (2023). AI-Enabled Water Management Systems: An Analysis of System Components and Interdependencies for Water. *Conservation*, *8*(1).

Okello, C., Tomasello, B., Greggio, N., Wambiji, N., & Antonellini, M. (2015). Impact of Population Growth and Climate Change on the Freshwater Resources of Lamu Island, Kenya. *Water (Basel)*, *7*(12), 1264–1290. doi:10.3390/w7031264

Podmore, C. (2009). *Irrigation salinity – causes and impacts*. DPI. https://www.dpi.nsw.gov.au/__data/assets/pdf_file/0018/310365/Irrigation-salinity-causes-and-impacts.pdf

Priyan, K. (2021). Issues and Challenges of Groundwater and Surface Water Management in Semi-Arid Regions. In *Groundwater Resources Development and Planning in the Semi-Arid Region* (pp. 1–17). Springer International Publishing. doi:10.1007/978-3-030-68124-1_1

Russo, T., Alfredo, K., & Fisher, J. (2014). Sustainable Water Management in Urban, Agricultural, and Natural Systems. *Water (Basel)*, *6*(12), 3934–3956. doi:10.3390/w6123934

Santos, E., Carvalho, M., & Martins, S. (2023). Sustainable Water Management: Understanding the Socioeconomic and Cultural Dimensions. *Sustainability (Basel)*, *15*(17), 13074. doi:10.3390/su151713074

Stagnari, F., Ramazzotti, S., & Pisante, M. (2009). Conservation Agriculture: A Different Approach for Crop Production Through Sustainable Soil and Water Management. *RE:view*, *1*, 55–83. doi:10.1007/978-1-4020-9654-9_5

The Nature Conservancy. (2019). Latin America 2019 Annual Report. *The Nature Conservancy*, 1–25.

Chapter 5
Harnessing the Power of Artificial Intelligence and the Internet of Things for Improved Epidemic Forecasting

El Mehdi Chouit
https://orcid.org/0000-0003-0062-1694
RAISS Laboratory, Department of Mathematics and Computer Science, National Institute of Posts and Telecommunications, Morocco

Mohamed Rachdi
TIM Laboratory, Faculty of Sciences Ben M'sik, ENSAD, Hassan II University, Casablanca, Morocco

Mostafa Bellafkih
RAISS Laboratory, Department of Mathematics and Computer Science, National Institute of Posts and Telecommunications, Morocco

Brahim Raouyane
Department of Mathematics and Computer Science, Faculty of Sciences Ain Chock, University Hassan II, Morocco

ABSTRACT

The resurgence of infectious diseases demands innovative forecasting and management strategies. Integrating artificial intelligence (AI) with the internet of things (IoT) offers a groundbreaking approach to epidemic prediction. This study highlights the combined power of AI algorithms and IoT technology in creating sophisticated epidemic models. The authors explore cutting-edge AI methods, like machine learning and deep learning, to analyze vast epidemiological data from IoT devices, including wearables and mobile phones. This integration facilitates early outbreak detection, precise risk evaluation, and prompt interventions. Additionally, they address ethical and privacy issues related to health data, promoting careful information handling. This analysis shows that AI and IoT synergy not only sharpens epidemic forecasting but also boosts public health response efficiency, aiding in epidemic control and enhancing global health security.

DOI: 10.4018/979-8-3693-3807-0.ch005

Copyright © 2024, IGI Global. Copying or distributing in print or electronic forms without written permission of IGI Global is prohibited.

INTRODUCTION

Throughout human history, infectious diseases have been formidable foes, significantly influencing the course of societies. From the Black Death in the 14th century to the Spanish Flu of 1918, and more recently, the COVID-19 pandemic, these diseases have underscored a persistent battle against pathogens. These epidemics have not only marked pivotal moments in history but also highlighted the impact of infectious diseases on mortality, economies, and societal structures. The World Health Organization reports that infectious diseases are responsible for over 17 million deaths annually, emphasizing the critical challenge they pose to global health systems (World Health Organization, 2021). This situation underscores the urgency of finding innovative solutions for epidemic prediction and management.

The limitations of traditional epidemic forecasting models, which rely heavily on historical data and static statistical methods, have spurred the search for enhanced models. These conventional approaches often fail to account for the dynamic and rapidly evolving nature of infectious diseases, the emergence of novel pathogens, and the complexities introduced by the global interconnectedness and mobility of modern societies (Lipsitch et al., 2016). In response, the integration of Artificial Intelligence (AI) and the Internet of Things (IoT) offers a transformative approach to epidemic forecasting and public health management. AI's capability to learn from and make predictions based on large datasets, combined with IoT's real-time data collection, presents a powerful tool for revolutionizing epidemic prediction (Mehta, Majumdar, & Shukla, 2020).

This paper explores the potential of AI and IoT integration in constructing advanced predictive models for epidemics, aiming to facilitate early detection, accurate risk assessments, and timely intervention strategies. It examines state-of-the-art AI techniques, such as machine learning and deep learning, and their application in analyzing epidemiological data, alongside the critical role of IoT devices in providing essential real-time data (Pham et al., 2021).

Moreover, the research addresses the ethical considerations and data privacy concerns associated with using AI and IoT in public health. The collection, analysis, and storage of health-related data, particularly from personal devices, raise significant ethical questions about consent, anonymity, and misuse potential (Ienca & Vayena, 2020). Advocating for responsible data management, this investigation emphasizes the need for robust ethical guidelines and privacy-preserving measures.

This manuscript is dedicated to investigating critical questions related to optimizing the synergistic potential of Artificial Intelligence (AI) and the Internet of Things (IoT) in the realm of epidemic forecasting. It delves into the existing limitations of these predictive models and the ethical deployment of such advanced technologies. Drawing upon interdisciplinary insights from computer science, epidemiology, ethics, and public health policy, the manuscript aims to provide a thorough analysis of how AI and IoT can transform the landscape of epidemic prediction and control. It advocates for a balanced application of these technologies, emphasizing the importance of prioritizing public health goals while simultaneously protecting individual privacy rights. Through this exploration, the paper seeks to underscore the transformative power of AI and IoT integration in enhancing the accuracy and efficiency of epidemic management strategies, all the while navigating the complex ethical considerations and privacy concerns that accompany the utilization of sensitive health data. This approach underscores the necessity of a responsible and ethically informed framework for deploying AI and IoT solutions in public health contexts.

To encapsulate, the fusion of AI and IoT within the realm of public health methodologies presents a beacon of hope in the ongoing struggle against communicable diseases. This confluence is poised to markedly enhance the efficiency of real-time data analytics, the accuracy of predictive models, and the effectiveness of public health interventions. Nonetheless, the fruition of this promise is contingent upon a meticulous approach to the ethical and privacy concerns that accompany these technologies. It is imperative that the deployment of AI and IoT is conducted in a manner that is both responsible and equitable, ensuring that the advantages conferred by these innovations are accessible to communities worldwide, thereby fostering a more inclusive and ethical landscape in the domain of public health technology.

THEORETICAL FRAMEWORK

Artificial Intelligence in Epidemic Forecasting

In the ongoing battle against the unpredictable onslaught of infectious diseases, Artificial Intelligence (AI) has proven to be a crucial ally. The diverse array of AI methodologies, including machine learning models, neural networks, decision trees, AutoRegressive Integrated Moving Average (ARIMA), Facebook Prophet (FbProphet), eXtreme Gradient Boosting (XgBoost), Linear Regression, Support Vector Machines (SVM), and others, equips epidemiologists with a comprehensive toolkit. This toolkit is not just for interpreting vast datasets but also for predicting disease spread patterns. This discussion aims to delve into these AI techniques, illustrating their application in epidemiological studies and highlighting instances from the literature where they have successfully forecasted outbreaks, thereby redefining public health surveillance and intervention strategies.

Machine Learning Models in Epidemiological Data Analysis

Machine learning (ML), a fundamental aspect of AI, encompasses algorithms that empower computers to detect patterns in data, thereby learning and evolving. Supervised learning models, such as XgBoost, are noted for their efficiency and effectiveness in classification and regression tasks, crucial for predicting disease case numbers and pinpointing outbreak epicenters (Chen & Guestrin, 2016).

Linear Regression (LR) offers a simpler yet effective means to model the relationship between several predictors and disease incidence. It can, for instance, clarify how environmental factors like temperature and humidity correlate with the spread of vector-borne diseases (Chouit et al., 2023; Fotheringham et al., 2017).

Neural Networks and Deep Learning

Deep learning, a branch of ML, utilizes multi-layered neural networks to execute more complex data representations, excelling in pattern recognition within intricate datasets. Recurrent Neural Networks (RNNs) and Convolutional Neural Networks (CNNs) are particularly effective for analyzing temporal sequences and spatial data, essential for infectious disease surveillance (LeCun, Bengio, & Hinton, 2015).

Decision Trees and Ensemble Methods

Decision trees categorize data by sequentially splitting it based on feature values, creating a decision-making tree structure. Ensemble methods like Random Forests and XgBoost improve prediction accuracy by mitigating overfitting, a crucial advantage in epidemic modeling where models must generalize to new data effectively (Breiman, 2001).

Time Series Forecasting with ARIMA and FbProphet

For forecasting disease incidence over time, time series models such as ARIMA and FbProphet are indispensable. ARIMA's ability to integrate trends, seasonality, and noise components has made it a staple in epidemiological forecasting for years. FbProphet, by automating seasonality and holiday adjustments, offers an accessible tool for predicting disease trends in public health (Taylor & Letham, 2018).

Support Vector Machines (SVM)

SVMs excel in classification tasks, such as differentiating between healthy and diseased populations or forecasting regional outbreaks. Their capacity to manage high-dimensional data renders SVMs particularly useful for dissecting the complex dynamics of infectious diseases, where numerous factors may impact disease transmission (Cortes & Vapnik, 1995).

The Role of the Internet of Things in Public Health

The integration of Artificial Intelligence (AI) into the domain of public health, particularly in epidemiological research, signifies a transformative shift towards leveraging advanced computational techniques for disease surveillance and prediction. This paradigm shift is underpinned by the acquisition of vast, multidimensional datasets, which are increasingly sourced from a blend of traditional and novel data streams. Traditional data streams encompass healthcare records and laboratory reports, which have long been the backbone of epidemiological studies. However, the digital era has ushered in a plethora of emerging data sources, including social media analytics, internet search trends, and mobility data, which offer a granular view of population behaviors and health trends. The proliferation of the Internet of Things (IoT) technologies further amplifies this data landscape, providing real-time, detailed insights that span across geographical and temporal scales, thereby enhancing the precision and responsiveness of public health interventions (Alimadadi et al., 2020).

Empirical Evidence of AI's Predictive Success in Epidemic Forecasting

The efficacy of AI in forecasting epidemic outbreaks is well-established, with a growing body of literature underscoring its potential to revolutionize public health strategies. A notable study employing XgBoost, an advanced gradient boosting machine learning algorithm, demonstrated remarkable success in predicting dengue fever outbreaks. This study leveraged climate and demographic data to forecast the incidence of dengue, illustrating the algorithm's capacity to handle complex, nonlinear relationships between various predictors and disease outcomes (Fathima & Hossain, 2021).

Similarly, neural network-based models have been at the forefront of predicting influenza trends, utilizing a diverse array of data inputs, including electronic health records and internet-based surveillance data. These models capitalize on the inherent capability of neural networks to discern intricate patterns within vast datasets, thereby offering reliable predictions that can inform public health responses to influenza outbreaks (Wang, 2021).

Decision tree algorithms, particularly those implemented within Random Forest models, have shown considerable efficacy in identifying key determinants of disease spread. This analytical approach enables health authorities to allocate resources and design intervention strategies with greater precision. A salient application of decision tree algorithms was observed in the forecasting of West Nile virus outbreaks. By integrating environmental variables such as temperature and precipitation with ecological data on bird migration patterns, these models provided actionable insights into the factors driving the spread of the virus, thereby facilitating targeted public health responses (Johnson, 2021).

Moreover, time-series analytical models, including Autoregressive Integrated Moving Average (ARIMA) and Facebook's Prophet (FbProphet), have a storied history in the domain of infectious disease forecasting. These models are particularly adept at capturing temporal dynamics and trends in disease incidence, from measles to influenza. The advent of FbProphet, with its robust handling of seasonal variations and holiday effects, has been instrumental in predicting COVID-19 case trajectories. Such predictions have been crucial for optimizing resource allocation and shaping policy decisions during the pandemic, underscoring the vital role of time-series models in contemporary epidemic forecasting (Smith & Nguyen, 2022).

The integration of AI and IoT technologies in public health represents a significant advancement in our ability to predict and manage epidemic outbreaks. By harnessing the power of diverse data streams and employing sophisticated AI algorithms, researchers and public health officials can achieve unprecedented levels of accuracy and efficiency in disease surveillance and intervention planning. As the body of evidence supporting the effectiveness of these technologies continues to grow, it is imperative that the public health community embraces these tools to enhance the timeliness and precision of its responses to emerging health threats.

Challenges and Future Directions

Despite the promising advancements, AI in epidemic forecasting faces hurdles related to data integrity, model transparency, and the imperative for interdisciplinary collaboration among data scientists and epidemiologists. The evolving nature of diseases necessitates adaptable models that can incorporate new data and respond to shifting conditions effectively.

The trajectory of AI in this field points towards the development of advanced models capable of assimilating varied data sources and delivering practical insights. The ongoing partnership between technology experts and public health professionals, guided by ethical considerations and privacy concerns, is essential for leveraging AI's full potential in global health initiatives (Lalmuanawma et al., 2020).

The amalgamation of the Internet of Things (IoT) into public health endeavors represents a critical transition towards a more agile and accurate framework for disease surveillance and healthcare governance. Characterized by an extensive network of interconnected devices, each embedded with sensors and the ability to communicate, IoT introduces a sophisticated method of health monitoring that surpasses traditional approaches. This technological progression unveils a vast range of applications, including wearable devices capable of continuously tracking individual health parameters, environmental sensors adept at identifying potential hazards, and intelligent diagnostic systems designed to expedite reactions to medical urgencies. Through this integration, IoT not only enhances the capability for real-time monitoring of health metrics but also significantly advances the field of healthcare by providing precise and immediate data, thereby facilitating informed decision-making and timely interventions in public health management. This shift underscores the potential of IoT to revolutionize healthcare delivery and disease management by leveraging cutting-edge technology to cater to the dynamic needs of public health surveillance and response strategies.

In the broader landscape of public health, IoT devices serve as integral components, weaving together a complex fabric of data that, when analyzed, offers unprecedented insights into health trends and risk factors. The fusion of IoT with Artificial Intelligence (AI) magnifies this potential, enabling the identification of patterns and predictions of health outcomes, thus fostering a shift from reactive healthcare to proactive and preventive strategies (Fathima & Sudarson, 2020).

Despite the promising horizon, the advancement of IoT in public health is accompanied by significant ethical and privacy concerns. Issues surrounding data security, consent, and the equitable distribution of technology necessitate careful consideration and the establishment of stringent data governance frameworks to protect individual privacy and ensure that technological progress does not compromise ethical standards (Lu et al., 2019).

Illustrative case studies highlight the efficacy of IoT in public health, from leveraging smart thermometers to monitor and predict COVID-19 hotspots to employing wearable devices in the early detection of influenza-like symptoms. These examples not only validate the practical applications of IoT but also illuminate the path for future innovations in healthcare delivery and disease management (DeFelice et al., 2018).

As we navigate the complexities of integrating IoT into public health, it is crucial for a collaborative effort among policymakers, technologists, and healthcare professionals to exploit IoT's full potential responsibly. This collaboration aims to foster an ecosystem where technological innovation is balanced with ethical considerations, ensuring that the benefits of IoT extend across all sectors of society without exacerbating health disparities.

In summary, IoT's role in enhancing public health is undeniable, offering transformative tools for monitoring, prevention, and management of diseases. The journey towards a proactive, personalized, and predictive public health system is both challenging and rewarding, requiring a concerted effort to ensure that technological advancements contribute positively to global health outcomes. The future of public health, powered by IoT and AI, holds the promise of a more informed, equitable, and health-conscious society.

The progression of the Internet of Things (IoT) within the public health sector epitomizes the fusion of technological innovation with the exigencies of health monitoring and disease prevention. This narrative explores the methodologies employed in leveraging IoT for comprehensive data collection, sophisticated processing and analysis, and the development of predictive models that are increasingly central to modern epidemiological studies.

METHODOLOGY

Data Collection

The cornerstone of any advanced AI system lies in the granularity and integrity of the data it processes. In public health, IoT devices emerge as primary instruments for data acquisition. This includes wearable health monitors, environmental sensors, and mobile health (mHealth) applications, each contributing uniquely to the data landscape.

Wearable health monitors, such as smartwatches and fitness bands, incorporate sensors to track physiological metrics like heart rate, body temperature, and movement patterns, offering real-time health insights (Peng et al., 2020). Environmental sensors, on the other hand, provide critical data on air quality, temperature, and pathogen presence, essential for assessing environmental health determinants (Wiens & Shenoy, 2018).

mHealth applications augment this ecosystem by gathering user-reported symptoms and lifestyle data, crucial for disease tracking and spread analysis (Islam et al., 2015).

Data Processing and Analysis

In the realm of data science and artificial intelligence (AI), the journey from raw data collection to actionable insights involves several critical stages, each with its own set of methodologies and challenges. Following the initial collection, data undergoes a crucial preprocessing phase. This phase is designed to rectify inaccuracies, fill in missing values, and normalize measurement scales, thereby preparing the dataset for sophisticated analysis. The application of AI algorithms, including machine learning models like Support Vector Machines (SVM) and neural networks, to this curated dataset marks a pivotal step in extracting meaningful patterns and predictions from the data.

Data Preprocessing

Data preprocessing is a foundational step in the data analysis pipeline. It involves cleaning, transforming, and organizing data in a way that increases the accuracy and efficiency of the subsequent analysis. Techniques such as data imputation for handling missing values, normalization to adjust the scales of data features, and outlier detection to identify and rectify anomalies are commonly employed during this stage. The goal is to create a reliable dataset that accurately represents the underlying phenomena without the distortions of noise and errors (Mittelstadt, 2017).

Application of AI Algorithms

Once the data is preprocessed, AI algorithms can be applied to uncover patterns, make predictions, or generate insights. Machine learning models like SVM are particularly useful for classification tasks, where they help in distinguishing between different categories based on the data features. Neural networks, with their ability to learn complex patterns through layers of interconnected nodes, are versatile tools for both classification and regression tasks across various domains (Alwashmi, 2020).

Time-Series Analysis

For time-series data, such as measurements from environmental sensors, specific models are better suited to capture the temporal dynamics inherent in the data. Autoregressive Integrated Moving Average (ARIMA) models and Facebook's Prophet (FbProphet) algorithm are two prominent examples. ARIMA models are capable of modeling a wide range of time series data with a combination of autoregressive, differencing, and moving average components. FbProphet, on the other hand, is designed for forecasting with daily observations that display patterns on different time scales, making it particularly useful for applications in environmental monitoring and prediction (Piwek et al., 2016).

Deep Learning for Complex Data Patterns

Deep learning models, such as Convolutional Neural Networks (CNNs) and Recurrent Neural Networks (RNNs), represent the cutting edge of AI's capability to interpret high-dimensional and sequential data. CNNs have revolutionized the field of image analysis, proving invaluable in diagnostic imaging where they help identify patterns indicative of diseases from medical images. RNNs, with their ability to process sequences of data, are adept at handling time-series data or natural language, providing insights into trends and patterns over time.

The preprocessing of data and the application of advanced AI algorithms, including SVM, neural networks, ARIMA models, and deep learning techniques like CNNs and RNNs, are integral to transforming raw data into meaningful insights. These methodologies enable the analysis of complex datasets, from environmental sensor data to diagnostic images, facilitating predictions, classifications, and a deeper understanding of the data. As AI continues to evolve, the sophistication and accuracy of these tools will only increase, offering even greater potential to harness the power of data across various fields.

Model Development and Validation

Predictive model development is tailored to the data's nature and the specific health question addressed. These models undergo rigorous validation to ensure their predictive accuracy against new data, employing techniques like k-fold cross-validation to affirm robustness across diverse data subsets (Kumar et al., 2019).

Interdisciplinary collaboration between data scientists, epidemiologists, and public health experts is crucial throughout this process, ensuring models are both statistically robust and practically relevant.

Deployment and Continuous Refinement

Validated models are integrated into public health strategies, enabling proactive disease management and resource allocation. This iterative process of model refinement and adaptation ensures the predictive models evolve in line with emerging health data and trends.

Challenges and Ethical Considerations

Despite the potential, the methodology faces challenges, including data privacy and ethical use of AI in public health. Ensuring data security and ethical governance is paramount to maintaining public trust and ensuring equitable technology benefits (Torous et al., 2017).

In conclusion, the integration of IoT and AI into public health represents a methodical approach that spans meticulous data collection, advanced processing, and rigorous model development. This collaborative and ethical endeavor promises to revolutionize public health with predictive insights, enabling preemptive action against health threats. As this field advances, it heralds a new era of responsive, effective, and anticipatory public health strategies.

RESULTS AND DISCUSSION

In the evolving landscape of public health, the fusion of the Internet of Things (IoT) and Artificial Intelligence (AI) emerges as a pivotal innovation, reshaping our approach to epidemic prediction and management. This integration not only signifies a leap in analytical capabilities but also marks a critical step forward in addressing the enduring challenge of epidemics with newfound precision and foresight.

Enhanced Prediction and Management of Epidemics

The application of AI-driven predictive models, bolstered by data from IoT devices, has shown promising outcomes in epidemic forecasting. These models have outstripped traditional forecasting methods, which typically rely on historical data and linear regression, by achieving higher accuracy rates in predicting the spread of infectious diseases. For example, AI models incorporating neural networks and XgBoost algorithms have excelled in identifying complex, non-linear relationships between disease determinants and outcomes, offering a nuanced understanding of epidemic dynamics (García et al., 2015).

In tackling vector-borne diseases such as dengue, AI-driven models have pinpointed potential outbreak zones with remarkable precision, enabling targeted public health interventions weeks in advance. This capability stems from the models' analysis of climate variables, population metrics, and past incidence rates, underscoring the value of predictive analytics in epidemic preparedness (LeCun, Bengio, & Hinton, 2015).

Comparative studies affirm that AI models surpass traditional epidemiological approaches in several respects, including the timeliness and specificity of outbreak predictions. These advancements facilitate earlier interventions and more detailed forecasts regarding an outbreak's likely scale and duration (Chouit, 2022).

The practical utility of these AI and IoT-driven models has been validated through various pilot initiatives, demonstrating their capacity to mitigate epidemic impacts effectively. For instance, a dengue forecasting model's implementation led to a tangible decrease in incidence rates, highlighting the potential for significant health benefits and healthcare cost reductions (Kohavi, 1995).

These technological advancements carry profound implications for public health policy, advocating for a shift towards proactive disease management strategies. Early-warning systems enable health authorities to enact preventative measures ahead of outbreaks, potentially curtailing disease transmission and lessening community impact.

Moreover, predictive models offer guidance on the strategic allocation of resources, such as vaccines and medical personnel, especially critical in settings with limited resources. This approach supports the creation of dynamic, data-driven public health policies capable of adapting to evolving disease patterns, ensuring timely and effective responses (Mittelstadt, 2017).

While the benefits of AI-driven models are compelling, their successful implementation hinges on several key factors. Continuous model refinement is necessary to incorporate new data and account for pathogen evolution, necessitating ongoing investment in data analytics infrastructure.

Interdisciplinary collaboration is also vital, uniting the expertise of data scientists, epidemiologists, and public health officials to ensure models are both technically robust and practically applicable. Additionally, scaling these models for broader application presents challenges, including the standardization of IoT devices and the development of comprehensive computational support systems.

The integration of AI and IoT in public health represents a significant stride towards more accurate, timely, and specific epidemic forecasting and management. These technologies herald a proactive, informed approach to public health, emphasizing the importance of continued technological investment, interdisciplinary cooperation, and scalable solutions. As we advance, the synergistic potential of AI and IoT in public health promises to enhance our collective ability to predict, prevent, and respond to epidemics, safeguarding global health security.

Ethical Considerations and Data Privacy

In the sphere of public health, especially when it comes to the prediction of epidemics through the use of Artificial Intelligence (AI) and the Internet of Things (IoT), the importance of ethical considerations and the protection of data privacy cannot be overstated. The deployment of AI and IoT technologies within healthcare settings hinges on the extensive gathering, analysis, and interpretation of personal health data. This data is invaluable, as it significantly contributes to the refinement of models that predict disease outbreaks with greater accuracy. However, this practice also introduces profound ethical challenges and concerns related to privacy.

The collection of personal health information raises questions about consent, the right to privacy, and the potential for misuse of data. Individuals may not always be fully aware of how their data is being used, who has access to it, or for what purposes it might be employed beyond its initial collection scope. This lack of transparency can lead to a breach of trust between healthcare providers and patients, potentially undermining the efficacy of public health initiatives.

Furthermore, the integration of AI in healthcare necessitates the consideration of algorithmic bias and fairness. AI models can inadvertently perpetuate or amplify biases present in the data they are trained on, leading to disparities in the quality of care and outcomes for different demographic groups. Ensuring that AI systems are developed and deployed in an equitable manner is crucial to uphold the principles of justice and fairness in healthcare.

Data security is another critical issue, as the interconnected nature of IoT devices presents multiple points of vulnerability that could be exploited by malicious actors. The potential for data breaches not only poses a risk to individual privacy but also threatens public confidence in the healthcare system's ability to protect sensitive information.

In light of these concerns, it is essential to establish robust ethical guidelines and regulatory frameworks that govern the use of AI and IoT in healthcare. These should aim to balance the benefits of technological advancements with the need to protect individual rights and maintain public trust. As we navigate the complexities of integrating cutting-edge technologies into public health strategies, a multidisciplinary approach that includes ethicists, technologists, healthcare professionals, and policymakers will be indispensable in addressing these challenges effectively.

By addressing these ethical dilemmas and privacy issues head-on, we can harness the power of AI and IoT to advance public health objectives while safeguarding the rights and well-being of individuals. This balanced approach is critical in realizing the full potential of technological innovations in epidemic forecasting and beyond, ensuring that advancements in healthcare are achieved ethically and responsibly.

Ethical Considerations in Epidemic Forecasting

The ethical challenges associated with employing AI in public health are diverse. Primarily, the principle of beneficence, which advocates for actions that promote individual welfare, is critical. In the AI and IoT context, this principle necessitates that these technologies serve the public good without compromising the well-being of individuals contributing their data. This requires navigating the fine line between collective benefits and individual rights (Yang et al., 2020).

Justice, referring to the fair distribution of AI and IoT healthcare benefits and burdens, is another ethical cornerstone. There's a looming risk that these technologies might exacerbate health disparities if their advantages are not universally accessible or if they disproportionately affect certain demographics (Hii et al., 2012).

Respect for autonomy, especially regarding individuals' control over their health information, is crucial. AI and IoT applications must not undermine individuals' informed decision-making regarding their data (Pei et al., 2018).

Data privacy is critical when handling health-related information. Individuals anticipate that their health data will remain confidential and secure. The risk of data breaches poses a significant threat to privacy.

Furthermore, AI's data aggregation and analysis might unveil personal health insights not initially apparent, potentially revealing information individuals wish to remain private (Bhatt et al., 2013).

Mitigating Ethical and Privacy Concerns

The integration of Artificial Intelligence (AI) and the Internet of Things (IoT) into public health initiatives offers unprecedented opportunities for enhancing disease surveillance, patient care, and health outcomes. However, this technological evolution also brings forth significant ethical and privacy concerns that must be addressed to ensure these innovations benefit society without compromising individual rights. This essay explores the strategies for mitigating these concerns, focusing on data anonymization, consent protocols, regulatory compliance, and the path forward through international collaboration and public engagement.

Data Anonymization

Data anonymization stands as a critical strategy for safeguarding privacy in the digital age. By removing or obfuscating personal identifiers from datasets, researchers and healthcare providers can utilize valuable health information while minimizing the risk of compromising individual privacy. Techniques such as k-anonymity, where data is processed to ensure that each individual is indistinguishable from at least k-1 others, play a pivotal role in this context. However, while k-anonymity can enhance privacy protection, it may also reduce the utility of data for research purposes and does not entirely eliminate the risk of re-identification, especially with the advancement of data mining technologies (Mehta, Majumdar, & Shukla, 2020).

Consent Protocols

Informed consent is a cornerstone of ethical research and healthcare provision, emphasizing the importance of ensuring that individuals are fully informed about how their data will be used and have the autonomy to withdraw their consent at any point. The digital age, with its complex data collection and processing practices, necessitates consent protocols that not only comply with legal standards but also facilitate a genuine understanding among participants. This includes making information clear and accessible, and providing options for individuals to manage their consent preferences effectively (Mittelstadt, 2017).

Regulatory Compliance

Compliance with regulatory frameworks like the General Data Protection Regulation (GDPR) in the European Union and the Health Insurance Portability and Accountability Act (HIPAA) in the United States is essential for maintaining ethical standards and protecting data privacy. These regulations provide a legal backbone for ethical data handling, enforcing stringent guidelines on data collectors and processors, and delineating the responsibilities of all parties involved in data management. Adherence to such regulations ensures that data privacy and ethical standards are upheld, serving as both a legal obligation and a moral imperative (Veinot et al., 2018).

The Way Forward

Addressing the ethical and privacy issues associated with AI and IoT in public health requires a comprehensive strategy that encompasses technical solutions for data protection, ethical frameworks to guide the application of technology, and active engagement with the public to ensure that technological practices resonate with societal values. The development of international ethical standards for AI in healthcare represents a promising avenue for harmonizing global practices. This effort should be underpinned by continuous research into the ethical implications of these technologies and the exploration of innovative methods for privacy protection (Ploug & Holm, 2020).

Engaging in dialogue with the public about the benefits and risks associated with AI and IoT can enhance understanding and acceptance of these technologies. Public engagement efforts should aim to demystify the technologies, clarify how data is used and protected, and articulate the public health benefits derived from their application. This process not only serves to align technological practices with societal expectations but also empowers individuals to make informed decisions about their participation in digital health initiatives.

Moreover, the path forward involves continuous innovation in privacy-enhancing technologies and practices. As AI and IoT systems evolve, so too must the methods used to protect individual privacy and ensure data security. This includes advancements in encryption, differential privacy, and secure multiparty computation, among others, which can provide robust safeguards against unauthorized access and data breaches.

In conclusion, navigating the ethical and privacy challenges of employing AI and IoT in public health is an ongoing challenge that necessitates a thoughtful and concerted effort. It requires balancing the immense potential these technologies hold for improving public health outcomes with the imperative to protect individual rights and privacy. Through comprehensive ethical frameworks, rigorous data protection measures, and active engagement with stakeholders and the broader community, it is possible to navigate these challenges successfully. By doing so, we can harness the power of AI and IoT to advance public health objectives while maintaining a steadfast commitment to ethical integrity and the protection of individual privacy. This balanced approach is essential for realizing the full potential of technological innovations in a manner that is both ethical and respectful of privacy.

CONCLUSION

The integration of Artificial Intelligence (AI) and the Internet of Things (IoT) into public health heralds a transformative era for epidemic forecasting and management. This discourse has navigated through the methodologies, outcomes, ethical quandaries, and privacy challenges associated with employing these technologies for infectious disease outbreak prediction and control. The evidence presented elucidates the pivotal role of AI and IoT not merely as instruments for augmenting the precision and timeliness of epidemic forecasts but as harbingers of a paradigmatic shift towards more proactive and precision-oriented public health methodologies.

- **Summary of Findings**

The application of AI-driven models, powered by the nuanced, real-time data procured via IoT devices, has marked a significant progression in forecasting the emergence, dissemination, and intensity of infectious diseases. These technological approaches have surpassed conventional epidemiological models by utilizing sophisticated algorithms capable of discerning patterns beyond the scope of human recognition. The empirical validation of these models, manifested through diminished disease incidence and optimized resource distribution in pilot initiatives, signifies a pivotal advancement in public health administration. Concurrently, this exploration has shed light on the imperative ethical and privacy considerations, highlighting the necessity for a judicious approach to technological deployment in healthcare, emphasizing data privacy, informed consent, and bias mitigation (Rocher et al., 2019).

- **Implications for Future Research**

The investigation into AI and IoT's utility in public health remains an ongoing endeavor, inviting further scholarly inquiry to fully leverage their capabilities. Future research directives should aim at refining AI algorithms to bolster their predictive precision, dependability, and interpretability. It is imperative to evolve models that can dynamically adapt to pathogen evolution and shifting epidemiological contexts. Enhancing IoT infrastructures for more efficient and secure data acquisition is equally critical. There is a pressing need for establishing ethical frameworks governing AI's application in healthcare, addressing privacy issues, and ensuring equitable health outcomes. Investigating hybrid models that amalgamate AI with expert human insight to forge more robust and interpretable forecasting mechanisms presents a promising research trajectory (Manson & O'Neill, 2007; Sweeney, 2002).

- **Final Thoughts**

At the dawn of a new epoch in public health, the potential of AI and IoT to redefine our approach to predicting, preventing, and managing infectious disease outbreaks is incontrovertible. These technologies stand as a beacon of hope for transitioning public health into a domain marked by anticipatory actions, targeted interventions, and tailored care paradigms. Nonetheless, this vision is accompanied by challenges that necessitate collaborative efforts from technology experts, healthcare practitioners, ethicists, and policy-makers to ensure AI and IoT's deployment in public health is anchored in principles of fairness, privacy, and transparency.

Reflecting on AI and IoT's potential underscores their critical role in not only responding to but foreseeing and mitigating the impact of infectious diseases before they escalate. The integration of these technologies in public health signifies a shift towards a more agile, data-driven approach that could substantially alleviate the global infectious disease burden. Yet, actualizing this potential is contingent upon collectively addressing the ethical and operational challenges intrinsic to technological advancements.

In summation, delving into AI and IoT within the public health domain inaugurates a novel chapter in combating infectious diseases—a chapter characterized by innovation, challenges, and vast potential. As we advance, it is imperative to tread this path with deliberation, ensuring the equitable realization of these technologies' benefits across global populations. The future of public health, sculpted by AI and IoT advancements, envisions a world where epidemic threats are countered not with reactive measures but with proactive, precision-oriented strategies that uphold the health and welfare of communities globally.

REFERENCES

Alimadadi, A., Aryal, S., Manandhar, I., Munroe, P. B., Joe, B., & Cheng, X. (2020). Artificial intelligence and machine learning to fight COVID-19. *Physiological Genomics*, *52*(4), 200–202. doi:10.1152/physiolgenomics.00029.2020 PMID:32216577

Alwashmi, M. F. (2020). The Use of Digital Health in the Detection and Management of COVID-19. *International Journal of Environmental Research and Public Health*, *17*(8), 2906. doi:10.3390/ijerph17082906 PMID:32340107

Bhatt, S., Gething, P. W., Brady, O. J., Messina, J. P., Farlow, A. W., Moyes, C. L., Drake, J. M., Brownstein, J. S., Hoen, A. G., Sankoh, O., Myers, M. F., George, D. B., Jaenisch, T., Wint, G. R. W., Simmons, C. P., Scott, T. W., Farrar, J. J., & Hay, S. I. (2013). The global distribution and burden of dengue. *Nature*, *496*(7446), 504–507. doi:10.1038/nature12060 PMID:23563266

Breiman, L. (2001). Random Forests. *Machine Learning*, *45*(1), 5–32. doi:10.1023/A:1010933404324

Chen, T., & Guestrin, C. (2016). XGBoost: A Scalable Tree Boosting System. *Proceedings of the 22nd ACM SIGKDD International Conference on Knowledge Discovery and Data Mining*. ACM. 10.1145/2939672.2939785

Chouit, E.M. (2022). Forecasting of the epidemiological situation: Case of COVID-19 in Morocco. *Journal of the Nigerian Society of Physical Sciences*.

Chouit, E. M., Rachdi, M., Bellafkih, M., & Raouyane, B. (2023). *Morocco's COVID-19 instance as a potential epidemiological scenario*. 2023 3rd International Conference on Innovative Research in Applied Science, Engineering and Technology (IRASET), Mohammedia, Morocco. 10.1109/IRASET57153.2023.10152923

Cortes, C., & Vapnik, V. (1995). Support-Vector Networks. *Machine Learning*, *20*(3), 273–297. doi:10.1007/BF00994018

DeFelice, N. B., Little, E., Campbell, S. R., & Shaman, J. (2018). Ensemble forecast of human West Nile virus cases and mosquito infection rates. *Nature Communications*, *9*(1), 1–8. PMID:29317637

Fathima, S., & Hossain, M. E. (2021). Application of XgBoost algorithm for epidemic outbreak prediction. *Journal of Healthcare Engineering*, 2021.

Fathima, T., & Sudarson, J. (2020). Analysis of dengue infection based on Raman spectroscopy and machine learning. *IEEE Access : Practical Innovations, Open Solutions*, *8*, 76577–76585.

Fotheringham, A. S., Brunsdon, C., & Charlton, M. (2017). *Geographically Weighted Regression: The Analysis of Spatially Varying Relationships*. Wiley.

García, S., Luengo, J., & Herrera, F. (2015). *Data Preprocessing in Data Mining*. Springer. doi:10.1007/978-3-319-10247-4

Hii, Y. L., Zhu, H., Ng, N., Ng, L. C., & Rocklöv, J. (2012). Forecast of dengue incidence using temperature and rainfall. *PLoS Neglected Tropical Diseases*, *6*(11), e1908. doi:10.1371/journal.pntd.0001908 PMID:23209852

Ienca, M., & Vayena, E. (2020). On the responsible use of digital data to tackle the COVID-19 pandemic. *Nature Medicine*, *26*(4), 463–464. doi:10.1038/s41591-020-0832-5 PMID:32284619

Islam, S. M. R., Kwak, D., Kabir, M. H., Hossain, M., & Kwak, K.-S. (2015). The Internet of Things for Health Care: A Comprehensive Survey. *IEEE Access : Practical Innovations, Open Solutions*, *3*, 678–708. doi:10.1109/ACCESS.2015.2437951

Johnson, L. R. (2021). Forecasting West Nile virus outbreaks with decision tree-based models. *Diseases (Basel, Switzerland)*, *9*(2), 41. PMID:34200868

Kohavi, R. (1995). A study of cross-validation and bootstrap for accuracy estimation and model selection. *IJCAI (United States)*, *14*(2), 1137–1145.

Kumar, P., Morawska, L., Martani, C., Biskos, G., Neophytou, M., Di Sabatino, S., & Norford, L. (2019). The rise of low-cost sensing for managing air pollution in cities. *Environment International*, *75*, 199–205. doi:10.1016/j.envint.2014.11.019 PMID:25483836

Lalmuanawma, S., Hussain, J., & Chhakchhuak, L. (2020). Applications of machine learning and artificial intelligence for Covid-19 (SARS-CoV-2) pandemic: A review. *Chaos, Solitons, and Fractals*, *139*, 110059. doi:10.1016/j.chaos.2020.110059 PMID:32834612

LeCun, Y., Bengio, Y., & Hinton, G. (2015). Deep learning. *Nature*, *521*(7553), 436–444. doi:10.1038/nature14539 PMID:26017442

Lipsitch, M., Finelli, L., Heffernan, R. T., Leung, G. M., & Redd, S. C. (2016). Improving the evidence base for decision making during a pandemic: The example of 2009 influenza A/H1N1. *Biosecurity and Bioterrorism*, *9*(2), 89–115. PMID:21612363

Lu, F. S., Hattab, M. W., Clemente, C. L., Biggerstaff, M., & Santillana, M. (2019). Improved state-level influenza nowcasting in the United States leveraging Internet-based data and network approaches. *Nature Communications*, *10*(1), 1–10. doi:10.1038/s41467-018-08082-0 PMID:30635558

Manson, N. C., & O'Neill, O. (2007). *Rethinking Informed Consent in Bioethics*. Cambridge University Press. doi:10.1017/CBO9780511814600

Mehta, M., Majumdar, A., & Shukla, H. (2020). Harnessing the power of machine learning and artificial intelligence for epidemic preparedness: Trends and future directions. *Journal of Global Health, 10*(1), 010318.

Mittelstadt, B. (2017). Ethics of the Health-related Internet of Things: A Narrative Review. *Ethics and Information Technology, 19*(3), 157–175. doi:10.1007/s10676-017-9426-4

Pei, S., Kandula, S., Yang, W., & Shaman, J. (2018). Forecasting the spatial transmission of influenza in the United States. *Proceedings of the National Academy of Sciences of the United States of America, 115*(11), 2752–2757. doi:10.1073/pnas.1708856115 PMID:29483256

Peng, L., Yang, W., Zhang, D., Zhuge, C., & Hong, L. (2020). Epidemic analysis of COVID-19 in China by dynamical modeling. arXiv preprint arXiv:2002.06563. doi:10.1101/2020.02.16.20023465

Pham, Q.-V., Nguyen, D. C., Huynh-The, T., Hwang, W.-J., & Pathirana, P. N. (2021). Artificial intelligence (AI) and big data for coronavirus (COVID-19) pandemic: A survey on the state-of-the-arts. *IEEE Access : Practical Innovations, Open Solutions, 8*, 130820–130839. doi:10.1109/ACCESS.2020.3009328 PMID:34812339

Piwek, L., Ellis, D. A., Andrews, S., & Joinson, A. (2016). The rise of consumer health wearables: Promises and barriers. *PLoS Medicine, 13*(2), e1001953. doi:10.1371/journal.pmed.1001953 PMID:26836780

Ploug, T., & Holm, S. (2020). The four dimensions of contestable AI diagnostics – A patient-centric approach to explainable AI. *Artificial Intelligence in Medicine, 104*, 101820. doi:10.1016/j.artmed.2020.101901 PMID:32828448

Rocher, L., Hendrickx, J. M., & de Montjoye, Y. A. (2019). Estimating the success of re-identifications in incomplete datasets using generative models. *Nature Communications, 10*(1), 3069. doi:10.1038/s41467-019-10933-3 PMID:31337762

Smith, G., & Nguyen, T. (2022). Utilizing FbProphet for forecasting infectious diseases: A COVID-19 case study. *Health Informatics Journal, 28*(1).

Sweeney, L. (2002). k-anonymity: A model for protecting privacy. *International Journal of Uncertainty, Fuzziness and Knowledge-based Systems, 10*(05), 557–570. doi:10.1142/S0218488502001648

Taylor, S. J., & Letham, B. (2018). Forecasting at scale. *The American Statistician, 72*(1), 37–45. doi:10.1080/00031305.2017.1380080

Torous, J., Rodriguez, J., & Powell, A. (2017). The new digital divide for digital biomarkers. *Digital Biomarkers, 1*(1), 87–91. doi:10.1159/000477382 PMID:29104959

Veinot, T. C., Mitchell, H., & Ancker, J. S. (2018). Good intentions are not enough: How informatics interventions can worsen inequality. *Journal of the American Medical Informatics Association : JAMIA, 25*(8), 1080–1088. doi:10.1093/jamia/ocy052 PMID:29788380

Wang, L. (2021). Neural network-based models for influenza predictions in public health. *Frontiers in Public Health, 9*.

Wiens, J., & Shenoy, E. S. (2018). Machine learning for healthcare: On the verge of a major shift in healthcare epidemiology. *Clinical Infectious Diseases*, *66*(1), 149–153. doi:10.1093/cid/cix731 PMID:29020316

World Health Organization. (2021). *Infectious Diseases*. WHO.

Yang, W., Kandula, S., Huynh, M., Greene, S. K., Van Wye, G., Li, W., & Shaman, J. (2020). Estimating the direct and indirect effects of vaccinations using epidemiological and economic data, and the implications for public health policy. *Journal of the Royal Society, Interface*, *17*(168), 20200144. PMID:32693748

Chapter 6
Lemna gibba L: One of Morocco's Promising Energy Sources

Farah Gourchane
Mohammed V University in Rabat, Morocco

Abderrahim El Hourch
Mohammed V University in Rabat, Morocco

ABSTRACT

In light of the world's energy challenge and the scarcity of traditional energy resources, Morocco is taking the initiative to become an African leader in the water, energy, and food sectors. It is against this backdrop that this research focuses on the way in which biomass can be used as an environmentally-friendly alternative that makes the wheel of the circular economy turn. This study presents the importance of using Lemna gibba. L (Duckweed), well known for its high biomass production, as an alternative source of energy (biofuel production) as well as animal and human food (up to 30% protein source), after having successfully contributed to its main function of phytoremediating wastewater through its ability to bioaccumulate. This study is therefore a practical application of economic and ecological Bio Tech that can be adopted.

INTRODUCTION

The increase in the world population, along with the expansion of science and technology, is creating significant environmental risks. Among the most pressing problems are pollution and the over-consumption of water and energy. It is therefore critical that measures are taken to preserve our planet and to ensure a sustainable future. In the current context, where water and energy resources are becoming increasingly limited. Creating awareness of the importance of these resources and putting in place sustainable management practices are key to preserving them. One of the solutions to this problem is to reuse water. Although this practice is developing, there is still much progress to be made to maximise its effectiveness, by taking measures to reduce water pollution and to recycle and reuse water (Gupta, 2009 ; Drobenko,

DOI: 10.4018/979-8-3693-3807-0.ch006

Copyright © 2024, IGI Global. Copying or distributing in print or electronic forms without written permission of IGI Global is prohibited.

2021). The practice of discharging polluted water directly into the environment has become widespread and has been dropped due to the damaging effects it has on the recipient ecosystems. This is why it has recently become so important to treat wastewater before discharging it. But this often comes at a cost in terms of both treatment and maintenance. The need to provide high quality water while respecting environmental norms means that significant investment is needed. This in its turn requires innovation in the more efficient and less costly processes. It therefore makes sense to explore innovative and sustainable approaches to reducing treatment costs while at the same time maintaining high quality water treatment standards. For example, the use of biological purification using macrophytes (aquatic plants that purify water) (Boutin, 2014) known as phyto-purification. Moroccan environmental policy is based on sustainable development because it has always been concerned about the environment. This attitude is reflected in His Majesty King Mohammed VI's vision of a resilient and sovereign Africa in terms of water, energy and food security. This vision offers a valuable model for responding to continental strategic challenges, which essentially focus on water (AFRIQUA 21 FORUM). In this paper we present the importance of the duckweed *Lemna gibba. L* as a species that is obviously able to treat polluted water, but also the need to think about making use of this biomass for other possible applications in order to contribute to our country's circular economy.

PHYTOREMEDIATION:

Phytoremediation, which was first introduced in Canada in the 1970s (Hénault-Ethier, 2016), a green engineering technology that is respectful of the environment. It is a strategy based on plants and solar energy. Phytoremediation is an action that removes pollutants from water or, in some cases, soil, using natural processes carried out by plants. A wide selection of floating, emergent and submerged aquatic plants known as macrophytes can be used for this process, most of which are capable of removing heavy metals from wastewater that are toxic to the environment. This method, both more effective and less costly, is used as a tertiary treatment and works perfectly with lagooning. The principle is simple: the bacteria, which need oxygen, transform the organic matter into mineral matter that can be assimilated by the plants, which in turn feed the bacteria through their roots, giving them the oxygen they need.

DUCKWEED

Duckweeds are one of the few known macrophytes, superior plants with 40 species identified so far around the world, belonging to the Lemnaceae family, subdivided into two subfamilies, being the Wolffioideae, comprising two genera (Wolffia and Wolfiella) and the Lemnoideae, comprising (*Lemna* and *Spirodella*). They are small, floating aquatic plants that can uptake large quantities of nutrients such as nitrogen and phosphorus (Paolacci, Stejskal, Toner, Jansen, 2022). For these reasons, duckweed is much used in the treatment of domestic and agricultural wastewater. Lemnaceae are therefore free-floating hydrophytes, not rooted to a substrate. Their structure is relatively simple, consisting of distinct roots, stems or leaves. The frond or leaf is elliptical with a slightly obtuse shape at the ends, between 1 and 3 millimetres long. It has a single, highly visible vein. Each frond has one or more unbranched roots up to 4 centimetres long (Duxbury, Dixon, Greenburg, 1997 ; Nunes, Veiga, Frankenbach, Serôdio, Pinto, 2019; Hattink, Wolterbeek, Goeii, 2001 ; Elshafai, Elgohary, Petervandersteen, Gijzen, 2007 ; Hillman,

1959). Duckweeds are fast-growing angiosperm plants that reproduce vegetatively asexually (by budding), forming colonies of very variable size in the form of solitary fronds or groups of 2 to 30. They reproduce from the basal part of the fronds, with new fronds emerging (daughter fronds) and eventually being detached from the plant that produced them (mother frond). Figure 1 provides a microscopic illustration of mother/daughter frond detachment.

Figure 1. Microscopic illustration of mother/daughter frond detachment's Lemna gibba

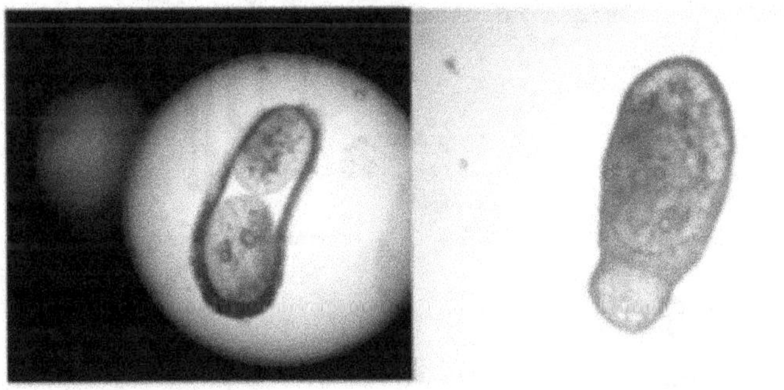

The geographical distribution of Lemnaceae depends on the physiological characteristics of each genus. We are particularly interested in the *Lemna* genus, which is generally adapted to the Mediterranean climate (Gourchane, Al-Jababi, Benrahmane, El Alami, Bouyahya, Laaouan, El Hourch, 2022).

VALORISATION OF BIOMASS AND APPLICATION

Duckweed biomass is considered a promising input material for biofuel production, due to the high starch rate between (31% and 45.8%) of the dry weight. Using the enzymatic hydrolysis approach for biomass between 25.8% and up to 94.75% of the starch could be converted to ethanol using yeast fermentation of the hydrolysate for fuel ethanol production (Fu, Ding, Kumpeangkeaw, Tan, Han, Sun, Zhang, 2020 ; Cheng, Stomp, 2009 ; Xu, Zhao, Stomp, Cheng, 2012 ; Calicioglu, Brennan, 2018). A more recent study found that hydrothermal pre-treatment of duckweed could increase ethanol yield (Souto, Silva, Ninow, Collins, Elliston, Waldron, 2019). Duckweed is therefore being considered as a new kind of energy source for the production of bioethanol and biofuel. In addition to this potential application, the economic value of duckweed has recently attracted attention. Because of its high protein content (Roman, Brennan, 2019). duckweed can be the perfect meal for livestock, poultry and even human food without any health risk, and this has been the subject of a number of research studies (Xu, 2023 ; Pagliuso, Grandis, Fortirer, Camargo, Iochevet, Floh, Buckeridge, 2022 ; Pradhan, Patel, Singh, 2019 ; Appenroth, 2016). Figure 2 presents a concrete example of recovering duckweed according to the concept of the circular economy, respecting the 3Rs (reduce, reuse, recycle).

Figure 2. Example of recovering duckweed according to the concept of the circular economy

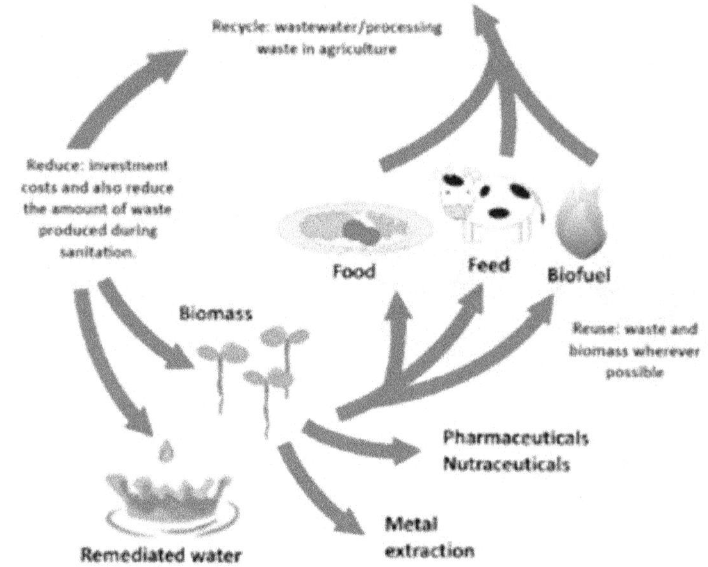

In the same context, studies and perspectives on the research and development of materials based on vegetal products are aimed not only at the environmental field but also at the industrial field. Duckweed can still be used, even as waste, after contributing to water purification. In one study, it was chosen to develop a powdered *Lemna gibba* plant biomaterial as a new alternative material to activated carbon, which was effective in eliminating dyes

(Gourchane, Bouyahya, Laaouan, Zaaboul, El Hourch, 2023). The use of aquatic plants has been extremely successful, mainly due to the phenolic compounds with antioxidant activity, and some very recent studies have proposed innovative products such as the addition of the ethanolic extract of *Lemna gibba* to paper interleaves to treat historic papers (Mohamed, Mansour, Salem, 2019). *Lemna minor* extracts have also been added to polymer systems for food packaging Luzi, 2022).

CONCLUSION

This paper defends the ecological and environmental objectives. Its main purpose is to highlight the process of phytoremediation by macrophytes as an inexpensive and environmentally friendly process, and to underline the practice of recovering plant waste, i.e. reusing and recycling it in order to obtain reusable materials or energy, as a new way of life and an economic solution.

REFERENCES

Appenroth, K.-J., Sree, K. S., Böhm, V., Hammann, S., Vetter, W., Leiterer, M., & Jahreis, G. (2017, February). Nutritional value of duckweeds (Lemnaceae) as human food. *Food Chemistry*, *217*, 266–273. doi:10.1016/j.foodchem.2016.08.116 PMID:27664634

Calicioglu, O., & Brennan, R. A. (2018, June). Sequential ethanol fermentation and anaerobic digestion increases bioenergy yields from duckweed. *Bioresource Technology*, *257*, 344–348. doi:10.1016/j.biortech.2018.02.053 PMID:29605287

Cheng, J. J., & Stomp, A. (2009, January). Growing Duckweed to Recover Nutrients from Wastewaters and for Production of Fuel Ethanol and Animal Feed. *Clean (Weinheim)*, *37*(1), 17–26. doi:10.1002/clen.200800210

Drobenko, B. (2021). *Introduction au droit de l'eau*. Johanet.

Duxbury, C. L., Dixon, D. G., & Greenberg, B. M. (1997, August). Effects of simulated solar radiation on the bioaccumulation of polycyclic aromatic hydrocarbons by the duckweed Lemna gibba. *Environmental Toxicology and Chemistry*, *16*(8), 1739–1748. doi:10.1002/etc.5620160824

Elshafai, S., Elgohary, F., Nasr, F., Petervandersteen, N., & Gijzen, H. (2007, March). Nutrient recovery from domestic wastewater using a UASB-duckweed ponds system. *Bioresource Technology*, *98*(4), 798–807. doi:10.1016/j.biortech.2006.03.011 PMID:16713255

Fu, L., Ding, Z., Kumpeangkeaw, A., Tan, D., Han, B., Sun, X., & Zhang, J. (2020, January). De novo assembly, transcriptome characterization, and simple sequence repeat marker development in duckweed Lemna gibba. *Physiology and Molecular Biology of Plants*, *26*(1), 133–142. doi:10.1007/s12298-019-00726-9 PMID:32158126

Gourchane, F., Al-Jadabi, N., Benrahmane, L., & El Alami, M. (2022). A.bouyahya, M.laaouan, A. EL hourch « Influence of climatic conditions and aquatic environment on the growth of Lemna gibba in the pilot lagoon treatment plant of Bouregreg RABAT-Morocco ». *International Journal of Chemical and Biochemical Science*, 176–182. https://www.iscientific.org/volume-22-2022/

Gourchane F. (2023). Bioadsorption of CI Reactive Blue 203 dye by duckweed Lemna gibba's powder. *Moroccan Journal of Chemistry*. doi:10.48317/IMIST.PRSM/morjchem-v11i3.40166

Gupta, S. V. K., & Suhas. (2009). Application of low-cost adsorbents for dye removal – A review. *Journal of Environmental Management*, *90*(8), 2313–2342. doi:10.1016/j.jenvman.2008.11.017

Hattink, J., Wolterbeek, H. Th., & De Goeij, J. J. M. (2001, May). Influence of salinity and eutrophication on bioaccumulation of 99 technetium in duckweed. *Environmental Toxicology and Chemistry*, *20*(5), 996–1002. doi:10.1002/etc.5620200509 PMID:11337889

Luzi, F., Del Buono, D., Orfei, B., Moretti, C., Buonaurio, R., Torre, L., & Puglia, D. (2022, June). Lemna minor aqueous extract as a natural ingredient incorporated in poly (vinyl alcohol)-based films for active food packaging systems. *Food Packaging and Shelf Life*, *32*, 100822. doi:10.1016/j.fpsl.2022.100822

Mohamed, W. A., Mansour, M. M. A., & Salem, M. Z. M. (2019, May). Lemna gibba and Eichhornia crassipes extracts: Clean alternatives for deacidification, antioxidation and fungicidal treatment of historical paper. *Journal of Cleaner Production*, *219*, 846–855. doi:10.1016/j.jclepro.2019.02.097

Nunes, B., Veiga, V., Frankenbach, S., Serôdio, J., & Pinto, G. (2019, November). Evaluation of physiological changes induced by the fluoroquinolone antibiotic ciprofloxacin in the freshwater macrophyte species Lemna minor and Lemna gibba. *Environmental Toxicology and Pharmacology*, *72*, 103242. doi:10.1016/j.etap.2019.103242 PMID:31473558

Oron, G., Porath, D., & Jansen, H. (1987, February). Performance of the duckweed species Lemna gibba on municipal wastewater for effluent renovation and protein production. *Biotechnology and Bioengineering*, *29*(2), 258–268. doi:10.1002/bit.260290217 PMID:18576384

Pagliuso, D., Grandis, A., Fortirer, J. S., Camargo, P., Iochevet, E., Floh, S., & Buckeridge, M. S. (2022). *Duckweeds as Promising Food Feedstocks Globally*., doi:10.3390/agronomy12040796

Paolacci, S., Stejskal, V., Toner, D., & Jansen, M. A. K. (2022). Wastewater valorisation in an integrated multitrophic aquaculture system; assessing nutrient removal and biomass production by duckweed species. *Environmental Pollution*, *302*, 119059. doi:10.1016/j.envpol.2022.119059 PMID:35227845

Pradhan, A., Patel, A. B., & Singh, S. K. (2019, June). Evaluation of live duckweed, Wolffia globosa as an allochthonous feed for Labeo rohita fry during nursery rearing. *Aquaculture Research*, *50*(6), 1557–1563. doi:10.1111/are.14025

Roman, B., & Brennan, R. A. (2019). A beneficial by-product of ecological wastewater treatment: An evaluation of wastewater-grown duckweed as a protein supplement for sustainable agriculture. *Ecological Engineering*, *142*, 100004. doi:10.1016/j.ecoena.2019.100004

Souto, L. R. F., Silva, I. F. D., Ninow, J. L., Collins, S. R. A., Elliston, A., & Waldron, K. W. (2019, August). Effect of hydrothermal pre-treatment on duckweed (Landoltia punctata) biomass for simultaneous saccharification and fermentation process. *Biomass and Bioenergy*, *127*, 105259. doi:10.1016/j.biombioe.2019.105259

Xu, J., Shen, Y., Zheng, Y., Smith, G., Sun, X. S., Wang, D., Zhao, Y., Zhang, W., & Li, Y. (2023, August). Duckweed (Lemnaceae) for potentially nutritious human food: A review. *Food Reviews International*, *39*(7), 3620–3634. doi:10.1080/87559129.2021.2012800

Xu, J., Zhao, H., Stomp, A., & Cheng, J. J. (2012). The production of duckweed as a source of biofuels. Future Science. doi:10.4155/bfs.12.31

Chapter 7
Methods and Technologies for the Treatment and Disposal of Brine in Seawater Desalination Plants

Driss Azdem
Mohammed V University in Rabat, Morocco

Jamal Mabrouki
https://orcid.org/0000-0002-3841-7755
Mohammed V University in Rabat, Morocco

Souad El Hajjaji
https://orcid.org/0000-0003-1467-704X
Mohammed V University in Rabat, Morocco

ABSTRACT

Seawater desalination is a water technology that is becoming increasingly important in meeting the water needs of many countries around the world. One of the constraints associated with the operation of desalination plants is the production of a concentrated solution, known as brine, which can cause damage to the marine environment due to its high salt content. This paper presents a review of contemporary technologies aimed at addressing the environmental challenges posed by concentrated solutions produced in desalination plants. Among these solutions are evaporation ponds, surface water discharge, and deep well injection. Nevertheless, these approaches encounter sustainability concerns, substantial initial investments, and constraints in terms of universal applicability. A systematic comparison of all these technologies is conducted, focusing on their individual merits and drawbacks, as well as evaluating their feasibility and current developmental status. The objective of this comprehensive analysis is to provide a structured decision-making tool for identifying the most appropriate technology for specific circumstances.

DOI: 10.4018/979-8-3693-3807-0.ch007

Copyright © 2024, IGI Global. Copying or distributing in print or electronic forms without written permission of IGI Global is prohibited.

INTRODUCTION

Numerous causes, such as population growth, economic expansion, the scarcity of water resources, and climate change, have contributed to the recent increase in freshwater supply and made water shortage issues worse globally (Gao et al., 2023; Ingrao et al., 2023). Based on recent estimates (Abou-Shady et al., 2023; Shemer et al., 2023), approximately 2.3 billion individuals, constituting 42% of the global populace, reside in areas facing water scarcity. Projections suggest this figure could escalate to 60% by the year 2025. In response to this challenge, many nations vulnerable to water scarcity have been exploring alternative approaches to traditional water sources. Strategies like constructing dams (Pradhan and Srinivasan, 2022) or water transportation (Dhakal et al., 2022) are insufficient and ineffective solutions to deal with the growing demand for water and the finite supply. Consequently, desalination is seen as a practical and viable remedy to tackle this global issue (Pontius and McIntosh, 2024). Desalination technologies leverage either phase change processes or semipermeable membrane techniques to effectuate the separation of solvent from solutes (de Nicolás et al., 2023; Politano et al., 2024). Phase change methods, such as distillation and multi-stage flash evaporation, exploit differences in boiling points to evaporate solvent, leaving behind concentrated brine (Lee et al., 2024). Alternatively, membrane-based processes, exemplified by reverse osmosis and electrodialysis, rely on selective permeability to allow solvent molecules to pass through while impeding the passage of solutes, thereby producing fresh water (Fig. 1). While desalination indeed yields freshwater, it also produces brine (Sirota et al., 2024; Tu et al., 2024), which presents a significant environmental hazard. Disposing of brine in the environment involves various methods, including surface water discharge, sewer discharge, deep-well injection, evaporation ponds, and land application (Khajah et al., 2023). Brine, comprising salt and water, boasts a complex composition beyond its salinity. It commonly contains an array of minerals and trace elements, influenced by its source and production process. Furthermore, organic matter such as algae and bacteria, along with dissolved gases like carbon dioxide, may be present in brine. Industrial activities and wastewater inputs can further affect its composition, introducing heavy metals, pretreatment chemicals, and other contaminants. Understanding the composition of brine is pivotal for evaluating its environmental impact and formulating effective disposal strategies. This paper provides an overview of the methods and technologies used to treat and dispose of brine generated from desalination processes. The aim is to evaluate their effectiveness, environmental implications, and potential for sustainable brine management. Through this exploration, we aim to offer insights and recommendations for optimizing brine disposal practices and minimizing environmental impact in the field of desalination

Figure 1. The classification of desalination methods based on their operational principles

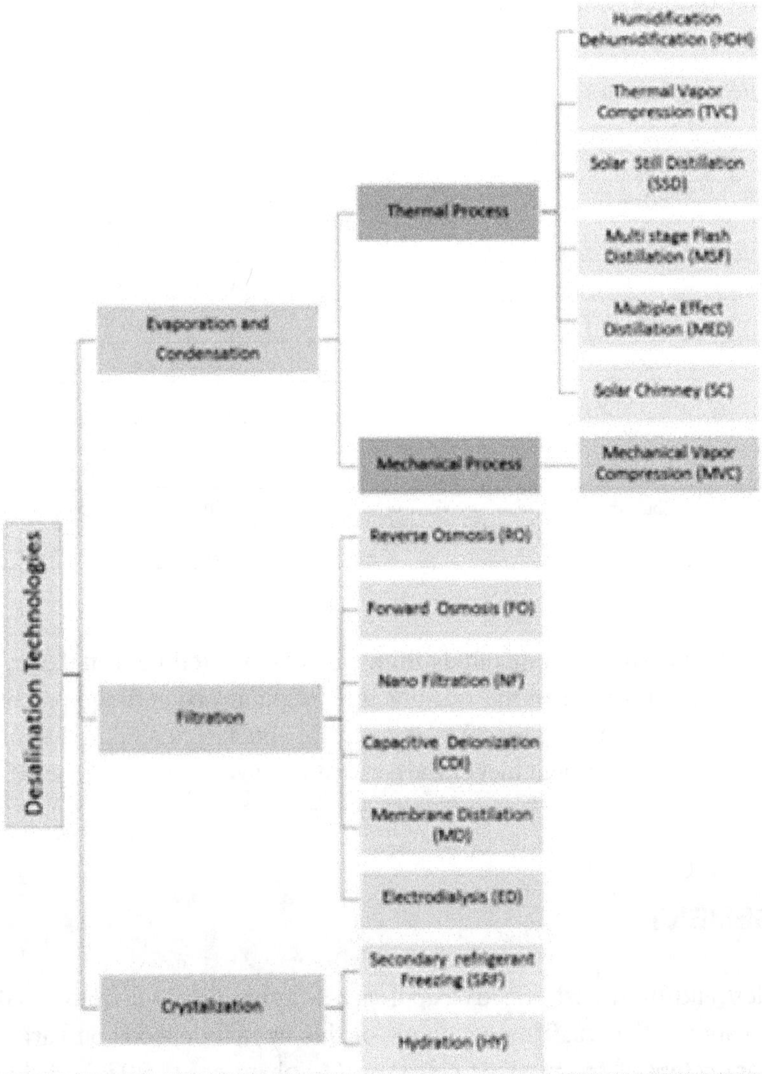

BRINE CHARACTERISTICS

Brine, an inevitable by-product of desalination (Rivero-Falcón et al., 2023), requires careful disposal strategies. The primary focus regarding brine discharge pertains to both the quantity released and its chemical composition, encompassing heavy metals, organic and inorganic compounds (Korak et al., 2023), as well as by-products resulting from pre-treatment and post-treatment procedures. Undoubtedly, the chemical composition of brine has a significant influence on the choice of disposal techniques (Linden, 2024). The characteristics of the brine are closely related to several factors, including the quality of the feed water, the specific desalination method used, the recovery rate, and any chemical additives used. Waste from membrane desalination can be divided into different types (Mohammadi et al., 2024)

(Chen et al., 2024): pre-treatment waste, membrane concentrate, cleaning waste and post-treatment waste (Poirier et al., 2023) (Abounahia et al., 2023)(see Table 1).

Table 1. Characteristics of the brine produced by seawater desalination plants

Process	*Chemical*	*Purpose of use*
Pretreatment	NaHSO3	Elimination of chlorine content to reach acceptable levels
	FeCl3/AlCl3	Removal of suspended solids
	NaOCl	Inactivation of microorganisms
	H2SO4/HCl	pH control
Posttreatment	Acids	Dissolving metals
	Detergents, surfactants, and caustics	Organic and silica dissolution
	Biocides	Microorganism inactivation
	Chelators	Removing scale
	Enzymes	Degradation of bacteria

Brine can also contain corrosive compounds from water flow (Vellore Vettalakar, 2023), dissolved gases and treatment chemicals used in the manufacture of pipework and equipment. Specific heavy metals (Rajamanickam and Selvasembian, 2024) are also of concern, including harmful elements such as copper (Cu), molybdenum (Mo) and nickel (Ni), as well as less harmful elements such as iron (Fe) and zinc (Zn).

BRINE MANAGEMENT

The optimal, simplest and most cost-effective approach to brine disposal from desalination plants is typically direct discharge to the sea. The magnitude of this impact depends on various environmental and hydrological parameters of the marine ecosystem (Aljohani et al., 2023), including bathymetric features (Hajiya Hasan et al., 2023), wave dynamics and water column depth. the brine management options are as follows:

(1) deep well injection;
(2) evaporation pond;
(3) spray irrigation;
(4) sewer;
(5) surface water,
(6) zero liquid discharge (ZLD).

Table 2. The main characteristics of brine treatment methods.

Disposal Methods	*Costs*	*Land requirements*	*Benefits*	*Constraints*
Deep well injection	High capital costs	Land required for injection wellfield	No anticipated effect on marine life	Pollution of nearby water aquifers and ground water contamination.
Evaporation ponds	High cost	Large	Potential for commercial salt extraction Requires minimal technological and managerial resources	Only viable in arid environments Requires a storage and distribution infrastructure Potential for soil and groundwater contamination
Sewer disposal	Economically viable with proximity to existing sewage and treatment facilities.	Small	Uses existing infrastructure Reduces the Biochemical Oxygen Demand (BOD) in the discharged wastewater.	Potential environmental risks arise from the high total dissolved solids (TDS) content in brine
Surface water discharge	Incurrs minimal capital and operational costs (Frequently the most economical choice)	Small	Capable of handling significant quantities.	Pollution of the marine ecosystem through changes in salinity, temperature and pH.
zero liquid discharge (ZLD).	Expensive	Large	Capable of commercially utilizing concentrate Anticipates no adverse impact on marine environments	Energy-intensive process
Land application	Low economy of scale	Large	Suitable for irrigating salt-tolerant species Does not pose a threat to marine ecosystems	Suitable for smaller discharge flows Has the potential to elevate the salinity of groundwater and the soil beneath.

ADAPTING BRINE FOR INDUSTRIAL APPLICATIONS

Utilizing Brine in the Chlor-Alkali Industry

Adapting brine for use in the chlor-alkali industry provides a cost-effective solution by selling concentrated brine to other sectors (Requena-Leal et al., 2024). In terms of energy efficiency, producing brine for the chlor-alkali industry from reverse osmosis (RO) concentrate requires less energy to concentrate sodium chloride compared to starting from seawater (Verma et al., 2023). Additionally, the hydrogen (H2) generated during this process could be utilized on-site for power generation, enabling self-sufficiency and rendering this approach suitable for regions with high energy costs. Adapting brine for the chlor-alkali sector reduces land requirements and helps to lower capital expenditures, particularly in regions where land availability is limited or expensive ("Comparative Analysis on Cost and Benefit of The Rejected Brine Valorization Technology for Sustainable Brine Management," 2022).

Producing HCl and NaOH Using Bipolar Membrane Electrodialysis

Membrane electrodialysis operates by harnessing enhanced ionic mobility triggered by an applied potential, while simultaneously controlling ion movement through ion-selective membranes. Bipolar membrane electrodialysis (BMED) employs this principle in conjunction with a bipolar membrane, enabling the division of water into protons and hydroxide ions. In designated chambers, the fusion of protons with anions leads to acid formation, whereas in other chambers, the combination of hydroxide ions with cations produces the respective base. Figure 2 illustrates the representation of ion diffusion in a typical BMED system.

Figure 2. Diagram illustrating the operating principle of a bipolar membrane electrodialysis (BMED) system

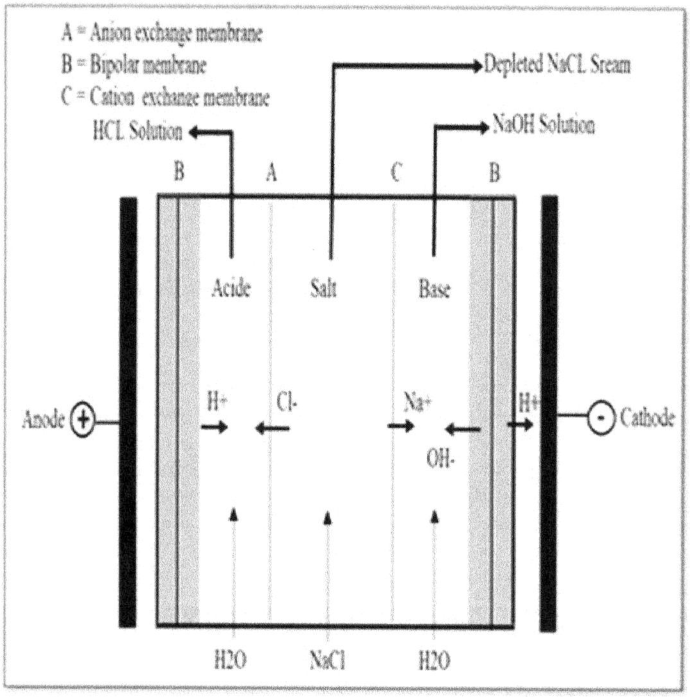

CONCLUSION

While desalination processes yield pure, high-quality drinking water, ensuring human health and safety, the associated environmental repercussions of managing and disposing of waste brine demand serious attention. Often, these reject brines are elevated in temperature and harbor chemicals that pose hazards to marine life. Brine management is essential, not optional, to protect the environment and maintain ecosystem integrity. This can be achieved by minimising the amount of brine discharged or by using Zero Liquid Discharge (ZLD) technologies, which allow for the responsible disposal of produced salts on land or in water, thereby reducing the environmental impact. Moreover, it's possible to extract valuable materials like salts and metals from reject brines, which can be sold for profit. The most sustainable approaches to brine management typically involve integrating various processes to recover water while simultaneously producing commercial products such as salts, chemicals, and precious metals.

One alternative strategy for managing brine involves its utilization in industrial processes, which requires intricate procedures to prepare brine suitable for feeding into industrial facilities. For example, adapting brine for use in the chlor-alkali industry necessitates concentration processes such as electrodialysis. Additionally, it's vital to eliminate divalent cations from the brine, as they surpass the specifications for membrane electrolysis. Although these treatments entail substantial costs, the products obtained from electrolysis can help offset them. It is advisable to install the electrolysis plant as an extension to the desalination plant. Bipolar membrane electrodialysis, which produces NaOH and HCl chemicals in high demand and easy to sell—shows promise. However, its industrial-scale implementation remains limited due to the established use of electrolytic cells in the chlor-alkali industry.

REFERENCES

Abou-Shady, A., Siddique, M. S., & Yu, W. (2023). A Critical Review of Recent Progress in Global Water Reuse during 2019–2021 and Perspectives to Overcome Future Water Crisis. *Environments (Basel, Switzerland), 10*(9), 159. doi:10.3390/environments10090159

Abounahia, N. M., El-Sayed, A. M. A., Saleem, H., & Zaidi, S. J. (2023). An overview on the progress in produced water desalination by membrane-based technology. *Journal of Water Process Engineering, 51*, 103479. doi:10.1016/j.jwpe.2022.103479

Aljohani, N. S., Kavil, Y. N., Al-Farawati, R. K., Aljohani, N. H., Orif, M. I., Ghandourah, M. A., Bahshwan, S. M., Aloufi, F. A., Halawani, R. F., & Salam, M. A. (2023). The assessment of environmental parameter along the desalination plants in the Kingdom of Saudi Arabia. *Open Chemistry, 21*(1), 20220274. doi:10.1515/chem-2022-0274

Chen, C., Yang, Y., Graham, N. J., Li, Z., Yang, X., Wang, Z., Farhat, N., Vrouwenvelder, J. S., & Hou, L. (2024). A comprehensive evaluation of the temporal and spatial fouling characteristics of RO membranes in a full-scale seawater desalination plant. *Water Research, 249*, 120914. doi:10.1016/j.watres.2023.120914 PMID:38007899

de Nicolás, A. P., Molina-García, Á., García-Bermejo, J. T., & Vera-García, F. (2023). Desalination, minimal and zero liquid discharge powered by renewable energy sources: Current status and future perspectives. *Renewable & Sustainable Energy Reviews, 187*, 113733. doi:10.1016/j.rser.2023.113733

Dhakal, N., Salinas-Rodriguez, S. G., Hamdani, J., Abushaban, A., Sawalha, H., Schippers, J. C., & Kennedy, M. D. (2022). Is Desalination a Solution to Freshwater Scarcity in Developing Countries? *Membranes, 12*(4), 381. doi:10.3390/membranes12040381 PMID:35448351

Gao, F., Luo, Y., & Zhao, C. (2023). Effects of Climate and Land-Use Change on the Supply and Demand Relationship of Water Provision Services in the Yellow River Basin. *Land (Basel), 12*(12), 2089. doi:10.3390/land12122089

Hajiya Hasan, A. H., Al-Bader, D. A., Woodward, S., Peters, A. F., & Küpper, F. C. (2023). Ecophysiology of Kuwaiti macroalgae with special emphasis on temperature and salinity tolerance related to the conditions at desalination plant outfalls. *Botanica Marina, 66*(5), 373–390. doi:10.1515/bot-2022-0049

Ingrao, C., Strippoli, R., Lagioia, G., & Huisingh, D. (2023). Water scarcity in agriculture: An overview of causes, impacts and approaches for reducing the risks. *Heliyon, 9*(8), e18507. doi:10.1016/j.heliyon.2023.e18507 PMID:37534016

Khajah, M., Ahmed, M.E., & Al-Matouq, A. (2023). *Characterization of reverse osmosis reject wastewater generated from Sulaibiya Wastewater Treatment Plant.*

Korak, J. A., Mungan, A. L., & Watts, L. T. (2023). Critical review of waste brine management strategies for drinking water treatment using strong base ion exchange. *Journal of Hazardous Materials, 441*, 129473. doi:10.1016/j.jhazmat.2022.129473

Lee, C., Ho, H., Chen, W., & Iizuka, A. (2024). Total Resource Circulation of Desalination Brine: A Review. *Advanced Sustainable Systems*, 2300460. doi:10.1002/adsu.202300460

Linden, M. & van der, (2024). *Environmental effects of brine disposal and seawater usage for offshore green hydrogen production and storage in the Dutch North Sea.*

Mohammadi, M., Fazeli, F., & Baghsheikhi, M. (2024). Enhancing water desalination and power generation in microbial desalination cells: A comprehensive review of effective parameters and structural types for optimal performance. *Journal of Power Sources, 602*, 234267. doi:10.1016/j.jpowsour.2024.234267

Poirier, K., Lotfi, M., Garg, K., Patchigolla, K., Anthony, E. J., Faisal, N. H., Mulgundmath, V., Sahith, J. K., Jadhawar, P., Koh, L., Morosuk, T., & Al Mhanna, N. (2023). A comprehensive review of pre-and post-treatment approaches to achieve sustainable desalination for different water streams. *Desalination, 566*, 116944. doi:10.1016/j.desal.2023.116944

Politano, A., Al-Juboori, R.A., Alnajdi, S., Alsaati, A., Athanassiou, A., Bar-Sadan, M., Naderi Beni, A., Campi, D., Cupolillo, A., & D'Olimpio, G. (2024). 2024 Roadmap on membrane desalination technology at the water-energy nexus. *Journal of Physics: Energy.*

Pontius, J., & McIntosh, A. (2024). Water Scarcity. In *Environmental Problem Solving in an Age of Climate Change, Springer Textbooks in Earth Sciences, Geography and Environment* (pp. 87–103). Springer International Publishing. doi:10.1007/978-3-031-48762-0_8

Pradhan, A., & Srinivasan, V. (2022). Do dams improve water security in India? A review of post facto assessments. *Water Security, 15*, 100112. doi:10.1016/j.wasec.2022.100112

Rajamanickam, R., & Selvasembian, R. (2024). A sustainable resource recovery approach through microalgae integrated brine management in minimal liquid discharge system. *Desalination, 578*, 117424. doi:10.1016/j.desal.2024.117424

Requena-Leal, I., Carvela, M., Fernández-Marchante, C. M., Lobato, J., & Rodrigo, M. A. (2024). On the use of chlor-alkali technology to power environmental electrochemical treatment technologies. *Current Opinion in Electrochemistry, 45*, 101461. doi:10.1016/j.coelec.2024.101461

Rivero-Falcón, Á., Peñate Suárez, B., & Melián-Martel, N. (2023). SWRO Brine Characterisation and Critical Analysis of Its Industrial Valorisation: A Case Study in the Canary Islands (Spain). *Water (Basel), 15*(8), 1600. doi:10.3390/w15081600

Shemer, H., Wald, S., & Semiat, R. (2023). Challenges and Solutions for Global Water Scarcity. *Membranes, 13*(6), 612. doi:10.3390/membranes13060612 PMID:37367816

Sirota, R., Winters, G., Levy, O., Marques, J., Paytan, A., Silverman, J., Sisma-Ventura, G., Rahav, E., Antler, G., & Bar-Zeev, E. (2024). Impacts of Desalination Brine Discharge on Benthic Ecosystems. *Environmental Science & Technology, 58*(13), 5631–5645. doi:10.1021/acs.est.3c07748 PMID:38516811

Tu, W. H., Zhao, Y., Chan, W. P., & Lisak, G. (2024). Reclaimed seawater discharge–Desalination brine treatment and resource recovery system. *Water Research, 251*, 121096. doi:10.1016/j.watres.2023.121096 PMID:38184912

Vellore Vettalakar, A.B. (2023). *THE PRODUCTION BRINE WATER PROBLEMS AND RESOURCES.*

Verma, S., Suresh, S., Mili, M., Diwedi, P., Bajpai, H., Khan, M., Hashmi, S., Rathore, S., & Srivastava, A. (2023). Future process options for brine sludge management from the chlor-alkali industry. *International Journal of Environmental Science and Technology, 20*(8), 9291–9310. doi:10.1007/s13762-022-04433-z

Chapter 8
Potential of Functional Foods in Diabetes Management:
A Review of Coriandrum sativum, Ficus carica, Camellia sinensis

Mounia Mezouara
Laboratory of Natural Resources and Sustainable Development, Ibn Tofaïl University, Morocco

Karima Hbyaj
Laboratory of Natural Resources and Sustainable Development, Ibn Tofaïl University, Morocco

Abdelaati Soufiani
Laboratory of Natural Resources and Sustainable Development, Ibn Tofaïl University, Morocco

Yassine Mouniane
https://orcid.org/0000-0002-6037-5301
Laboratory of Natural Resources and Sustainable Development, Ibn Tofaïl University, Morocco

Ahmed Chriqui
Laboratory of Natural Resources and Sustainable Development, Ibn Tofaïl University, Morocco

Khadija Ben Ramdane
Laboratory of Natural Resources and Sustainable Development, Ibn Tofaïl University, Morocco

Ismail Ouhammou
https://orcid.org/0000-0003-2340-3438
Laboratory of Natural Resources and Sustainable Development, Ibn Tofaïl University, Morocco

Jamal Mabrouki
https://orcid.org/0000-0002-3841-7755
Laboratory of Spectroscopy, Molecular Modelling, Materials, Nanomaterials, Water, and Environment, Morocco

Driss Hmouni
https://orcid.org/0000-0001-7598-6204
Laboratory of Natural Resources and Sustainable Development, Ibn Tofaïl University, Morocco

ABSTRACT

This synthesis delves into the potential of natural foods, particularly coriander, fig, and green tea, in the management of diabetes. These foods are abundant in bioactive compounds such as polyphenols and flavonoids, which possess antioxidant, anti-inflammatory, and hypoglycemic properties. Research indicates that they play a vital role in regulating blood glucose levels, enhancing insulin sensitivity, and mitigating complications associated with diabetes. For instance, coriander has been shown to exert hypoglycemic

DOI: 10.4018/979-8-3693-3807-0.ch008

Copyright © 2024, IGI Global. Copying or distributing in print or electronic forms without written permission of IGI Global is prohibited.

effects in animal models of diabetes, while fig extracts demonstrate the ability to reduce hyperglycemia and normalize lipid profiles. Green tea, with its rich content of catechins, has been linked to improved glucose tolerance and insulin sensitivity. Incorporating these functional foods into daily diets offers a comprehensive approach to diabetes management, complementing traditional therapies. Nonetheless, promoting the consumption of these natural foods presents a promising avenue for combating diabetes and improving overall metabolic health.

INTRODUCTION

The worldwide diabetes pandemic, characterized by insulin resistance and elevated blood sugar levels, is defined by a persistent endocrine condition that disrupts the metabolism of carbohydrates, proteins, fats, electrolytes, and water, primarily stemming from irregularities in insulin production and/or function (Farzaei et al., 2017; Qian et al., 2020). Type 1 and type 2 diabetes are the two primary forms of the disease that have been identified by studies (Chukwuma et al., 2017; Hu et al., 2019). The primary cause of Type 1 diabetes is insufficient insulin production, which results in persistent hyperglycemia (Chukwuma et al., 2017). On the other hand, Type 2 diabetes is on the rise globally, with projections indicating a significant increase to 552 million cases by 2030 (Pratley, 2013; Whiting et al., 2011). It is caused by inefficient metabolic processes, organ damage, or dysfunction, and it affects about 90% of the world's population. This escalating prevalence of diabetes has fueled a burgeoning interest in alternative methods for its control and prevention. Insulin resistance stands as a pivotal factor in the onset of type 2 diabetes and atherosclerosis. It prompts an increase in insulin secretion by pancreatic β cells; however, this capacity diminishes over time, leading to hyperglycemia. Inflammatory signaling pathways and oxidative stress are activated, accelerating protein glycation. Moreover, insulin resistance is linked to endothelial dysfunction and chronic vascular complications, underscoring the necessity to uncover effective hypoglycemic agents (Goh & Cooper, 2008).

Improving diets on a global scale is crucial for promoting healthy and sustainable nutrition worldwide. Incorporating nutrient-rich foods like legumes, grains, vegetables, and fruits into daily meals represents a practical and effective strategy to combat micronutrient deficiencies (Maoka, 2020; Meléndez-martínez et al., 2022). Furthermore, there's a growing emphasis on utilizing food plants with medicinal properties in the fight against diabetes. *Coriandrum sativum* L. (coriander), *Ficus carica* (fig), and *Camellia sinensis* (green tea) emerge as promising candidates for regulating blood sugar levels and enhancing metabolic health. With their widespread consumption as part of daily diets, these plants offer a natural and potentially impactful approach to complement traditional diabetes treatments.

In this review, we will explore recent scientific evidence supporting the use of these plants in diabetes management, highlighting their potential as functional dietary supplements in a holistic approach to health and wellness.

Role of Nature's Food in Diabetes Management

Diabetes is recognized as a major global health problem in the 21st century. Therefore, the exploration of therapeutic agents derived from medicinal plants for managing these conditions is of paramount importance (Asafo-Agyei et al., 2023). Consequently, the World Health Organization has advocated for its integration into the mainstream healthcare system, which includes the inclusion of certain herbal products on the essential medicine list (World Health Organization, 2013). Significant efforts have been made to enhance and strengthen herbal medicine, supported by the availability of scientific evidence validating its efficacy (Abdollahi et al., 2010; Christodoulou et al., 2019; Dineshkumar et al., 2010). Research has investigated the significance of medicinal plants in managing diabetes (Torres-León & Rebolledo, 2023).

Antidiabetic Activity of *Coriandrum sativum* L.

Known by many names as Chinese parsley or coriander, *Coriandrum sativum* L. is an annual or biennial herbaceous plant that belongs to the Apiaceae family (Wei et al., 2019). Originating in the Euro-Mediterranean area, coriander has been used widely in traditional medicine in many different nations to treat a wide range of conditions, including diabetes, inflammation, sleeplessness, anxiety, seizures, and digestive, respiratory, and urinary issues (Grieve, 1971; Ugulu et al., 2009). Its roots, leaves, and seeds are all edible and full of minerals, fiber, vitamins B and C, and carotene, among other benefits (Bhat et al., 2014). Coriander is widely used in Asian cookery; its leaves and shoots are used as seasonings for soups and cold foods, while its seeds and roots are frequently used for cooking meat. Dried coriander seeds are a popular spice in India and the Mediterranean region (Gastón et al., 2016).

Since ancient times, the medicinal virtues of Coriandrum sativum have been acknowledged and valued. Numerous biological properties, including as antioxidant, anticancer, neuroprotective, anxiolytic, anticonvulsant, analgesic, anti-migraine, hypolipidemic, hypoglycemic, hypotensive, antimicrobial, and anti-inflammatory effects, are attributed to the bioactive phytochemicals found in it. Coriander contains a variety of bioactive components that can be extracted, with the specific plant sections and environmental circumstances influencing the extraction of essential oils, lipids, and polyphenols, among other substances (Laribi et al., 2015; Punetha et al., 2018).

Studies have demonstrated the hypoglycemic effects of coriander and its bioactive phytochemicals in animal models of diabetes induced by streptozotocin. For example, ethanol extracts of coriander seeds significantly reduce blood glucose levels at doses of 200 and 250 mg/kg, enhance pancreatic β-cell activity, and promote insulin release in diabetic rats (Eidi et al., 2009). Similar hypoglycemic effects were observed with aqueous extracts of coriander seeds at a dose of 20 mg/kg in obese, hyperglycemic, and hyperlipidemic rats (Aissaoui et al., 2011).

Moreover, long-term oral administration of coriander seed powder has shown hypoglycemic effects in rats fed a high-fat diet. In a study by Chithra & Leelamma (1999), a 10% coriander seed powder administered for 90 days resulted in normalized plasma glucose levels, comparable to the reference drug glibenclamide administered at a daily dose of 2.5 mg/kg. However, a rebound effect was observed after discontinuation of the coriander extract for one week, indicating a temporary and acute hypoglycemic effect of coriander extracts (Prachayasittikul et al., 2018).

The mechanisms underlying the hypoglycemic effects of coriander involve improvements in glucose utilization by peripheral tissues, enhanced hepatic glucose absorption, decreased glucose production, inhibition of gastrointestinal enzymes such as α-glucosidase and α-amylase, and reduction of glucose absorption in the gastrointestinal tract through glucose transporter inhibition.

Figure 1. Syndrome management mechanism, with hypoglycemic in Coriandrum sativum (Prachayasittikul et al., 2018)

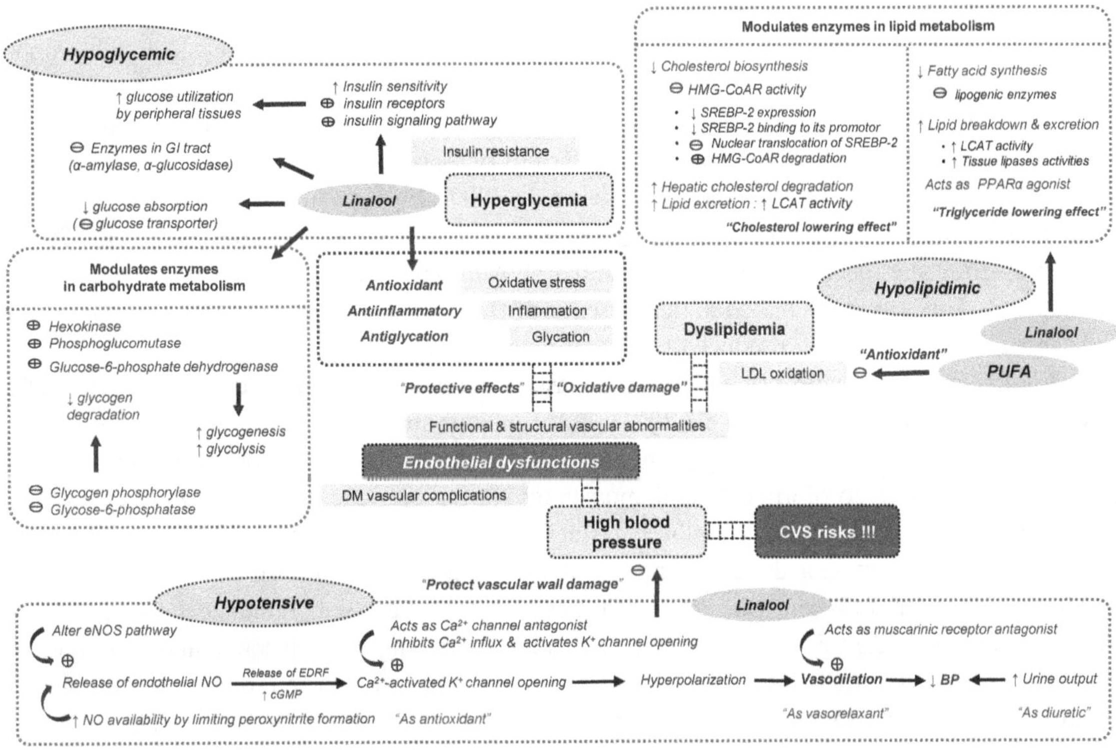

Antidiabetic Activity of *Ficus carica*

Fig (*Ficus carica*) is considered one of the earliest cultivated plants, boasting a long history of human interaction and utilization (Valdeyron & Lloyd, 1979). Belonging to the Moraceae family, which comprises over 800 species of trees, shrubs, and epiphytes cultivated primarily in subtropical regions worldwide (Singh et al., 2011), nearly half of the Ficus species are monoecious, while the rest are functionally dioecious (Berg, 2003). Various genetically diverse varieties of Ficus species exhibit remarkable pharmacological properties and are commercially exploited (Woodland, 1997). The common fig has been a staple crop, particularly in the eastern Mediterranean region since ancient times, although its native range includes parts of the southern Arabian Peninsula, Italy, the Balkan Peninsula, and Russia. Cultivation of the fig began approximately 3000 years ago in the southern Arabian Peninsula before spreading to Iraq, Turkey, Syria, and other Mediterranean countries. The peel of the fig fruit serves to protect it from environmental factors and maintain its structural integrity. Secondary metabolites found in the peel, such as pigments, aroma compounds, and tannins, contribute to the fruit's appearance, quality, and shelf life (Li et al., 2013). Structurally, fig trees vary in size from small to medium-sized, typically reaching heights of 5–8 meters, with a short trunk and irregular branches. Fig fruits come in various colors, including green, yellow-brown, purplish, or black, depending on the cultivar, and are generally pear-shaped with either a velvety or smooth skin. Botanically known as syconium, the edible fig is a compound fruit comprising a hollow fleshy receptacle with a hole at the tip and numerous small flowers lining the inner surface.

Tolerant to subtropical climates, mature fig trees can withstand temperatures as low as -12 to -9.5 °C, depending on the cultivar. However, young trees require protection during winter as temperatures around 1.5–4.5 °C can be damaging when the tree is not in a dormant state (Hajam & Saleem, 2022).

Figs are rich in carbohydrates, minerals, lipids, phenolic compounds, and enzymes. They are an important source of dietary fiber and primarily contain glucose, fructose, and sucrose (Kolesnik et al., 1986). Dried figs are particularly high in iron, with a content higher than that of beef liver, and also contain calcium and potassium. The lipids present include various types of fatty acids such as myristic, palmitic, stearic, oleic, linoleic, and linolenic acids (Marwat et al., 2009). Figs provide a defense against diabetes by regulating carbohydrate metabolism, thereby counteracting hyperglycemia (Barkaoui et al., 2017).

The oral administration of fig-leaf decoction has shown hypoglycemic effects in individuals with type I diabetes. This activity is linked to the presence of flavonoids such as quercetin, kaempferol, and chrysin in fig leaves, along with reduced levels of transaminases. Furthermore, the protective effects observed in diabetic neuropathy are believed to be associated with the hypoglycemic properties of fig leaves against type I diabetes (Zhao et al., 2021). Both aqueous and chloroform extracts from fig leaves exhibit the potential to mitigate hyperglycemia in diabetic rats, normalizing fatty acid and vitamin E levels. Particularly, the chloroform extract demonstrates superior efficacy in reducing fatty acid levels. Moreover, fig leaf extracts are shown to lower fasting blood glucose levels and enhance the concentrations of superoxide dismutase and glutathione peroxidase in serum, as well as superoxide dismutase and glutathione in the liver of diabetic rats. Recent investigations have underscored the beneficial effects of fig latex and leaves on carbohydrate enzymes, indicating promising hypoglycemic and hypolipidemic properties in type 2 diabetic rats (Meziant et al., 2021). Fig pulp can be used to control improper metabolism of carbohydrates, which is linked to diabetes and hyperglycemia (Ramadan, 2023). Fig leaf extract inhibits PEPCK and G6Pase expression in the liver and HepG2 cells of diabetic mice, potentially through AMPK signaling pathway activation, leading to reduced expression of PGC-1α, HNF4α, and FOXO1. Consequently, gluconeogenesis is hindered, suggesting a potential therapeutic avenue for diabetes management (Zhang et al., 2019). Fig extracts possess significant antidiabetic properties by inhibiting glucose absorption in the intestines through the inhibition of α-glucosidase or α-amylase, enhancing glucose uptake via activation of GLUT4, PI3K, or serine/threonine protein kinase, and regulating glucose homeostasis through AMPK activation. These findings highlight the potential of figs as a preventive measure against diabetes (Ayuso, Carpena, Taofiq, Albuquerque, Simal-Gandara, Oliveira, & Barros, 2022).

Figure 2. The antidiabetic efficacy of fig: mechanism of action (Ayuso, Carpena, Taofiq, Albuquerque, Simal-Gandara, Oliveira, Prieto et al, 2022)

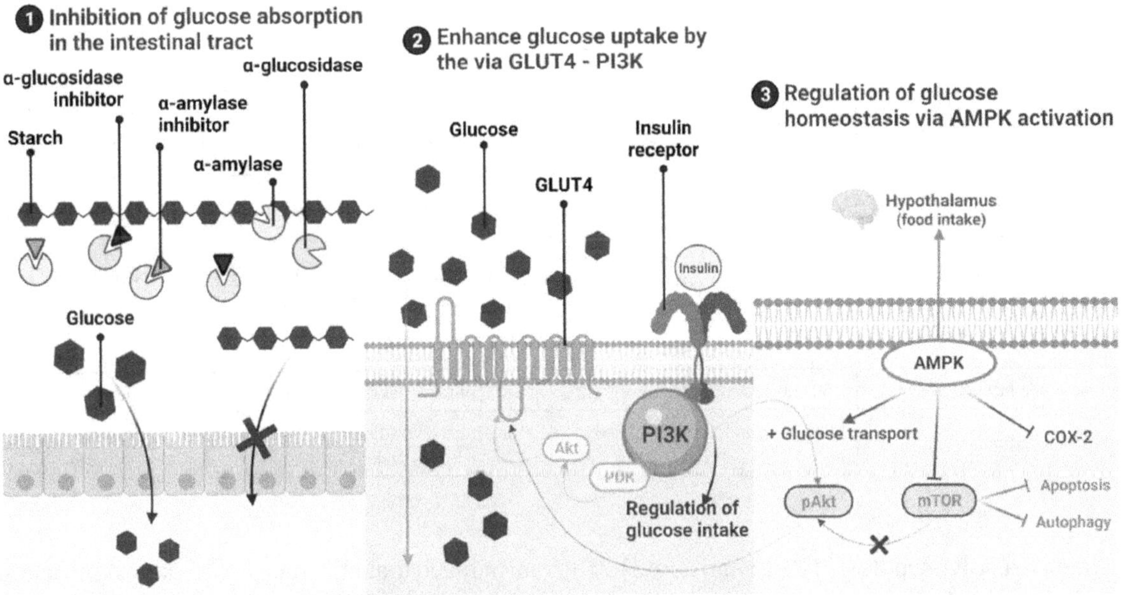

Antidiabetic Activity of *Camellia sinensis*

Green tea, scientifically known as *Camellia sinensis* (L), is a member of the Teaceae family and is cultivated and consumed in over 160 countries worldwide due to its aromatic, flavorful, and medicinal characteristics (Xing et al., 2019). The primary bioactive components found in green tea are polyphenols, with flavonols being a prominent class among them. These compounds play a significant role in imparting antioxidant, anticancer, and anti-inflammatory properties to green tea (Rashidi et al., 2017). Therefore, extracts from green tea offer a promising alternative for the development of functional foods and novel medications, potentially replacing conventional chemical agents commonly utilized (Silovská et al., 2020). Tea stands as one of the most popular beverages globally, offering a myriad of health advantages, including antioxidant, anticancer, anti-cardiovascular, and antidiabetic properties (Yu et al., 2020). Tea seed, sourced from the *Camellia Sinensis* (L.) O. Kuntze plant, represents the mature fruit essential for plant propagation and growth. Following a growth period of 4–6 years, the tea plant reaches maturity (Zhang et al., 2023).

Tea flowers have been studied for their potential antidiabetic effects, which are attributed to various functional metabolites. In vitro and in vivo investigations demonstrate their ability to exert antidiabetic effects (Table 1).

Table 1. Studies on the antidiabetic effects of tea flowers

<table>
<tr><th>Method</th><th>The inhibitory concentration</th><th>Main functional components</th><th>References</th></tr>
<tr><td>α-glycosidase inhibitory activity</td><td>5 mg/mL</td><td>Polysaccharide</td><td>(Wang et al., 2012)</td></tr>
<tr><td>α-glucosidase inhibitory activity</td><td>2.0 mg/mL (inhibitory rate of 83.3%)</td><td>Polysaccharides (31 kDa; 5 kDa)</td><td>(Wei et al., 2010)</td></tr>
<tr><td>Alloxan-treated Sprague-Dawley rats</td><td>200 and 800 mg/kg·body weight</td><td>Polysaccharide</td><td>(Cai et al., 2011)</td></tr>
<tr><td>α-glucosidase inhibitory activity</td><td>2.0 mg/mL (inhibitory rate of 57.2%)</td><td>Polysaccharide (1.12 × 104 g/mol)</td><td rowspan="3">(Han et al., 2011)</td></tr>
<tr><td>α-amylase inhibitory activity</td><td>2.0 mg/mL (inhibitory rate of 29.5%)</td><td>Polysaccharide (1.12 × 104 g/mol)</td></tr>
<tr><td>Alloxan-induced diabetic mice</td><td>75, 150, and 300 mg/kg·body weight</td><td>Polysaccharide (1.12 × 104 g/mol)</td></tr>
<tr><td>Sucrose-loaded mice</td><td>50 and 100 mg/kg·body weight</td><td>Saponins (chakasaponins I-III)</td><td>(Matsuda et al., 2012)</td></tr>
<tr><td>Sucrose-loaded rats</td><td>50, 100, and 200 mg/kg·body weight</td><td>Saponins (floratheasaponins A, B and C)</td><td>(Yoshikawa et al., 2008)</td></tr>
</table>

Green tea polyphenols (GTP) administered to alloxan-induced diabetic mice for 15 days significantly reduced serum glucose levels by 44%, while also decreasing renal and hepatic enzymes and serum lipid peroxidation (Chacko et al., 2010). GTP altered insulin tolerance in peripheral tissues, increased glucose transporter IV levels, and improved glucose tolerance in animal models of obesity, diabetes, and dyslipidemia (Nunes et al., 2014). Additionally, epigallocatechin gallate (EGCG) treatment reduced glucose production in rat hepatoma cells by enhancing insulin receptor phosphorylation and reducing the expression of the gluconeogenic enzyme phosphoenolpyruvate carboxykinase (Waltner-Law et al., 2002).

CONCLUSION

In conclusion, this synthesis highlights the remarkable potential of functional foods, such as coriander, fig and green tea, in managing diabetes. The studies reviewed revealed the multiple mechanisms of action of these plants, including their ability to regulate blood sugar, improve insulin sensitivity and protect against complications associated with diabetes. Their richness in bioactive compounds, such as polyphenols, flavonoids and polysaccharides, gives these foods antioxidant, anti-inflammatory and hypoglycemic properties. By integrating these foods into the daily diet, it is possible to complement conventional diabetes treatments and promote better metabolic health. Ultimately, this review emphasizes the significance of promoting a diet high in functional foods to prevent and manage diabetes, providing an all-natural holistic approach to improving metabolic health.

REFERENCES

Abdollahi, M., Zuki, A. B. Z., Goh, Y. M., Rezaeizadeh, A., & Noordin, M. M. (2010). The effects of Momordica charantia on the liver in streptozotocin-induced diabetes in neonatal rats. *African Journal of Biotechnology*, *9*(31), 5004–5012.

Aissaoui, A., Zizi, S., Israili, Z. H., & Lyoussi, B. (2011). Hypoglycemic and hypolipidemic effects of Coriandrum sativum L. in Meriones shawi rats. *Journal of Ethnopharmacology*, *137*(1), 652–661. doi:10.1016/j.jep.2011.06.019 PMID:21718774

Asafo-Agyei, T., Appau, Y., Barimah, K. B., & Asase, A. (2023). Medicinal plants used for management of diabetes and hypertension in Ghana. *Heliyon*, *9*(12), e22977. doi:10.1016/j.heliyon.2023.e22977 PMID:38076168

Ayuso, M., Carpena, M., Taofiq, O., Albuquerque, T. G., Simal-Gandara, J., Oliveira, M. B. P. P., & Barros, L. (2022). Fig "Ficus carica L." and its by-products: A decade evidence of their health-promoting benefits towards the development of novel food formulations. *Trends in Food Science & Technology*, *127*, 1–13. doi:10.1016/j.tifs.2022.06.010

Ayuso, M., Carpena, M., Taofiq, O., Albuquerque, T. G., Simal-Gandara, J., Oliveira, M. B. P. P., Prieto, M. A., Ferreira, I. C. F. R., & Barros, L. (2022). Fig "*Ficus carica* L." and its by-products: A decade evidence of their health-promoting benefits towards the development of novel food formulations. *Trends in Food Science & Technology*, *127*, 1–13. doi:10.1016/j.tifs.2022.06.010

Barkaoui, M., Katiri, A., Boubaker, H., & Msanda, F. (2017). Ethnobotanical survey of medicinal plants used in the traditional treatment of diabetes in Chtouka Ait Baha and Tiznit (Western Anti-Atlas). *Journal of Ethnopharmacology*, *198*, 338–350. doi:10.1016/j.jep.2017.01.023 PMID:28109915

Berg, C. C. (2003). Flora Malesiana precursor for the treatment of Moraceae 1: the main subdivision of Ficus: the subgenera Blumea-Biodiversity. *Blumea*, *48*(1), 166–177. doi:10.3767/000651903X686132

Bhat, S., Kaushal, P., & Sharma, H. K. (2014). Coriander (*Coriandrum sativum* L.): Processing, nutritional and functional aspects. *African Journal of Plant Science*, *8*(1), 25–33. doi:10.5897/AJPS2013.1118

Cai, X., Wang, Y. F., Mao, F. F., Lan, Y. U., Liu, C. X., Zhang, H., & Wei, X. L. (2011). Hypoglycemic and hyperglycemia-prevention effects of crude tea flower polysaccharide. Modern [in Chinese]. *Food Science and Technology (Campinas)*, *27*(3), 262–266.

Chacko, S. M., Thambi, P. T., Kuttan, R., & Nishigaki, I. (2010). Beneficial effects of green tea: A literature review. *Chinese Medicine*, *5*(1), 13. doi:10.1186/1749-8546-5-13 PMID:20370896

Chithra, V., & Leelamma, S. (1999). *Coriandrum sativum* — Mechanism of hypoglycemic action. *Food Chemistry*, *67*(3), 229–231. doi:10.1016/S0308-8146(99)00113-2

Christodoulou, M., Tchoumtchoua, J., Skaltsounis, A., Scorilas, A., & Halabalaki, M. (2019). Natural alkaloids intervening the insulin pathway: New hopes for anti-diabetic agents? *Current Medicinal Chemistry*, *26*(32), 5982–6015. doi:10.2174/0929867325666180430152618 PMID:29714135

Chukwuma, C. I., Mopuri, R., Nagiah, S., Chuturgoon, A. A., & Islam, M. S. (2017). Erythritol reduces small intestinal glucose absorption, increases muscle glucose uptake, improves glucose metabolic enzymes activities and increases expression of Glut-4 and IRS-1 in type 2 diabetic rats. *European Journal of Nutrition*. PMID:28770335

Dineshkumar, B., Mitra, A., & Manjunatha, M. (2010). A comparative study of alpha amylase inhibitory activities of common anti-diabetic plants at Kharagpur 1 block Int. *J. Green Pharm*, *4*(2).

Eidi, M., Eidi, A., Saeidi, A., Molanaei, S., Sadeghipour, A., Bahar, M., & Bahar, K. (2009). Effect of coriander seed (Coriandrum sativum L.) ethanol extract on insulin release from pancreatic beta cells in streptozotocin-induced diabetic rats. *Phytotherapy Research*, *23*(3), 404–406. doi:10.1002/ptr.2642 PMID:19003941

Farzaei, F., Morovati, M. R., Farjadmand, F., & Farzaei, M. H. (2017). A mechanistic review on medicinal plants used for diabetes mellitus in traditional Persian medicine. *Journal of Evidence-Based Complementary & Alternative Medicine*, *22*(4), 944–955. doi:10.1177/2156587216686461 PMID:29228789

Gastón, M. S., Cid, M. P., Vázquez, A. M., Decarlini, M. F., Demmel, I. G., Rossi, L. I., Aimar, M. L., & Salvatierra, N. A. (2016). Sedative Effect of Central Administration of Coriandrum Sativum Essential Oil and Its Major Component Linalool in Neonatal Chicks. *Pharmaceutical Biology*, *54*(10), 1954–1961. doi:10.3109/13880209.2015.1137602 PMID:26911626

Goh, S. Y., & Cooper, M. E. (2008). Clinical review: The role of advanced glycation end products in progression and complications of diabetes. *The Journal of Clinical Endocrinology and Metabolism*, *93*(4), 1143–1152. doi:10.1210/jc.2007-1817 PMID:18182449

Grieve, M. (1971). *A modern herbal Dover Publications.*

Hajam, T. A., & Saleem, H. (2022). Phytochemistry, biological activities, industrial and traditional uses of fig (*Ficus carica*): A review. *Chemico-Biological Interactions*, *368*, 110237. doi:10.1016/j.cbi.2022.110237 PMID:36288779

Han, Q., Yu, Q.Y., Shi, J., Xiong, C.Y., Ling, Z.J., & He, P.M. (2011). Molecular characterization and hypoglycemic activity of a novel water-soluble polysaccharide from tea (Camellia sinensis) flower. *Carbohydrate Polymers, 86* (2), 797-805.

Hu, Z., Gao, F., Qin, L., Yang, Y., & Xu, H. (2019). A case-control study on risk factors and their Interactions with prediabetes among the elderly in rural communities of yiyang city, hunan province. *Journal of Diabetes Research*, *2019*, 1386048. doi:10.1155/2019/1386048 PMID:30911549

Kolesnik, A. A., Kakhniashvili, T. A., Zherebin, Y. L., Golubev, V. N., & Pilipenko, L. N. (1986). Lipids of the fruit of Ficus carica. *Chemistry of Natural Compounds*, *22*(4), 394–397. doi:10.1007/BF00579808

Laribi, B., Kouki, K., M'Hamdi, M., & Bettaieb, T. (2015). Coriander (Coriandrum sativum L.) and its bioactive constituents. *Fitoterapia*, *103*, 9–26. doi:10.1016/j.fitote.2015.03.012 PMID:25776008

Li, P., Ma, F., & Cheng, L. (2013). Primary and secondary metabolism in the sun-exposed peel and the shaded peel of apple fruit. *Physiologia Plantarum*, *148*(1), 9–24. doi:10.1111/j.1399-3054.2012.01692.x PMID:22989296

Maoka, T. (2020). Carotenoids as natural functional pigments. *Journal of Natural Medicines*, *74*(1), 1–16. doi:10.1007/s11418-019-01364-x PMID:31588965

Marwat, S. K., Khan, M. A., Khan, M. A., Ahmad, M., Zafar, M., Rehman, F., & Sultana, S. (2009). Fruit plant species mentioned in the Holy Qura'n and Ahadith and their ethno medicinal importance. *American-Eurasian Journal of Agricultural & Environmental Sciences*, *5*, 284–295.

Matsuda, H., Hamao, M., Nakamura, S., Kon'i, H., Murata, M., & Yoshikawa, M. (2012). Medicinal Flowers. XXXIII. Anti-hyperlipidemic and anti-hyperglycemic effects of chakasaponins I-III and structure of chakasaponin IV from flower buds of Chinese tea plant (Camellia sinensis). *Chemical & Pharmaceutical Bulletin*, *60*(5), 674–680. doi:10.1248/cpb.60.674 PMID:22689407

Meléndez-martínez, A. J., Mandić, A. I., Bantis, F., Borge, G. I. A., Brnčić, M., Bysted, A., Cano, M. P., Dias, G., Elgersma, A., Fikselová, M., García-alonso, J., Gonçalves, V. S. S., Hornero-méndez, D., Lavelli, V., Manganaris, G. A., Mapelli-brahm, P., Olmedilla-alonso, B., Periago-castón, M. J., Sheehan, J. J., & Šaponjac, V. T. (2022). A comprehensive review on carotenoids in foods and feeds: Status quo, applications, patents, and research needs. *Critical Reviews in Food Science and Nutrition*, *62*(8), 1999–2049. doi:10.1080/10408398.2020.1867959 PMID:33399015

Meziant, L., Bachir-bey, M., Bensouici, C., Saci, F., Boutiche, M., & Louaileche, H. (2021). Assessment of inhibitory properties of flavonoid-rich fig (Ficus carica L.) peel extracts against tyrosinase, α-glucosidase, urease and cholinesterases enzymes, and relationship with antioxidant activity. *European Journal of Integrative Medicine*, *43*, 101272. doi:10.1016/j.eujim.2020.101272

Nunes, A. R., Alves, M. G., Moreira, P. I., Oliveira, P. F., & Silva, B. M. (2014). Can tea consumption be a safe and effective therapy against diabetes mellitus-induced neurodegeneration? *Current Neuropharmacology*, *12*(6), 475–489. doi:10.2174/1570159X13666141204220539 PMID:25977676

Prachayasittikul, V., Prachayasittikul, S., Ruchirawat, S., & Prachayasittikul, V. (2018). Coriander (*Coriandrum sativum*): A promising functional food toward the well-being. *Food Research International*, *105*, 305–323. doi:10.1016/j.foodres.2017.11.019 PMID:29433220

Pratley, R. E. (2013). The early treatment of type 2 diabetes. *The American Journal of Medicine*, *126*(9), S2–S9. doi:10.1016/j.amjmed.2013.06.007 PMID:23953075

Punetha, D., Tewari, G., & Pande, C. (2018). Compositional variability in inflorescence essential oil of Coriandrum sativum from North India. *Journal of Essential Oil Research, 30*(2), 113-119.

Qian, X., Liu, X., Mao, Z., Abdulai, T., Dong, X., Tu, R., Wang, Y., Liu, X., Luo, Z., Qiao, D., Wang, C., & Li, Y. (2020). Fruit and vegetable intake modifies the associations between suppressor of cytokine signaling 3 genetic variants and type 2 diabetes. *European Journal of Nutrition*, *59*(8), 3441–3449. doi:10.1007/s00394-020-02178-4 PMID:31927672

Ramadan, M.F. (2023). *Fig* (Ficus carica): *Production, processing, and properties*. Springer.

Rashidi, B., Malekzadeh, M., Goodarzi, M., Masoudifar, A., & Mirzaei, H. (2017). Green tea and its anti-angiogenesis effects. *Biomedicine and Pharmacotherapy*, *89*, 949–956. doi:10.1016/j.biopha.2017.01.161 PMID:28292023

Silovská, T., Matoušek, J., Fajstavr, D., Švorčík, V., & Kolská, Z. (2020). Antimicrobial effect of polymers grafted with cinnamaldehyde. *Materials Letters*, *277*, 128274. doi:10.1016/j.matlet.2020.128274

Singh, D., Singh, B., & Goel, R. K. (2011). Traditional uses, phytochemistry and pharmacology of Ficus religiosa: A review. *Journal of Ethnopharmacology*, *134*(3), 565–583. doi:10.1016/j.jep.2011.01.046 PMID:21296646

Torres-León, C., & Rebolledo, R. F. (2023). Article. *Saudi Pharmaceut. J., 31*(1), 21–28.

Ugulu, I., Baslar, S., Yorek, N., & Dogan, Y. (2009). The investigation and quantitative ethnobotanical evaluation of medicinal plants used around Izmir province. *Turkey Journal of Medicinal Plant Research, 3*, 345–367.

Valdeyron, G. & Lloyd, D.G. (1979). Sex differences and flowering phenology in the common fig. *Ficus carica L Evolution, 33*, 673-685.

Waltner-Law, M. E., Wang, X. L., Law, B. K., Hall, R. K., Nawano, M., & Granner, D. K. (2002). Epigallocatechin gallate, a constituent of green tea, represses hepatic glucose production. *The Journal of Biological Chemistry, 277*(38), 34933–34940. doi:10.1074/jbc.M204672200 PMID:12118006

Wang, Y. F., Yu, L., & Wei, X. L. (2012). Monosaccharide composition and bioactivity of tea flower polysaccharides obtained by ethanol fractional precipitation and stepwise precipitation. *CYTA: Journal of Food, 10*(1), 1–4. doi:10.1080/19476337.2010.523901

Wei, J.-N., Liu, Z.-H., Zhao, Y.-P., Zhao, L.-L., Xue, T.-K., & Lan, Q.-K. (2019). Phytochemical and bioactive profile of Coriandrum sativum L. *Food Chemistry, 286*(juillet), 260–267. doi:10.1016/j.foodchem.2019.01.171 PMID:30827604

Wei, X. L., Chen, M. A., Xiao, J. B., Liu, Y., Yu, L., Zhang, H., & Wang, Y. F. (2010). Composition and bioactivity of tea flower polysaccharides obtained by different methods. *Carbohydrate Polymers, 79*(2), 418–422. doi:10.1016/j.carbpol.2009.08.030

Whiting, D. R., Guariguata, L., Weil, C., & Shaw, J. (2011). J. Shaw IDF diabetes atlas: Global estimates of the prevalence of diabetes for 2011 and 2030. *Diabetes Research and Clinical Practice, 94*(3), 311–321. doi:10.1016/j.diabres.2011.10.029 PMID:22079683

Woodland, D. W. (1997). *Contemporary Plant Systematics* (2nd ed.). Andrews University Press.

World Health Organization. (2013). *WHO Traditional Medicine Strategy*. World Health Organization.

Xing, L., Zhang, H., Qi, R., Tsao, R., & Mine, Y. (2019). Recent advances in the understanding of the health benefits and molecular mechanisms associated with green tea polyphenols. *Journal of Agricultural and Food Chemistry, 67*(4), 1029–1043. doi:10.1021/acs.jafc.8b06146 PMID:30653316

Yoshikawa, M., Wang, T., Sugimoto, S., Nakamura, S., & Nagatomo, A. (2008). Functional saponins in tea flower (Flower buds of Camellia sinensis): Gastroprotective and hypoglycemic effects of floratheasaponins and qualitative and quantitative analysis using HPLC Yakugaku. *Zasshi-Journal of the Pharmaceutical Society of Japan, 128*(1) 141-151.

Yu, X., Xiao, J., Chen, S., Yu, Y., Ma, J., Lin, Y., Li, R., Lin, J., Fu, Z., Zhou, Q., Chao, Q., Chen, L., Yang, L., & Liu, R. (2020). Metabolite signatures of diverse Camellia sinensis tea populations. *Nature Communications, 11*(1), 5586–5600. doi:10.1038/s41467-020-19441-1 PMID:33149146

Zhang, Y., Chen, J., Zeng, Y., Huang, D., & Xu, Q. (2019). Involvement of AMPK activation in the inhibition of hepatic gluconeogenesis by Ficus carica leaf extract in diabetic mice and HepG2 cells. *Biomedicine and Pharmacotherapy, 109*, 188–194. doi:10.1016/j.biopha.2018.10.077 PMID:30396076

Zhang, Z. B., Xiong, T., Chen, J. H., Ye, F., Cao, J. J., Chen, Y. R., Zhao, Z. W., & Luo, T. (2023). Understanding the origin and evolution of tea (*Camellia sinensis* [L.]): Genomic advances in tea. *Journal of Molecular Evolution*, *91*(2), 156–168. doi:10.1007/s00239-023-10099-z PMID:36859501

Zhao, C., Li, S., Li, C., Wang, T., Tian, Y., & Li, X. (2021). Flavonoids from Fig (*Ficus carica* Linn.) Leaves: The development of a new extraction method and identification by UPLC-QTOF-MS/MS. *Applied Sciences (Basel, Switzerland)*, *11*(16), 7718. doi:10.3390/app11167718

Chapter 9
Rainfall Prediction Using Machine Learning

A. V. Senthil Kumar
https://orcid.org/0000-0002-8587-7017
Hindusthan College of Arts and Science, India

S. Antony Roshan
Hindusthan College of Arts and Science, India

Amit Dutta
All India Council for Technical Education, India

Samrat Ray
Indian Institute of Management, Pune, India

Shadi R. Masadeh
Isra University, Jordan

G. Prasanna Lakshmi
Sandip University, India

Dimitris Michalopoulos
https://orcid.org/0000-0001-7496-8872
University of the People, Greece

Rakesh Nyayapati
UBS Bank, USA

Ismail Bin Musirin
Universiti Teknologi MARA, Malaysia

Gaganpreet Kaur
https://orcid.org/0000-0002-3322-1315
Chitkara University, India

ABSTRACT

Rainfall prediction is a crucial aspect of weather forecasting and plays a significant role in various fields, including agriculture, water resource management, and disaster preparedness. In this chapter, the authors explore the application of two machine learning algorithms, random forest and cat boost, for predicting rainfall events. They utilize historical weather data from a specific location to train and evaluate the performance of both models. The evaluation metrics employed include accuracy, precision, recall, and F1-score. The findings suggest that incorporating additional features, such as humidity, can enhance the predictive capabilities of both random forest and cat boost. Overall, this project demonstrates the effectiveness of random forest and cat boost in predicting rainfall events.

DOI: 10.4018/979-8-3693-3807-0.ch009

Copyright © 2024, IGI Global. Copying or distributing in print or electronic forms without written permission of IGI Global is prohibited.

INTRODUCTION

In Ruslan et al. (2020) rainfall prediction is a pivotal aspect of meteorological science with far-reaching implications across various sectors, including agriculture, water resource management, and disaster preparedness. Understanding and accurately forecasting rainfall patterns contribute significantly to informed decision-making and planning in these fields. In recent years, the integration of advanced machine learning techniques has emerged as a promising avenue for enhancing the precision and reliability of rainfall predictions. By leveraging historical meteorological data and sophisticated algorithms, researchers and practitioners aim to develop models that can discern intricate patterns and relationships within the atmospheric conditions, leading to more accurate and timely predictions (Widiasari et al., 2019). This intersection of meteorology and machine learning represents a frontier in scientific research, seeking to address the challenges posed by the dynamic and complex nature of weather systems. In this context, this study explores the application of machine learning algorithms, specifically Random Forest and Cat Boost, in the realm of rainfall prediction, with the objective of advancing our ability to anticipate and adapt to changing weather patterns. Through the synthesis of historical data and cutting-edge computational approaches, the aim is to contribute to the development of robust and efficient rainfall prediction models, ultimately fostering resilience and preparedness in the face of evolving climatic conditions.

Rainfall, a fundamental component of Earth's climate system, plays a crucial role in shaping ecosystems, sustaining agriculture, and influencing various aspects of human life (Ruslan et al., 2016). Understanding and predicting rainfall patterns are central to effective water resource management, agricultural planning, and disaster preparedness. Traditional methods of weather forecasting have evolved, and contemporary approaches increasingly incorporate advanced technologies, including machine learning, to enhance prediction accuracy. Machine learning algorithms analyze historical meteorological data, discern intricate patterns, and make predictions that aid in proactive decision-making. As climate change introduces greater variability in weather patterns, the demand for more accurate and timely rainfall predictions becomes imperative. This study explores the intersection of meteorology and machine learning in the context of rainfall prediction, aiming to contribute to the development of more robust and efficient models. By delving into historical data and leveraging computational advancements, this research seeks to improve our ability to anticipate and adapt to changing rainfall patterns, ultimately fostering resilience and preparedness in a world characterized by dynamic climatic conditions.

RAINFALL PREDICTION

Rainfall, a fundamental component of Earth's climate system, plays a crucial role in shaping ecosystems, sustaining agriculture, and influencing various aspects of human life (Hiroi & Kawaguchi, 2016). Understanding and predicting rainfall patterns are central to effective water resource management, agricultural planning, and disaster preparedness. Traditional methods of weather forecasting have evolved, and contemporary approaches increasingly incorporate advanced technologies, including machine learning, to enhance prediction accuracy. Machine learning algorithms analyze historical meteorological data, discern intricate patterns, and make predictions that aid in proactive decision-making. As climate change introduces greater variability in weather patterns, the demand for more accurate and timely rainfall predictions becomes imperative. This study explores the intersection of meteorology and machine learning in the context of rainfall prediction, aiming to contribute to the development of more robust and efficient models. By delving into historical data and leveraging computational advancements, this research seeks to improve our ability to anticipate and adapt to changing rainfall patterns, ultimately fostering resilience and preparedness in a world characterized by dynamic climatic conditions.

MACHINE LEARNING

In Boni (2016), machine learning, at the intersection of computer science and statistical modeling, represents a transformative paradigm that empowers systems to learn patterns from data and make informed decisions without explicit programming. In the digital era, the explosive growth of data and computational power has propelled machine learning to the forefront of technological innovation. This field encompasses a diverse range of algorithms and techniques that enable machines to recognize patterns, classify information, and make predictions or decisions autonomously. From recommendation systems and natural language processing to image recognition and predictive analytics, machine learning applications have permeated various industries, revolutionizing the way we approach problem-solving and decision-making. As we delve into an era increasingly dominated by data, the capacity of machine learning to extract meaningful insights and drive advancements across disciplines makes it a cornerstone of contemporary technological progress. This introduction sets the stage for exploring the multifaceted landscape of machine learning, its applications, and its transformative impact on how we interact with and harness the power of information.

CLIMATE MODELING

Climate modeling stands as a cornerstone in our pursuit to comprehend and anticipate the complex dynamics of Earth's climate system. In the face of global environmental changes, understanding the intricacies of climate patterns becomes paramount for scientific inquiry, policy formulation, and proactive adaptation strategies. Climate models, sophisticated numerical representations of atmospheric, oceanic, and terrestrial processes, enable scientists to simulate and analyze the behavior of our planet's climate under various conditions (Adnan et al., 2021). These models serve as invaluable tools for unraveling the interconnected web of climatic factors, projecting future scenarios, and assessing the potential impacts of anthropogenic activities. As technology advances and observational datasets expand, climate modeling evolves, offering increasingly accurate insights into the drivers of climate change. This introduction sets the stage for an exploration into the significance of climate modeling, emphasizing its role in unraveling the complexities of our planet's climate system and contributing essential knowledge to global efforts addressing the challenges posed by a changing climate.

ADVANTAGES OF RAINFALL PREDICTION

- The model has been shown to achieve high accuracy in predicting the rainfall range for a given location.
- The model can predict rainfall in real time, which allows for early warning and response.

- The model can be used for a variety of applications, including flood early warning systems, agricultural water management, hydropower generation, and urban drainage planning.
- The model is relatively affordable to implement and maintain.
- MLR assumes that the relationship between the input variables and the target variable is linear. This assumption may not hold true in reality, especially for complex phenomena like rainfall.
- It is sensitive to outliers. A single outlier can have a significant impact on the model's coefficients and predictions.
- This can be prone to over fitting. This occurs when the model learns the training data too well and is unable to generalize to new data.
- It can be computationally expensive, especially when using a large number of input variables (Abbot & Marohasy, 2011).

FEASIBILITY STUDY

In Akhtar (2019), preliminary investigation examines project feasibility; the likelihood the system will be useful to the organization. The main objective of the feasibility study is to test the Technical, Operational and Economical feasibility for adding new modules and debugging old running system. All system is feasible if they are unlimited resources and infinite time. There are aspects in the feasibility study portion of the preliminary investigation:

- Technical Feasibility
- Operation Feasibility
- Economical Feasibility

Data Pre-processing is pivotal for refining the acquired dataset before it is fed into our machine learning model variables, and scaling numerical features. By rectifying data inconsistencies and ensuring uniformity, this module enhances the overall quality of the dataset.

TECHNICAL FEASIBILITY

In Kanan et al. (2010), the technical issue usually raised during the feasibility stage of the investigation includes the following:

- Does the necessary technology exist to do what is suggested?
- Do the proposed equipment have the technical capacity to hold the data required to use the new system?
- Will the proposed system provide adequate response to inquiries, regardless of the number or location of users?
- Can the system be upgraded if developed?
- Are there technical guarantees of accuracy, reliability, ease of access and data security?

Earlier no system existed to cater to the needs of 'Secure Infrastructure Implementation System'. The current system developed is technically feasible. It is a web based user interface for audit workflow at DB2 Database. Thus it provides an easy access to the users. The database's purpose is to create, establish and maintain a workflow among various entities in order to facilitate all concerned users in their various capacities or roles. Permission to the users would be granted based on the roles specified.

Therefore, it provides the technical guarantee of accuracy, reliability and security. The software and hard requirements for the development of this project are not many and are already available in-house at NIC or are available as free as open source. The work for the project is done with the current equipment and existing software technology. Necessary bandwidth exists for providing a fast feedback to the users irrespective of the number of users using the system.

OPERATIONAL FEASIBILITY

In Geetha and Selvaraj (2010), proposed projects are beneficial only if they can be turned out into information system. That will meet the organization's operating requirements. Operational feasibility aspects of the project are to be taken as an important part of the project implementation. Some of the important issues raised are to test the operational feasibility of a project includes the following: -

- Is there sufficient support for the management from the users?
- Will the system be used and work properly if it is being developed and implemented?
- Will there be any resistance from the user that will undermine the possible application benefits?

This system is targeted to be in accordance with the above-mentioned issues. Beforehand, the management issues and user requirements have been taken into consideration. So there is no question of resistance from the users that can undermine the possible application benefits.

The well-planned design would ensure the optimal utilization of the computer resources and would help in the improvement of performance status.

ECONOMIC FEASIBILITY

In Guhathakurta (2008), a system can be developed technically and that will be used if installed must still be a good investment for the organization. In the economic feasibility, the development cost in creating the system is evaluated against the ultimate benefit derived from the new systems. Financial benefits must equal or exceed the costs.

The system is economically feasible. It does not require any addition hardware or software. Since the interface for this system is developed using the existing resources and technologies available at NIC, there is nominal expenditure and economical feasibility for certain.

TYPES OF MAINTENANCE

- Corrective maintenance

- Adaptive maintenance
- Perfective maintenance
- Preventive maintenance

CORRECTIVE MAINTENANCE

In Tun and Sein (2018), changes made to a system to repair flows in its design coding or implementation. The design of the software will be changed. The corrective maintenance is applied to correct the errors that occur during that operation time. The user may enter invalid file type while submitting the information in the particular field, then the corrective maintenance will displays the error message to the user in order to rectify the error.

Maintenance is a major income source. Nevertheless, even today many organizations assign maintenance to unsupervised beginners, and less competent programmers.

The user's problems are often caused by the individuals who developed the product, not the maintainer. The code itself may be badly written maintenance is despised by many software developers unless good maintenance service is provided, the client will take future development business elsewhere. Maintenance is the most important phase of software production, the most difficult and most thankless.

ADAPTIVE MAINTENANCE

In Kanna and Ghosh (2010), it means changes made to system to evolve its functionalities to change business needs or technologies. If any modification in the modules the software will adopt those modifications. If the user changes the server then the project will adapt those changes. The modification server work as the existing is performed.

PERFECTIVE MAINTENANCE

Perfective maintenance means made to a system to add new features or improve performance. The perfective maintenance is done to take some perfect measures to maintain the special features. It means enhancing the performance or modifying the programs to respond to the users need or changing needs. This proposed system could be added with additional functionalities easily. In this project, if the user wants to improve the performance further then this software can be easily upgraded.

PREVENTIVE MAINTENANCE

Preventive maintenance involves changes made to a system to reduce the changes of features system failure. The possible occurrence of error that might occur are forecasted and prevented with suitable preventive problems. If the user wants to improve the performance of any process then the new features can be added to the system for this project.

Methodology

This chapter carries out an in-depth review of relevant literature to reveal the different practices authors take to predict rainfall. The review covers several aspects which relate to the input into, output from, and methods used in the various system devised in the literature for this purpose. The review specifically focuses on studies that use supervised learning for both regression and classification problems.

Literature on Rainfall Prediction Using Machine Learning Model

Figure 1. PicChart

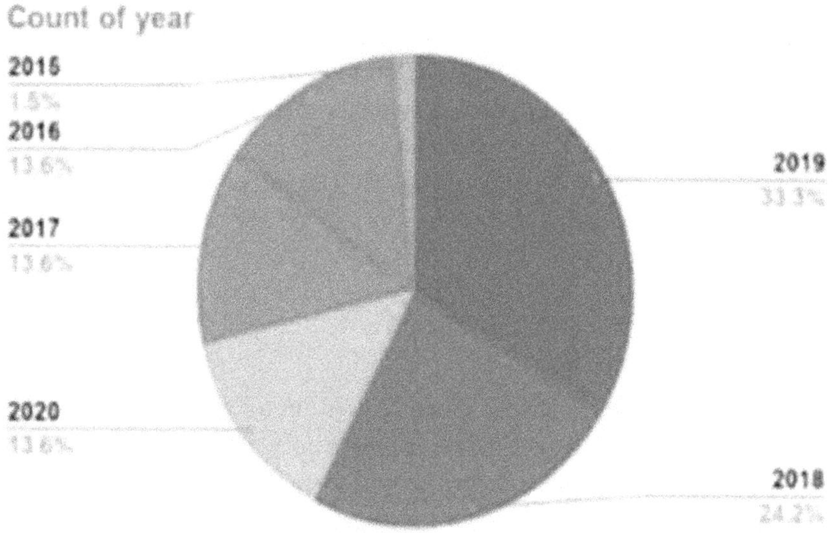

In Kumar et al. (2012), rainfall is a key geophysical parameter that is essential for many applications water resource management, especially in the agriculture sector. Predicting rainfall can help managers in various sectors to make decisions regarding a range of important activities such as crop planting, traffic control, the operation of sewer systems, and managing disasters like droughts and floods.

The his chapter serves as an addition to the field by surveying recent relevant studies focusing on the use of ML in rainfall prediction in a variety of geographic locations from 2016–2020. After detailing the methods used to forecast rainfall, one of the important contributions of this chapter is to demonstrate various pitfalls that lead to an overestimation in model performance of the ML models in various papers. Thisin turn leads to unrealistic hype and expectations surrounding ML in the current literature. It also leads to an unrealistic understanding of the advancements in, and gains by, ML research in this field. It is therefore important to clearly state and demonstrate these pitfalls in order to help researchers avoid them.

This chapter carries out an in-depth review of relevant literature to reveal the different practices authors take to predict rainfall. The review covers several aspects which relate to the input into, output from, and methods used in the various systems devised in the literature for this purpose. The review specifically focuses on studies that use supervised learning for both regression and classification problems (Htike Khalifa, 2010).

Figure 2. Experimental methodology

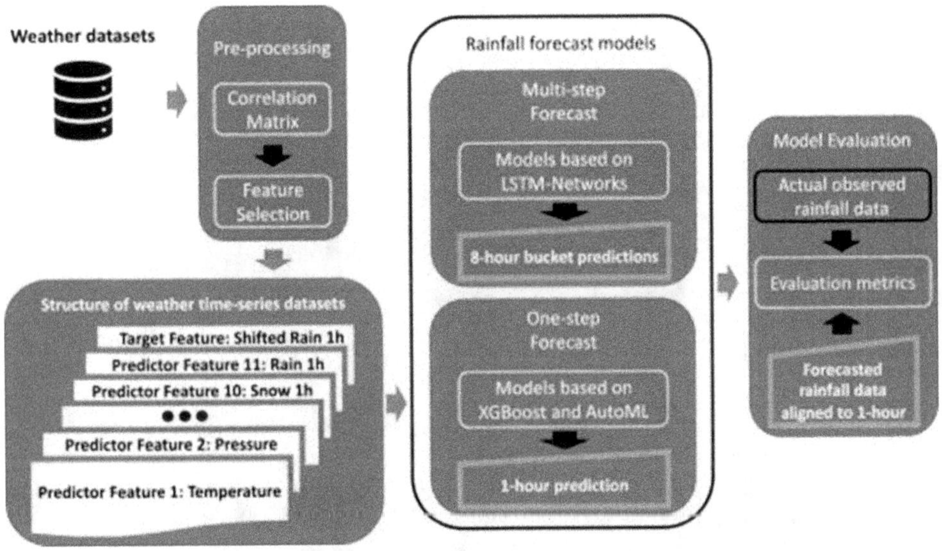

Rainfall Prediction Approach

- The datasets as a baseline. Perform a non-exhaustive hyperparameter search to obtain the best hyperparameter values to train each model.
- Use an AutoML tool to explore the performance of different well-known regression algorithms in the literature. For each of the five datasets, train a model recommended by the AutoML tool.
- From the literature review, build two models based on LSTM-Networks and one model based on Stacked-LSTM Networks. For each of the five datasets, train an individual version of the three models built. During the training process of each individual model, perform a non-exhaustive hyperparameter grid search to obtain the best hyperparameter values that boost its performance.
- From the previous models based on LSTM-Networks and Stacked-LSTM Networks, identify the model that achieves the best performance and accuracy across all five datasets. Based on the architecture of this model, build two different model versions of Stacked-LSTM Networks and one Bidirectional-LSTM Networks model.

Figure 3. Correlation matrix of the city of the both

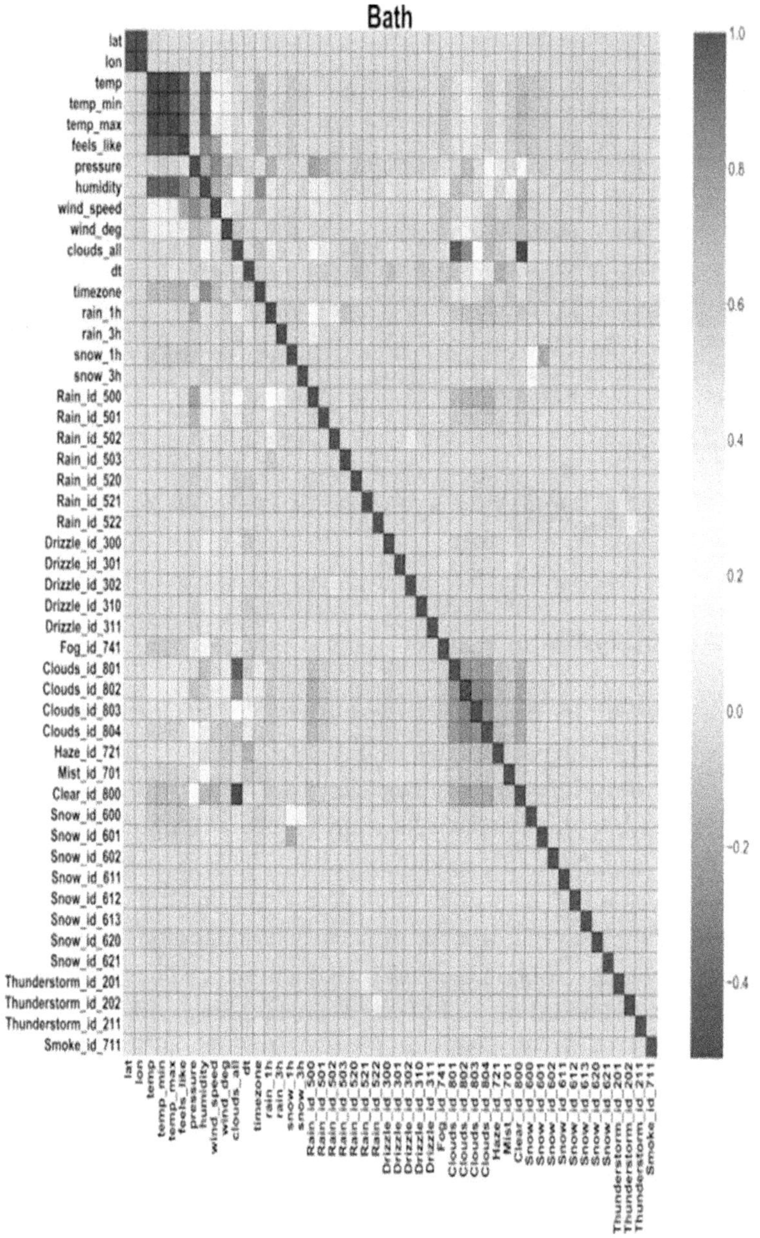

Table 1. Tuples of features

Tuples of features			
Number	**Tuple**	**Number**	**Tuple**
1	Latitude - Longitude	5	Min Temp - Max Temp
2	Temp - Min Temp	6	Min Temp - Feels Like
3	Temp - Max Temp	7	Max Temp - Feels Like
4	Temp - Feels Like	8	Snow 1h - Snow ID 601

- Latitude or Longitude features cannot be considered independent features since they both represent a geographic coordinate, making a proper interpretation of one without the other impossible. Considering that the datasets are divided by each city, and the previous statement, both features are removed from the datasets.
- All features related to temperature measurements are highly correlated one to another. Thus, the main feature is retained (temperature) and the rest features that concentrates the maximum, minimum, and feels like values are removed
- The weather code of snow ID 601 is highly correlated with the feature that measures the volume of snow in one hour. The main weather feature is Snow 1 h and the Snow ID 601 feature represents an internal code used by the OpenWeather services. Consequently, the feature Snow ID 601 is removed (Nikhil Binoye et al., 2019).

Data Preparation

- In Kanna et al. (2010), the problem solved at this paper is that of predicting the accumulated rainfall precipitation for the next day, by using data gathered in the previous days.
- More specifically, we employed data from a meteorological station located in a central area of Manizales. The station makes transmissions every five minutes, which are grouped into a daily time series covering from 2002 to 2013.
- The dataset used for model validation contains a total of forty-seven explanatory characteristics, including temperature, relative humidity, barometric pressure, sun brightness, speed and direction of wind, and other derivatives of the reset.
- Multiple network architectures were implemented with different configurations for the autoencoder and the MLP. The trainings and tests were carried out in parallel on a cluster consisting of 72 nodes.

- Initially, several combinations of value parameters were tested in order to look for the best possible architecture configuration. More specifically, we carried out a grid search for the following set of parameters, the Table 2 presents the approved values for each parameter.

Figure 4. Rainfall volume in the dataset

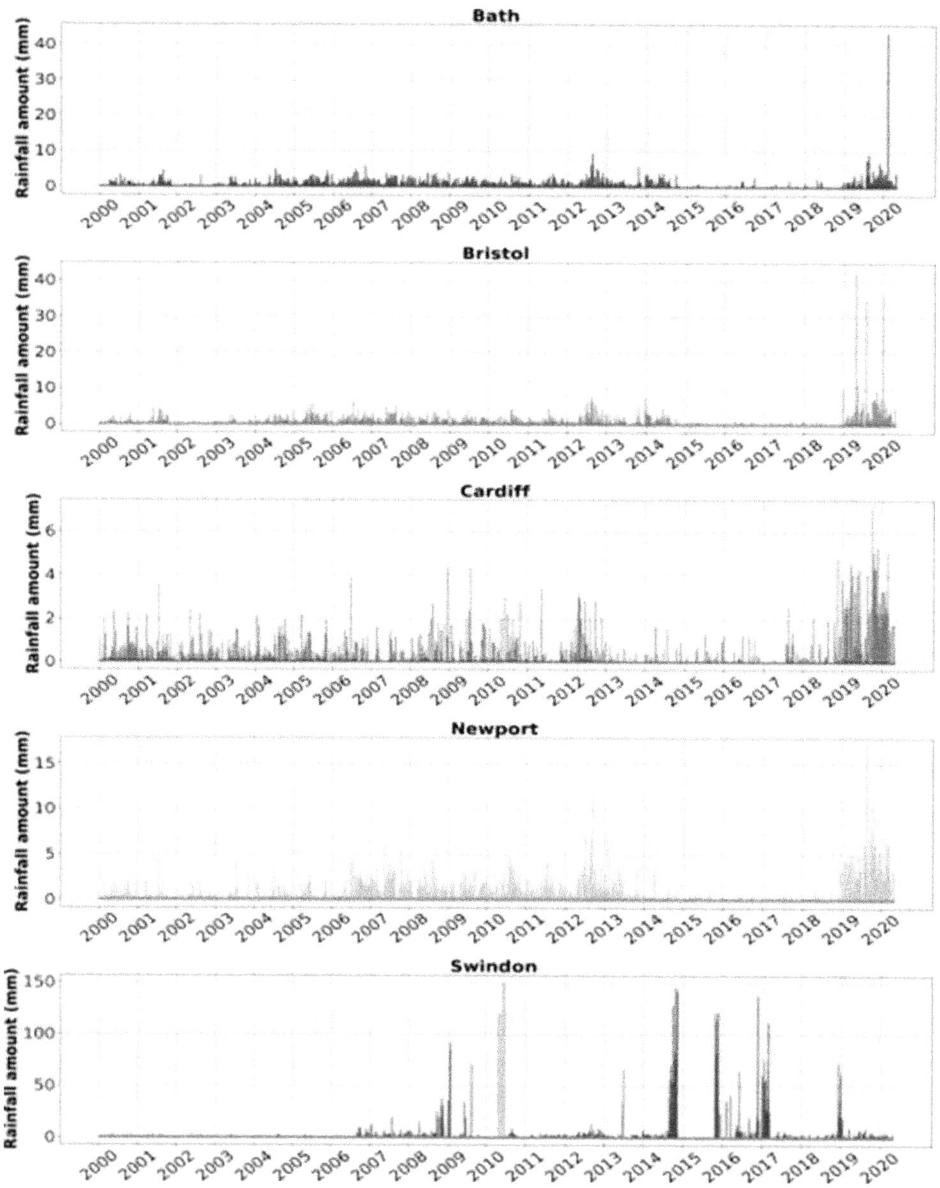

Dataset Pre-Processing

- In Wu et al. (2010), separate weather data by city. This step aims to preserve the weather singularities of each city during the training process.

- Obtain the nullness percentage of each feature. This step help in the decision of removing or filling up missing values of each column vector. The nullness percentage values of each feature in the five major UK cities are concentrated in AppendixRemove the sea and ground-level measurements. Both features present 100% of nullness therefore, these features cannot be used in the experimental phaseRemove the city name and date-time iso features. The city name is no longer required due to the process of dividing the data for each city. The date and time values of the records are duplicated in the date-time feature; then, the date-time iso column is removed. Replenish the missing values of rain 1 h, rain 3 h, snow 1 h, and snow 3 h features. These features refer to the volume of rain and snow that fell during one and three hours. Consequently, missing values of these features were filled in with zeros. This approach follows the rationale that for any hour that any of these values are missing, there was no precipitation.

CONCLUSION

This study set out to compare the prediction performance of rainfall forecasting models based on LSTM-Networks architectures with modern Machine Learning algorithms. To achieve this objective, 2 models based on LSTM-Networks, 3 models based on Stacked-LSTM, and 1 Bidirectional-LSTM Networks model were compared with an XGBoost (baseline model) and an ensemble model that resulted from carrying out an Automated Machine Learning approach.

In order to evaluate the performance of the implemented rainfall forecasting models, historical weather data of two decades from the UK cities of Bath, Bristol, Cardiff, Newport, and Swindon were used. Good practices were implemented when conducting Machine Learning experiments such as (a) a pre-processing procedure to eliminate categorical and incomplete data within the datasets, (b) a feature selection process carried out through a Correlation Matrix analysis, (c) implementation of a non-exhaustive hyperparameters grid search to obtain the best performance of models based on XGBoost and LSTM-Networks architectures.

To the authors' knowledge, this study is the first to present a comparative analysis of the performance of rainfall forecasting models based on modern Machine Learning algorithms in predicting hourly rainfall volume using weather time-series data from cities in the United Kingdom. Specifically, the main contributions of this study are highlighted below:

- From the literature reviewed, three models based on the LSTM and Stacked-LSTM Networks were adapted for the task of forecasting hourly rainfall using time-series data from five major UK cities.
- A model based on Bidirectional-LSTM Networks was proposed for the task of forecasting rainfall on an hourly basis using time-series data from five major UK cities.
- A comparison of the performance of models based on LSTM-Networks, Stacked-LSTM Networks, Bidirectional-LSTM Networks, XGBoost, and an ensemble model resulted from performing an AutoML approach was performed in the task of forecasting the amount of rainfall per hour using time-series data from five major UK cities.

Results showed that a Bidirectional-LSTM Network can be used as a rainfall forecast model with comparable performance to a Stacked-LSTM Network with two hidden layers. Among all the rainfall prediction models tested in the comparison, which included three models adapted from the literature, the Stacked-LSTM (Model 4) and Bidirectional-LSTM (Model 6) models achieved, in general, lower values in the evaluation metrics RMSE, MAE, and RMSLE. Model 4 achieved Loss values between 0.0014–0.0001, RMSE values in the range of 0.0375–0.0084, MAE values between 0.0071–0.0013, and RMSLE values ranging between 0.0157–0.0037. On the other hand, Model 6 obtained values ranging between 0.0014–0.0001, 0.0377–0.0099, 0.0072–0.0015, and 0.0111–0.0044 for Loss, RMSE, MAE, and RMSLE, respectively.

REFERENCES

Abbot, J., & Marohasy, J. (2012). Application of Artificial Neural Networks to rainfall forecasting in Queensland, Australia. *Advances in Atmospheric Sciences*, *29*(4), 717–730. doi:10.1007/s00376-012-1259-9

Adnan, R. (2016). *A 3-hours river water level rainfall prediction model using NNARX with improves modelling strategy*. 2021 6th International Conference on System Engineering and Technology (ICSET), Bandung. 10.1109/ICSEngT.2016.7849617

Akhtar, N., Rehman, A., Hussnain, M., Rohail, S., Missen, M. S., Nasir, M., Hayder, A., Salamat, N., & Pasha, M. (2019). Hierarchical Coloured Petri-Net Based Multi-Agent System for Rainfall Monitoring, Prediction, and Rescue (FMPR). *IEEE Access : Practical Innovations, Open Solutions*, *7*, 180544–180557. doi:10.1109/ACCESS.2019.2958258

Boni, G. (2016, June). A Prototype System for Rainfall Monitoring Based on Rainfall Forecast Combined With COSMO-SkyMed and Sentinel-1 Data. *IEEE Journal of Selected Topics in Applied Earth Observations and Remote Sensing*, *9*(6), 2794–2805. doi:10.1109/JSTARS.2016.2514402

Deshpande, R. R. (2012). On the rainfall time series prediction using Multilayer Perceptron Artificial Neural Network. *International Journal of Emerging Technology and Advanced Engineering*, *2*(1), 148–153.

Geetha & Selvaraj. (2011). Prediction of monthly rainfall in Chennai using Back Propagation Neural Network model. *Int. J. of Eng. Sci. and Technology, 3*(1), 211-213.

Guhathakurta, P. (2008). Long lead monsoon rainfall prediction using deterministic Artificial Neural Network model. *Meteorology and Atmospheric Physics*, *101*, 93–108. doi:10.1007/s00703-008-0335-2

Hiroi & Kawaguchi. (2016). RainfallEye: Real-time flash rainfall prediction system for urban complex water flow. 2021 IEEE SENSORS, 1-3. . doi:10.1109/ICSENS.2016.7808626

Htike, K. K., & Khalifa, O. O. (2010). Rainfall forecasting models using Focused Time-Delay Neural Networks. Comput. and Commun. Eng. (ICCCE), Int. Conf. on IEEE. 10.1109/ICCCE.2010.5556806

Kannan, M., Prabhakaran, S., & Ramachandran, P. (2010). Rainfall Forecasting Using Data Mining Technique. *IACSIT International Journal of Engineering and Technology*, *2*(6), 397–401.

Kannan & Ghosh. (2010). Prediction of daily rainfall state in a river basin using statistical downscaling from GCM output. Springer-Verlag.

Kumar, A., Kumar, A., Ranjan, R., & Kumar, S. (2012). *A rainfall prediction model using artificial neural network. In Control and Syst. Graduate Research Colloq*. ICSGRC.

Li, C., Peng, J., Wang, H., & Yang, S. X. (2017). A rainfall prediction method based on streaming big data processing. *2022 IEEE International Conference on Information and Automation (ICIA)*, 898-902. 10.1109/ICInfA.2017.8079030

Nikhil Binoy, C., Arjun, N., Keerthi, C., Sreerag, S., & Nair, A. H. (2019). Rainfall Prediction Using Flow and Depth Measurement with Artificial Neural Network in Canals. 2019 3rd International Conference on Computing Methodologies and Communication (ICCMC), 798-801. 10.1109/ICCMC.2019.8819807

Ruslan, F. A., Samad, A. M., & Adnan, R. (2020). Modelling of rainfall prediction system using hybrid NNARX and Extended Kalman Filter. *2017 IEEE 13th International Colloquium on Signal Processing & its Applications (CSPA)*, 149-152. 10.1109/CSPA.2017.8064941

Ruslan, F. A., Samad, A. M., Tajjudin, M., & Adnan, R. (2016). 7 hours rainfall prediction modelling using NNARX structure: Case study Terengganu. 2019 IEEE 12th International Colloquium on Signal Processing & Its Applications (CSPA), 263-268. 10.1109/CSPA.2016.7515843

Sharma, M. A., & Singh, J. B. (2011). *Comparative Study of rainfall forecasting models*. New York Sci. J.

Tun, P. P. P., & Sein, M. M. (2018). Rainfall Prediction System for Middle Region of Myanmar. *2018 IEEE 7th Global Conference on Consumer Electronics (GCCE)*, 769-770. 10.1109/GCCE.2018.8574757

Widiasari, Nugroho, & Widyawan. (2017). Deep learning multilayer perceptron (MLP) for rainfall prediction model using wireless sensor network based hydrology time series data mining. 2019 International Conference on Innovative and Creative Information Technology (ICITech), 1-5. . doi:10.1109/INNOCIT.2017.8319150

Wu, C. L., Chau, K. W., & Fan, C. (2010). Prediction of rainfall time series using Modular Artificial Neural Networks coupled with data preprocessing techniques. *Journal of Hydrology (Amsterdam)*, *389*(1), 146–167. doi:10.1016/j.jhydrol.2010.05.040

Chapter 10
Recent Approaches for Mitigating Climate Change in Agriculture

Saddam Hussain
https://orcid.org/0000-0001-7706-6778
University of Florida, USA

Umer Sharif
Muhammad Nawaz Shareef University of Agriculture, Multan, Pakistan

Muhammad Zakwan Hadeed
University of Veterinary and Animal Sciences, Lahore, Pakistan

Aamir Raza
https://orcid.org/0009-0001-1867-2660
University of Agriculture, Faisalabad, Pakistan

Muhammad Kashif Aman
https://orcid.org/0000-0003-1169-3132
University of Agriculture, Faisalabad, Pakistan

Muhammad Safdar
https://orcid.org/0009-0006-1779-6967
University of Agriculture, Faisalabad, Pakistan

Nasir Habib
The Islamia University of Bahawalpur, Pakistan

Muntaha Munir
https://orcid.org/0009-0005-3404-3636
University of the Punjab, Lahore, Pakistan

ABSTRACT

Climate change is posing significant challenges to global agriculture, necessitating the development and adoption of sustainable practices. This chapter explores innovative approaches to mitigate climate change's impact on agriculture, including precision agriculture, agroforestry, conservation agriculture, and sustainable livestock management. It emphasizes the role of organic farming in improving soil health and enhancing carbon sequestration in agricultural soils. The chapter also discusses the importance of efficient water use in agriculture and the integration of renewable energy sources like solar, wind, and biomass. It also discusses the role of policy and economic instruments in facilitating the transition to sustainable practices, supported by case studies and success stories. The chapter concludes by identifying challenges faced in adopting these practices and providing insights into future directions and emerging technologies in this field.

DOI: 10.4018/979-8-3693-3807-0.ch010

Copyright © 2024, IGI Global. Copying or distributing in print or electronic forms without written permission of IGI Global is prohibited.

INTRODUCTION

Aquaponics and Hydroponics

Hydroponics, the practice of growing plants in water that has a high concentration of nutrients, has been in existence for more than a hundred years (Sharma , *et al.*, 2018). The system employs several closed-loop mechanisms in which the roots of plants are suspended in a fluid that contains vital minerals and dissolved oxygen. By circumventing the constraints of soil absorption, this method of directly delivering nutrients results in accelerated growth rates and increased yields in comparison to conventional agriculture.

Aquaponics enhances hydroponics by integrating it with aquaculture, which involves the cultivation of aquatic creatures such as fish (Rakocy, 2012). The nutrient-rich effluent from the fish tank serves as a fertilizer for the plants cultivated in the hydroponic system in this symbiotic arrangement. Beneficial microorganisms transform ammonia derived from fish waste into nitrates that can be readily absorbed by plants, creating a self-sustaining environment that saves water and reduces waste. Aquaponic systems have the capacity to use up to 90% less water compared to conventional agriculture, making them very suitable for dry regions or countries experiencing water constraint (Obirikorang , *et al.*, 2021).

Aquaponics and hydroponics are modern agricultural methodologies that provide a departure from conventional soil-based farming methods. Aquaponics is a self-contained system that combines aquaculture with hydroponics, creating a mutually beneficial connection between aquatic creatures and plants. This system utilizes fish excrement as a source of nutrients for plant development, while the plants in turn cleanse the water for aquatic life.

Historical Background and Evolution

Aquaponics and hydroponics are agricultural technologies that have transitioned from old approaches to modern solutions. The origins of these technologies may be traced back to the Aztec Chinampas circa 1000 AD, when fish played a vital role as fertilizers, producing a mutually beneficial system akin to contemporary aquaponics (Henkel, 2015). The Hanging Gardens of Babylon, built around 600 BC, serve as early illustrations of hydroponics, demonstrating the practice of growing plants using water supplied via complex networks. The Floating Rice Gardens in Southeast Asia showcased a sophisticated comprehension of nutrient fluxes in aquatic ecosystems by combining the development of rice paddies with fish. The use of "water of dung" for plant nourishment, arising from modernization and technological progress, anticipated the principles of wastewater usage in present-day aquaponics.

In the 1970s, the New Alchemy Institute was founded with a specific emphasis on combining aquaculture with hydroponics. This laid the groundwork for the development of contemporary aquaponic systems. The efficiency and potential of aquaponics and hydroponics have been further boosted by technological developments in automation, sensor technology, and vertical farming. The historical evolution of various agricultural technologies demonstrates humanity's ongoing quest for sustainable and efficient practices. The development of aquaponics and hydroponics, from historical experimentation to contemporary scientific methods, demonstrates our continuous commitment to enhancing food production while reducing environmental harm. Analyzing this historical backdrop offers a more profound comprehension of humanity's persistent endeavor to develop a more sustainable and productive relationship with the earth.

PRINCIPLES

Hydroponics

Hydroponics, a revolutionary technique, involves growing plants without soil by manipulating their basic requirements for survival. It provides a structure for achieving the most favorable conditions for plant development in this soil-free environment. The fundamental tenets of hydroponics include the provision of nutrients, the provision of root support and oxygenation, the maintenance of regulated conditions, and the implementation of effective water management as shown in figure 1. Efficient nutrient delivery is vital in hydroponic systems, since it directly supplies necessary minerals to the plant's roots via a nutrient solution, hence optimizing absorption and promoting maximum growth potential. Root support and oxygenation are crucial in hydroponic systems, which use various substrates like perlite, rockwool, or recycled materials like coconut coir to anchor roots and facilitate oxygen passage.

Controlled settings, such as greenhouses, provide meticulous control over light, temperature, and humidity, therefore reducing plant stress and maximizing growth potential. Some crops, such as lettuce, may be harvested with a much higher speed of up to 50% when using hydroponic systems as opposed to traditional soil-based approaches (Wang , *et al.*, 2023). Effective water management is crucial in hydroponics, since closed-loop systems that recirculate the nutrient solution greatly decrease water use in comparison to traditional agriculture. Effective control of microbial populations inside the system may facilitate the decomposition of nutrients and safeguard plants against detrimental diseases.

Hydroponic systems have the capacity to reduce water use compared to conventional agriculture, making them well-suited for locations facing water scarcity. Utilizing hydroponics, vertical farming may greatly enhance the food production capacity in metropolitan regions that have limited land resources. Hydroponics offers a method for achieving sustainable and effective food production by controlling the plant's surroundings and accurately supplying necessary supplies.

Aquaponics

Aquaponics is an environmentally-friendly agricultural technique that integrates aquaculture with hydroponics to create a mutually beneficial and efficient system (Pantanella, 2018). The system utilizes a closed-loop recirculating water system, resulting in significantly reduced water use, making it more water-efficient compared to conventional soil-based agriculture as shown in figure 2 below. The fundamental principle of aquaponics is in the symbiotic interdependence between fish and plants, where fish waste acts as a valuable food supply for plants, while plants effectively cleanse and purify the water. Biological filtration, also known as the nitrogen cycle, is a vital process in aquaponics. During this process, beneficial bacteria transform ammonia, which is present in fish waste, into nitrites and nitrates. These compounds are critical nutrients that promote plant development. This mechanism guarantees an equilibrium and enduring ecology. Aquaponics also integrates systems for pH and water quality regulation, guaranteeing the ideal pH levels for the well-being of both fish and plants.

The efficacy of aquaponics in promoting plant development, resulting in greater basil harvests in aquaponic systems as compared to conventional hydroponics. In addition, aquaponics enhances resource efficiency by using fish waste as a fertilizer supply for plants, therefore reducing the environmental consequences linked to conventional farming. Aquaponics tackles issues like water shortage and nutrient contamination by fostering a symbiotic relationship between aquatic and plant organisms. The fundamentals of aquaponics include a mutually beneficial interaction between fish and plants, hydroponic farming, the recycling of nutrients, the management of water, the nutrition of plants, the chemistry of water, and the ecology of microorganisms.

Plant nutrition is crucial for achieving optimum development, and aquaponic plants require the same vital nutrients as plants cultivated on soil. The chemical composition of water has a vital role in determining the availability of nutrients and the activity of microorganisms. Additionally, water quality and the health of plants are influenced by factors such as temperature and the amount of dissolved oxygen present. An assortment of advantageous bacteria is crucial for the processes of nitrification, waste breakdown, and disease prevention. It is important to provide favorable circumstances for these microorganisms in order to maintain the general well-being of the aquaponic system. Aquaponics systems may attain agricultural yields that are 2-3 times greater than those of conventional agriculture for certain crops. Additionally, it mitigates soil erosion, fertilizer runoff, and pesticide use, so promoting a more sustainable food production system.

SYSTEM DESIGN AND COMPONENTS

Hydroponics

The design of a hydroponics system is a crucial factor that directly impacts the effectiveness and achievement of soilless growing. For successful implementation and optimization of hydroponic systems, it is crucial to have a thorough understanding of the related components as discussed below (Conn , *et al.*, 2013,Yanes , *et al.*, 2020).

Designs of the System

The Nutrient Film Technique (NFT) is a widely used method for cultivating leafy greens and herbs. In NFT systems, a constant flow of a thin film of nutrient solution is circulated over the roots of the plants. Channels or troughs serve as containers for the plants, enabling their roots to hang and directly receive nutrients from the flowing fluid. This method guarantees optimal root contact and effective absorption of nutrients.

Deep Water Culture (DWC) is a suitable method for cultivating bigger plants such as tomatoes and peppers. In DWC systems, the plants are suspended in floating rafts or baskets inside a deep reservoir filled with nutritional solution. The roots penetrate the fluid, consistently immersed in an abundant supply of nutrients and oxygen. Close observation of water temperature and levels of dissolved oxygen is essential for the proper functioning of this system.

The Ebb and Flow system, also known as Flood and Drain, is a flexible method that cyclically inundates the cultivation substrate with a nutrient solution, followed by letting it to recede back into a reservoir. The alternating rise and fall of water levels generates a cycle of aeration, which supplies the roots with new oxygen as the water is drained. It is versatile and may be used with many types of growth soil and crops.

Aeroponics is an advanced method of gardening that involves suspending plants in mist-filled chambers. The roots of the plants are sprayed with a nutrient solution at regular intervals, allowing them to grow without being rooted in soil. This technique completely removes the need for a substrate, so optimizing the flow of air and perhaps expediting the growth process. Nevertheless, it requires careful regulation of nutrition administration and the frequency of misting.

Vertical farming utilizes hydroponics to optimize space by using vertical systems. Utilizing stacking grow beds, hanging towers, or multi-tiered racks optimizes land use, making it well-suited for urban agriculture and locations with limited land availability.

Components

The system comprises a reservoir, pump, timer, Nutrient Dosing System, optional Growing Medium, Lighting, and Environmental Control Systems. The reservoir functions as a container for nutrient solution and serves as the central hub, its dimensions determined by the number of plants and the design. The pump enables the movement of the solution, ensuring even distribution and oxygenation. The Nutrient Dosing System dispenses concentrated nutrients to ensure optimal nutrient levels are maintained. The pH and EC meter measure the acidity and electrical conductivity of the solution, enabling adjustments to ensure optimal nutrient availability. Perlite, rockwool, and coconut coir are types of growing mediums that offer structural support to plant roots. Lighting enhances or replaces natural sunlight, ensuring the most favorable conditions for photosynthesis. Environmental Control Systems employ sensors to regulate temperature, humidity, and ventilation to achieve optimal plant growth.

Appropriate System Selection

The decision-making process of an appropriate system design and components is contingent upon several aspects, such as the selection of plants, the allocated money, space limitations, and technical expertise. Various crops possess distinct requirements and root structures, which impact the appropriateness of specific systems. The expenses are contingent upon the intricacy of the system, level of automation, and the components used.

The success of hydroponic systems is characterized by the tight coordination of components and thorough design. Every component, from the dynamic core of the reservoir to the regulated dispersion of mist in aeroponics, serves a crucial function in establishing a harmonious and supportive setting for the cultivation of plants without soil. Gaining knowledge about the system designs and fundamental components enables individuals to effectively traverse the dynamic field of hydroponics and harness its capacity for sustainable and efficient food cultivation.

Aquaponics

Aquaponics is a complex combination of aquaculture and hydroponics that depends on a well-planned system to enable the mutually beneficial interaction between fish and plants. The interplay among components is crucial for establishing a sustainable and efficient environment. Exploring the intricacies of system design uncovers the captivating interaction between technology and biology, highlighting the cleverness behind this sustainable food production technique.

Components

The aquaponic system comprises a fish tank, grow beds, pumps, piping, biofilter, sump tank, and monitoring and control devices. The fish tank accommodates many freshwater species such as tilapia, catfish, and koi, and its dimensions and structure may vary depending on its size. Grow beds provide both structural reinforcement and a spacious surface area for efficient nutrient absorption, while also allowing for customizable water flow and system intricacy. Pumps and pipes enable the movement of water containing nutrients from the fish tank to plant beds and filters.

The biofilter houses microorganisms that facilitate the conversion of ammonia, derived from fish waste, into nitrates that are used by plants. Additional designs include media filters, moving bed filters, and plant-based systems. The sump tank functions as a storage unit for accumulated water and houses submersible pumps, enabling the force of gravity to return the water to the fish tank. Advanced systems may include sensors and controllers to monitor water quality parameters like pH, temperature, and nutrient levels. Automated adjustments may be made to guarantee optimal conditions for aquatic life and plants.

System Explanation

Aquaponic systems use a synergistic blend of technology and biology to achieve sustainable food production. The systems consist of many components such as horizontal or vertical grow beds, bell-siphon, media filters, and plant-based filters. These systems need a substantial space for efficient functioning and necessitate periodic maintenance. Biofilters, such as moving bed filters, media filters, and plant-based filters, provide a substantial surface area for bacterial colonization, but they need a constant flow of water.

Surveillance and management systems are crucial for ensuring the optimal operation of the system. By using sensors, it is possible to monitor various parameters like pH, temperature, dissolved oxygen, and nutrient concentrations. Controllers may then make automated adjustments to pumps, lights, and ventilation systems according on the data collected. Aquaponic systems may be customized to accommodate diverse specifications and circumstances. Typical variations of aquaponics include recirculating aquaponics, which employs a closed-loop system with continuous water circulation, non-recirculating aquaponics, which involves periodic water changes, and integrated aquaponics, which combines hydroponics with a small component.

Understanding the system architecture and components of aquaponics unveils the intricate interaction between technology and biology, facilitating the feasibility of this sustainable method of food production. Every element, ranging from the fish tank sanctuary to the grow beds, has a vital function in fostering the mutually beneficial connection between plants and fish. Through a thorough analysis of these factors, we can fully comprehend the capacity of aquaponics to revolutionize the practice of food growing and provide a more ecologically sustainable future.

BENEFITS OF AQUAPONICS AND HYDROPONICS

Environmental Benefits

Water Conservation

Aquaponics and hydroponics represent a fundamental change in how we use water, making a substantial contribution to worldwide water conservation initiatives. Traditional agriculture is well recognized for its significant water requirements, with estimates suggesting that it may use more than 2,000 liters of water to produce a single kilogram of tomatoes (Haq , *et al.*, 2021). On the other hand, aquaponic and hydroponic systems exhibit exceptional water efficiency by using recirculating systems that limit water waste. Aquaponic systems exhibit a water use reduction of up to 90% compared to traditional soil-based agriculture (Goddek , *et al.*, 2015). Hydroponic systems, on average, use 70-90% less water in comparison to conventional agricultural techniques (Elmulthum , *et al.*, 2023).

Reduced Land Use

Aquaponic and hydroponic systems provide a compact and regulated structure, enabling the practice of vertical farming and optimal usage of space. Conventional farming requires extensive areas of cultivable land, leading to deforestation and the destruction of natural habitats. Hydroponic towers and aquaponic systems, on the other hand, enable the cultivation of crops in urban areas and the conversion of non-arable ground for agricultural purposes. Vertical hydroponic systems have the ability to enhance crop productivity per unit area, making them well-suited for urban agriculture. Aquaponics allows for the efficient combination of fish farming with crop production, resulting in a mutually beneficial relationship that maximizes the use of space (Pantanella, 2012).

Lower Chemical Usage

Aquaponic and hydroponic systems mitigate the need for synthetic fertilizers and pesticides, hence reducing chemical dependency. Conventional agricultural methods often include the use of chemical inputs, resulting in soil deterioration and water contamination. Hydroponics involves growing plants in water solutions that are rich in nutrients, which removes the need for fertilizers found in soil. On the other hand, aquaponics uses the waste produced by fish to provide nourishment for plants. Hydroponic systems intrinsically decrease the need for chemical fertilizers, since nutrient solutions are customized to meet the specific needs of plants (Velazquez-Gonzalez , *et al.*, 2022). Aquaponics obviates the need of synthetic fertilizers by depending on the nutrient conversion process enhanced by beneficial microorganisms .

Economic Benefits

Increased Yield

One of the main economic benefits of aquaponics and hydroponics is their ability to provide greater crop yields in comparison to conventional soil-based agriculture. The regulated conditions and exact distribution of nutrients in hydroponic systems, along with the mutually beneficial connection in aquaponics, result in ideal circumstances for plant development. Hydroponic systems have been shown to accelerate plant development by 30-50%, resulting in higher crop yields (Yep , *et al.*, 2020). Aquaponic systems often demonstrate superior productivity in comparison to conventional farming, which may be due to the presence of nutrient-dense water originating from the fish tanks (Colt , *et al.*, 2022).

Lower Transportation Costs

Aquaponics and hydroponics, especially when used in urban or local environments, have the benefit of being close to markets. This minimizes the need for elaborate transportation and distribution networks, leading to cost reductions and a reduced carbon impact. Localized hydroponic and aquaponic farms in close proximity to metropolitan areas may substantially decrease transit distances, leading to reduced carbon emissions and transportation expenses.

Market Potential

Aquaponics and hydroponics have a promising economic feasibility that translates into significant market potential. Aquaponics and hydroponics are well-positioned in the market because to the rising consumer demand for fresh, locally produced products and the growing interest in sustainable agricultural techniques. Worldwide hydroponics industry is expected to see significant expansion because to its rising popularity in both commercial and residential environments. Aquaponics has become more popular in specialized markets, thanks to its sustainable and efficient production techniques. There is also possibility for further growth and development. To summarize, aquaponics and hydroponics provide economic advantages that go beyond higher crop yields. These advantages include cost-efficient local production, potential for market growth, and the generation of job prospects. As these systems progress and win more widespread approval, their favorable economic influence is set to become more substantial.

PLANT GROWTH AND NUTRITION IN HYDROPONIC SYSTEMS

Nutrient Solutions and Their Management

Hydroponic systems depend on nutrient solutions, which are a mixture of vital components that directly provide plants with the nutrients they need. The solutions consist of 16 vital elements, comprising macronutrients such as nitrogen (N), phosphorus (P), potassium (K), calcium (Ca), magnesium (Mg), and sulfur (S), as well as micronutrients like iron (Fe), manganese (Mn), zinc (Zn), copper (Cu), boron (B), molybdenum (Mo), chlorine (Cl), and nickel (Ni) (Trejo-Téllez and Gómez-Merino, 2012). Each element has a crucial function in plant physiology, including nitrogen, phosphorus, potassium, calcium, magnesium, sulfur, iron, manganese, zinc, copper, boron, molybdenum, chlorine, and nickel.

The absorption of nutrients is influenced by several parameters including pH, electrical conductivity (EC), temperature, oxygen levels, and light exposure. The ideal pH range for most plants is between 5.5 and 6.5. Any variations from this range might impede the absorption of nutrients and lead to shortages. Consistent monitoring, fine-tuning, substitution, and thorough cleaning of the system are crucial for preserving the ideal nutritional solution. Other factors to take into account include the kind of plant, the stage of development it is in, and the quality of water. Based on research conducted by the International

Society of Soilless Culture, hydroponically cultivated crops have the potential to provide a 20-50% higher yield compared to conventional soil-based approaches.

Effective fertilizer management may enhance water use efficiency by 70-80% in comparison to traditional agricultural practices. Advanced hydroponic systems have the capability to automate the monitoring and modification of nutrient solutions, resulting in decreased labor costs and assuring constant plant development. Ultimately, hydroponic cultivators have the ability to establish an optimal setting that enhances both crop output and quality, while simultaneously limiting the use of resources. Effective fertilizer management may significantly enhance water usage efficiency by 70-80% in comparison to traditional agricultural practices.

Factors Affecting Plant Growth

Hydroponic systems use light, temperature, humidity, air movement, and carbon dioxide (CO2) to maximize plant development. The intensity of light plays a vital role in the process of photosynthesis, while the length of light exposure controls the growth of plants in terms of both vegetative and reproductive development. The shape and yield of plants are influenced by the quality of light, with blue and red wavelengths playing a crucial role in photosynthesis and blooming.

Temperature is crucial for achieving optimum development, since the majority of plants flourish within a narrow range of 65-75°F (18-24°C). Optimal nutrient absorption and root growth are promoted by maintaining lower root temperatures. Keeping the relative humidity at a modest level of 50-70% helps to maintain a balance between transpiration and avoid excessive water loss. Proper air circulation minimizes the accumulation of still air and decreases moisture levels, hence lowering the likelihood of sickness and facilitating the interchange of gases.

Physical aspects include the choice of a suitable growth medium, such as inert media like rock wool, perlite, or coconut coir, as well as providing enough support for climbing plants or heavy fruits. Water quality should be pristine and devoid of any pathogens, while the system design should guarantee optimal water circulation and frequent examination to avert blockages and equipment malfunction.

Additional elements include plant diversity, insect infestations and diseases, as well as the practice of monitoring and analyzing data. Well-regulated hydroponic systems have the potential to enhance productivity by 20-50% in comparison to conventional soil-based agriculture. By using LED lights that have been fine-tuned to emit certain light spectra, it is possible to improve both the speed and quality of plant development. Additionally, ensuring that the temperature of the root zone remains at an ideal level may accelerate the absorption of nutrients and result in a yield increase of up to 15%. Through comprehending and controlling these pivotal elements, hydroponic cultivators may establish an optimal setting for their crops, hence enhancing efficiency and reducing resource use.

FISH PRODUCTION IN AQUAPONIC SYSTEMS

A new way to grow food is aquaponics, which combines fish farming and plant growth in a way that works well for both. In this complicated environment, plants eat the nutrients that fish waste, and the plants clean the water for the fish. To get the most out of aquaponics, you need to understand how fish generation works in this mutual dance.

Selection of Suitable Fish Species

Picking the right fish species is important for aquaponic output because it helps with food cycle, keeping the system stable, and making money. Temperature tolerance, pH tolerance, ammonia tolerance, nutrient conversion efficiency, feed conversion ratio (FCR), system compatibility, market demand, and revenue are some of the most important things to think about when choosing fish.

Tilapia are often used in aquaponics because they are hardy, eat a variety of foods, convert nutrients quickly, and are omnivore. Channel catfish grow faster, but they like water that is cooler. Rainbow trout, on the other hand, do better in colder areas. Yields of 1-2 kg per square meter of water area per year are possible with tilapia farming, which is about the same as with standard aquaculture methods.

Disease resistance is another thing to think about. This helps the system stay healthy and reduces the need for action. Also, fish that are nice to look at, like koi carp, can make your aquaponic system more appealing, especially if it's in a public or teaching space. You can make the best choice for your aquaponic system in terms of fish health, plant growth, and overall success by carefully considering these factors and doing study on specific fish species. Remember that picking the right fish is the first and most important thing you need to do to make an aquaponic food production system that works and lasts.

Fish Health and Nutrition Management

For an aquaponic system to work well, it is very important to keep the fish healthy and fed properly. A healthy fish population not only gives plants the nutrients they need to grow, but it also helps keep the environment in balance and the water quality stable. Let's look at the science side of controlling these important parts:

Water Quality and Fish Health

- **pH:** Most fish do best in a range from 6.5-7.8. Changes outside of this range can be stressful for fish, which can hurt their growth and defense systems. Check the pH often and change it as needed by adding natural bases or acidifiers.
- **Temperature:** The best temperature for different types of fish is different. Tropical fish like water that is 75 to 82°F (24 to 28°C), while colder fish do better in water that is 60 to 70°F (16 to 21°C). Keep the temperature ranges stable so as not to stress the fish.
- **Dissolved Oxygen (DO):** Fish need dissolved oxygen levels of 5 to 8 mg/L in order to breathe. Not getting enough DO can cause breathing, tiredness, and eventually death. Check the DO level often and use air pumps if needed.
- Ammonia and nitrite are byproducts of fish waste that are harmful to fish even in small amounts. If the biofilter works, it can change them into nitrate, which is less dangerous. Keep a close eye on the amounts of ammonia and nitrite and fix the problem if they get too high. You can do this by changing some of the water or making changes to the biofiltration.
- **Pathogens and Parasites:** Using good cleanliness, putting in place quarantine rules, and choosing fish breeds that don't get diseases help lower the chance of outbreaks. If problems happen, they need to be found quickly and treated with natural or organic means.

Nutrition Management

- **Diet:** Pick full, high-quality fish food made for aquaponics systems, taking into account the nutritional needs of the fish type and stage of life.
- **Feeding Rate:** Don't feed them too much, because that can damage the water quality and cause nutrition problems. Feed the fish at the right rate for their size, temperature, and amount of movement.
- **Feed Type:** Try different types of feed, like pellets, flakes, or real food, to make sure the animal likes them and gets all the nutrients they need.
- **Feeding:** To keep things consistent and avoid overfeeding, use automatic feeds or set feeding plans.
- **Monitoring:** Watch how the fish behave and what they eat to see if they might be having stomach problems or not getting enough nutrients. Check the vitamin content of fish waste to see how well the feed is being used and make changes to the diet if needed.

Considerations

Aquaponic systems can significantly reduce fish deaths by up to 50% compared to traditional farming methods. Biodiverse fish groups can enhance resilience and improve water quality through biofiltration. Fish fed hydroponic meals efficiently use nutrients, reducing waste and requiring fewer chemicals for plants. Stress management is crucial for fish health and immune system function. Monitoring fish growth, feeding rates, and water quality trends helps identify trends and make necessary adjustments. Continuous learning ensures long-term success in aquaponic systems. By controlling water quality, providing the right food, and prioritizing fish health, a healthy population of fish is essential for the longevity and success of an aquaponic system. A healthy fish population is the foundation of an effective and long-lasting aquaponic system.

INTEGRATED PEST MANAGEMENT IN SOILLESS AGRICULTURE

Soilless agriculture, including hydroponics and aquaponics, has many benefits compared to conventional soil-based techniques. Nevertheless, pest and disease problems may still occur in these regulated ecosystems. Integrated Pest Management (IPM) is a scientifically grounded approach that aims to avoid pest issues and reduce the need for strong chemical pesticides. Integrated Pest Management (IPM) plays a vital role in soilless agriculture because it addresses the specific difficulties associated with this method, including restricted biodiversity, demanding management practices, and environmental considerations (Barzman , *et al.*, 2015).

The main principles of Integrated Pest Management (IPM) in soilless agriculture are implementing preventive measures, maintaining cleanliness and hygiene, conducting regular monitoring, using biological and physical controls, and optimizing environmental conditions. Preventative techniques include the selection of pest-resistant types, frequent cleaning and sterilization of equipment, tanks, and growth systems, vigilant monitoring for early indications of pest activity, introduction of beneficial insects or predators, application of bacteria or fungus, and use of insect barriers and traps. Physical control is manipulating variables such as temperature, humidity, and ventilation to hinder the growth of pests and provide an inhospitable environment for their survival.

Other factors to take into account include the maintenance of records, ongoing education, and available local support. Recording pest sightings, techniques for managing them, and the efficacy of these tactics aids in shaping future Integrated Pest Management (IPM) approaches. Remaining informed about novel pest management methodologies and developing risks guarantees the ability to adjust and achieve efficient pest control. The use of indigenous beneficial organisms and management strategies adds to the long-term sustainability and environmental compatibility.

Ultimately, implementing an Integrated Pest Management (IPM) strategy in soilless agriculture may effectively address pest-related obstacles while simultaneously enhancing plant well-being, ensuring food safety, and upholding environmental stewardship. To ensure successful crop growth and a sustainable future for soilless agriculture, it is crucial to prioritize prevention and implement preventive measures together with a diverse management plan.

Challenges of Pest and Disease Management

The control of pests and diseases in soilless systems, such as aquaponics and hydroponics, has distinct issues. The aforementioned components consist of a closed-loop system, regulated environmental conditions, and water serving as a transmission medium. The self-contained structure of these systems restricts the occurrence of natural predation and the spread of pest populations, potentially resulting in a fast increase in outbreaks if not controlled. Moreover, the absence of soil-dwelling predators and decreased competition for resources in these environments might amplify the prevalence of certain pest species.

In the controlled environment, it is crucial to maintain continual surveillance in order to detect early indications of pest activity, since outbreaks have the potential to rapidly propagate within the limited area. Proficiency in technical knowledge is essential for taking proactive measures in pest management. However, the availability of intervention methods is restricted owing to worries about water pollution and fish health, which adds further difficulties.

Notable difficulties in aquaponics include fungal infections, proliferation of algae, bacterial pathogens, insect pests, nutritional imbalances, and root illnesses. Pythium and root rot, fungal diseases, may be worsened in aquaponics because to the warm and humid conditions, as well as the presence of organic materials from fish waste. Aphids, thrips, whiteflies, and fungus gnats are prevalent pests in hydroponics. Additionally, imbalances in nitrogen levels may compromise plant health, rendering them more vulnerable to infestations and diseases.

By using proactive integrated pest management (IPM) tactics in soilless agriculture, it is possible to achieve a reduction in pesticide when compared to traditional approaches. Successful pest and disease management in soilless systems necessitates a comprehensive strategy that encompasses timely identification, proactive intervention, and the use of a variety of biological, physical, and cultural control techniques.

Strategies for Sustainable Pest Control

Implementing sustainable pest management in aquaponics and hydroponics entails taking preemptive actions to mitigate pest issues prior to their occurrence. This encompasses the process of choosing pest-resistant cultivars, optimizing the environmental conditions for growth, ensuring cleanliness and hygiene, and closely monitoring for first indications of pest presence. Biological control includes the use of indigenous predators like as ladybugs, lacewings, and predatory mites, together with naturally existing microbes like Bacillus thuringiensis and Trichoderma species.

Physical control approaches include several techniques, including the use of insect barriers, water flow management, light manipulation, cultural control practices, crop rotation, intercropping, and nutrient management. These methods are quite easy and have little influence on the environment. Utilizing local resources may help reduce environmental impact and enhance system resilience. Keeping records provides valuable information for developing future Integrated Pest Management (IPM) plans and enhances the efficiency of pest control measures in the long run.

Empirical evidence demonstrates that the use of Integrated Pest Management (IPM) techniques in soilless agriculture may result in a remarkable reduction in pesticide usage, reaching up to 80% when compared to traditional approaches. The introduction of beneficial insects, like as ladybugs, may result in a 50% decrease in aphid populations in hydroponic systems. Additionally, the use of certain light spectra can interfere with pest egg-laying and lower pest pressure by up to 30%. Ultimately, it is crucial to use a wide range of sustainable pest management methods in order to establish a flourishing and resilient ecosystem in soilless agriculture. To establish a flourishing and resilient ecosystem in an aquaponic or hydroponic system, it is crucial to prioritize prevention and use proactive measures, biological control, and a holistic approach.

TECHNOLOGICAL ADVANCEMENTS AND AUTOMATION

Technology's Role in Aquaponic and Hydroponic Systems Optimization

The progress in technology is transforming the enhancement of aquaponic and hydroponic systems, providing more accuracy, automation, and insights based on data. These technological developments result in improved efficiency, production, and sustainability. Precision monitoring and control include the utilization of sensors and data recorders to oversee critical characteristics such as pH, temperature, dissolved oxygen, nutrient concentrations, and water flow. Environmental control systems use sensor data to provide a steady growing environment, while nutrient dosing systems administer vital nutrients according to plant requirements and water analysis. Advanced automation and robotics include the use of automated feeding systems, robotic plant manipulation, and automated water management systems.

Machine learning algorithms, cloud-based platforms, and decision support tools are used to provide data analysis and decision assistance. Machine learning algorithms process past data and current sensor readings to enable predictive maintenance, proactive control of pests and diseases, and improvement of system settings. Cloud-based solutions provide remote monitoring and control of systems in real-time, hence reducing geographical constraints. AI-driven systems assess data and provide suggestions for modifying system parameters, nutrient concentrations, and pest management tactics.

The advantages of these technologies include heightened efficacy and output, diminished labor expenses, augmented sustainability, and enhanced food safety and quality. Research indicates that the use of automation in hydroponic systems may enhance crop production by a maximum of 20% while concurrently decreasing water consumption by 30%. Utilizing data-driven analysis from sensor networks may result in a 15% decrease in expenses related to fertilizers, while employing artificial intelligence-based models for predicting pests and diseases can avert up to 50% of possible crop losses.

Ultimately, technology is swiftly revolutionizing the terrain of soilless agriculture, providing a potent set of tools for enhancing aquaponic and hydroponic systems. However, comprehending the fundamental scientific ideas and customizing these innovations to particular systems are essential for attaining best outcomes.

Automation and Monitoring Technologies

The field of soilless agriculture, including hydroponics and aquaponics, is undergoing a significant transformation as a result of technical breakthroughs and more automation. These instruments facilitate environmental monitoring and control, nutrient management, and automation. Sensors and data collecting systems oversee crucial elements such as pH, temperature, conductivity, dissolved oxygen, and humidity, guaranteeing prompt modifications. Automated control systems regulate the distribution of nutrients, control the temperature, and manage water flow in order to sustain ideal circumstances for plant development. Remote monitoring and management enable farmers to remotely observe the functioning of their systems and get notifications about possible problems.

Nutrient management and automation include the use of automated dosing systems, nutrient level monitors, and closed-loop nutrient recycling systems. These systems enhance resource efficiency, mitigate wastage, and facilitate nutrient recycling while lowering water use. Robotics and automation in soilless agriculture include many technologies like as automated sowing and transplanting, autonomous harvesting robots, and AI-driven plant monitoring and analysis.

Research indicates that the use of automation in hydroponics might result in a significant decrease in water consumption, up to 90%, in comparison to conventional soil-based farming methods. Utilizing precision nutrient dosing devices may enhance crop yields by up to 20% while also reducing fertilizer waste. Algorithms using artificial intelligence may effectively detect and identify pests and illnesses, resulting in a reduction of crop losses by up to 30%. To summarize, technical progress and automation have the capacity to greatly transform soilless agriculture. Through the use of sensors, robotics, and AI, farmers may enhance their operations by achieving higher levels of accuracy, productivity, and environmental consciousness. This allows for the optimization of resource utilization, waste reduction, and increased output.

SUSTAINABILITY AND ENVIRONMENTAL IMPACT

Assessment of the Sustainability

Aquaponic and hydroponic systems are environmentally-friendly techniques of producing food that provide notable benefits in terms of efficient resource use, regulated conditions, and the possibility for enhanced output. Hydroponics may achieve a substantial reduction in water use compared to conventional agriculture. This is mostly due to the implementation of closed-loop systems and efficient delivery techniques, resulting in a potential decrease of up to 90% in water usage. Aquaponics further repurposes fish waste into fertilizers for plants, therefore significantly reducing the water impact.

Both systems have a smaller land footprint, making them well-suited for urban settings or regions with limited land resources. Additionally, they maintain a regulated environment, therefore eliminating the need for chemical fertilizers and pesticides. This practice fosters environmental conservation and diminishes the likelihood of contamination hazards. Vertical farming allows for a more efficient use of land compared to traditional agriculture, since it may be conducted in a vertical manner.

Nevertheless, they also possess vulnerabilities, such as excessive energy consumption, reliance on technology, waste disposal, and economic feasibility. Hydroponic systems significantly depend on artificial lighting and temperature control, resulting in a substantial energy consumption, especially when non-renewable energy sources are used. In addition, aquaponics produces fish waste that requires appropriate management and disposal to prevent water contamination and maintain nutritional equilibrium.

Evaluating the sustainability of aquaponic and hydroponic systems is a multifaceted undertaking. Although they have benefits in terms of resource efficiency, regulated settings, and potential for higher production, it is important to carefully assess their reliance on technology, energy usage, and waste management difficulties. Maximizing the beneficial environmental effect and long-term sustainability of these systems requires optimizing system design, leveraging renewable energy sources, and applying sustainable waste management methods.

Environmental Impact and Carbon Footprint Analysis

Soilless agriculture, including hydroponics and aquaponics, is an innovative method of food production that provides substantial yields, optimal resource use, and little land requirements. It has various beneficial effects, including as water preservation, reduced land use, recycling of organic waste, enhanced crop production and food safety, and less reliance on chemicals. Nevertheless, there are certain adverse consequences associated with it, including the consumption of energy caused by the operation of pumps, climate control systems, and automation technologies, the generation of electronic waste, the contamination of nutrients and water, the restriction of crop diversity, and the higher initial investment costs for small-scale farmers.

To evaluate the carbon footprint of soilless agriculture, a comprehensive analysis must be conducted using a lifecycle assessment methodology. This involves examining every step of the process, including system building, operation, product transportation, and disposal. Key determinants of the carbon footprint include the use of sustainable energy sources such as solar or wind power, the selection of energy-efficient machinery and sustainable materials, the mitigation of emissions related to product transportation, and the implementation of responsible waste management practices. Research indicates that soilless systems might potentially reduce environmental effect by up to 80% in comparison to traditional agriculture, taking into account factors such as water use, soil erosion, and pesticide usage.

By using renewable energy sources and utilizing efficient equipment, the carbon footprint of soilless agriculture may be further reduced via energy consumption optimization. Life cycle evaluations demonstrate that the carbon footprint of soilless agriculture may be equivalent to or even lower than that of conventional agriculture when it is responsibly maintained. Soilless agriculture may become a really sustainable and ecologically conscious method of food production by embracing renewable energy sources, emphasizing local production, and practicing appropriate waste management.

CHALLENGES AND FUTURE DIRECTIONS

Current Limitations and Challenges

Soilless agriculture, while it has the potential for sustainable food production, encounters several limits and problems that impede its widespread implementation. The challenges include substantial upfront capital, specialized knowledge, continuous system fine-tuning and upkeep, reliance on energy resources, restricted variety of cultivated crops, control of pests and diseases, and addressing nutritional imbalances and inadequacies.

From a technological standpoint, soilless systems need specialized equipment, sensors, and automation technologies, resulting in greater upfront expenses in comparison to conventional farming. Proficiency in technical skills is necessary for the management of intricate systems that include precise nutrition control, temperature regulation, and automation. Continuous attention and skill are necessary for system optimization and maintenance, which may be challenging for large-scale operations or resource-limited environments.

From a biological standpoint, soilless systems may encounter difficulties when it comes to cultivating some crops, especially root vegetables and others that depend on intricate soil interactions. Efficient pest and disease control methods are essential for soilless systems, since they have unique difficulties arising from the enclosed habitat and the scarcity of natural predator populations. Plant health may be rapidly impacted by imbalances and deficits in nutrients, requiring ongoing monitoring and modification of nutrient solutions.

From an economic standpoint, market acceptability and consumer perception exhibit significant variation, but long-term profitability may be substantial as a result of enhanced yields and resource efficiency. Nevertheless, the initial expenses associated with investment and the intricacy of operations might initially impede the financial feasibility, particularly for farmers operating on a small scale. Access to inexpensive equipment, training programs, and information resources pertaining to soilless agriculture is essential for the acceptance and development of this practice, especially in poor nations.

Evidence suggests that the initial capital required to establish a hydroponic system might be 2-5 times more than that of conventional soil-based agriculture. Additionally, up to 20% of crop losses in hydroponic systems can be linked to problems related to pests and diseases. Insufficient knowledge among consumers and limited acceptability in the market might result in higher prices for soilless-produced products when compared to traditionally grown alternatives.

Notwithstanding these difficulties, soilless agriculture continues to be a hopeful remedy for tackling global food security issues and advancing sustainable food production. By surmounting these obstacles and promoting cooperative endeavors, we may establish the groundwork for a future in which soilless agriculture flourishes, guaranteeing the sustenance of food, the optimization of resources, and the preservation of the environment for future generations.

Future Trends and Potential Areas

Soilless agriculture, including aquaponics and hydroponics, is undergoing fast development and has significant promise for future food cultivation. One emerging trend is precision agriculture, which utilizes artificial intelligence, robots, and data analytics to optimize fertilizer delivery, light spectrum, and climatic conditions tailored to the particular requirements of each plant. This approach may increase yields by up to 30% compared to conventional approaches. Vertical and rooftop soilless systems are becoming popular in urban farming since they provide fresh food while using less land and reducing transportation emissions.

The integration of circular economy entails the use of waste heat from buildings or renewable energy sources for the operation of systems, as well as the utilization of plant biomass from processing plants for the production of biofuels or compost. The advanced study of aquaponics is on the integration of various fish species, the optimization of biofiltration systems, and the use of alternate water sources such as saltwater or treated wastewater.

Possible research areas encompass comprehending the intricate interconnections among plant roots, beneficial microorganisms, and nutrient absorption, formulating novel probiotic supplements, investigating sustainable growth substrates, establishing closed-loop water management systems, employing advanced sensors and AI algorithms for instantaneous monitoring, and breeding new plant varieties tailored for soilless environments.

The projected value of the worldwide soilless agriculture market is anticipated to reach USD 12.7 billion by 2027. Furthermore, it is estimated that expenditures in research and development would have a yearly growth rate of 15% in the near future. Enhancing resource efficiency and achieving greater closure in soilless systems may result in a 50% reduction in their environmental effect compared to existing methods. The potential of aquaponics and hydroponics in addressing pressing global issues such as food hunger, resource scarcity, and climate change is very promising.

CONCLUSION

Aquaponics and hydroponics are cutting-edge, adaptable, and effective methods of food production that tackle urgent agricultural challenges such as water shortages, land constraints, and environmental degradation. They decrease water use, need less acreage, and diminish dependence on artificial fertilizers and pesticides. Aquaponics and hydroponics facilitate diversification by allowing the growing of a broader range of crops, especially in difficult regions or metropolitan areas. The future prospects for aquaponics and hydroponics include the use of AI, robots, and sophisticated sensors, which will result in highly tailored growth conditions and improved plant well-being, productivity, and resource efficiency. Soilless systems provide a great opportunity for sustainable and localized food production in urban environments, which helps promote healthy cities and decrease transportation emissions.

Advancements in the study of microbe-plant interactions, development of new growth medium, and implementation of closed-loop water management will enhance efficiency and broaden the potential of these systems. Genomics and plant breeding may promote productivity and profitability by creating cultivars specifically adapted to soilless conditions, with increased disease resistance, greater nutrient absorption, and higher yields. Nevertheless, the key obstacles to widespread adoption still revolve around surmounting the upfront financial expenses, ensuring proper handling of electronic trash, and facilitating the diffusion of information on a larger scale. By adopting these systems, promoting research and innovation, and implementing them responsibly, we can unleash a future that is both sustainable and food-secure for future generations.

REFERENCES

Barzman, M., Bàrberi, P., Birch, A. N. E., Boonekamp, P., Dachbrodt-Saaydeh, S., Graf, B., Hommel, B., Jensen, J. E., Kiss, J., Kudsk, P., Lamichhane, J. R., Messéan, A., Moonen, A.-C., Ratnadass, A., Ricci, P., Sarah, J.-L., & Sattin, M. (2015). Eight principles of integrated pest management. *Agronomy for Sustainable Development, 35*(4), 1199–1215. doi:10.1007/s13593-015-0327-9

Colt, J., Schuur, A. M., Weaver, D., & Semmens, K. (2022). Engineering design of aquaponics systems. *Reviews in Fisheries Science & Aquaculture, 30*(1), 33–80. doi:10.1080/23308249.2021.1886240

Conn, S. J., Hocking, B., Dayod, M., Xu, B., Athman, A., Henderson, S., Aukett, L., Conn, V., Shearer, M. K., Fuentes, S., Tyerman, S. D., & Gilliham, M. (2013). Protocol: Optimising hydroponic growth systems for nutritional and physiological analysis of Arabidopsis thaliana and other plants. *Plant Methods, 9*(1), 1–11. doi:10.1186/1746-4811-9-4 PMID:23379342

Elmulthum, N. A., Zeineldin, F. I., Al-Khateeb, S. A., Al-Barrak, K. M., Mohammed, T. A., Sattar, M. N., & Mohmand, A. S. (2023). Water Use Efficiency and Economic Evaluation of the Hydroponic versus Conventional Cultivation Systems for Green Fodder Production in Saudi Arabia. *Sustainability (Basel), 15*(1), 822. doi:10.3390/su15010822

Goddek, S., Delaide, B., Mankasingh, U., Ragnarsdottir, K. V., Jijakli, H., & Thorarinsdottir, R. (2015). Challenges of sustainable and commercial aquaponics. *Sustainability (Basel), 7*(4), 4199–4224. doi:10.3390/su7044199

Haq, S., Boz, I., & Shahbaz, P. (2021). Adoption of climate-smart agriculture practices and differentiated nutritional outcome among rural households: A case of Punjab province, Pakistan. *Food Security, 13*(4), 913–931. doi:10.1007/s12571-021-01161-z

Obirikorang, K. A., Sekey, W., Gyampoh, B. A., Ashiagbor, G., & Asante, W. (2021). Aquaponics for improved food security in Africa: A review. *Frontiers in Sustainable Food Systems, 5*, 705549. doi:10.3389/fsufs.2021.705549

Pantanella, E. (2012). Nutrition and quality of aquaponic systems.

Pantanella, E. (2018). Aquaponics production, practices and opportunities. *Sustainable Aquaculture*, 191-248.

Rakocy, J. E. (2012). Aquaponics—integrating fish and plant culture. *Aquaculture production systems*, 344-386.

Sharma, N., Acharya, S., Kumar, K., Singh, N., & Chaurasia, O. P. (2018). Hydroponics as an advanced technique for vegetable production: An overview. *Journal of Soil and Water Conservation, 17*(4), 364–371. doi:10.5958/2455-7145.2018.00056.5

Trejo-Téllez, L. I. (2012). Nutrient solutions for hydroponic systems. *Hydroponics-a standard methodology for plant biological researches.*

Velazquez-Gonzalez, R. S., Garcia-Garcia, A. L., Ventura-Zapata, E., Barceinas-Sanchez, J. D. O., & Sosa-Savedra, J. C. (2022). A review on hydroponics and the technologies associated for medium-and small-scale operations. *Agriculture*, *12*(5), 646. doi:10.3390/agriculture12050646

Wang, L., Ning, S., Zheng, W., Guo, J., Li, Y., Li, Y., Chen, X., Ben-Gal, A., & Wei, X. (2023). Performance analysis of two typical greenhouse lettuce production systems: commercial hydroponic production and traditional soil cultivation. Frontiers in Plant Science, 14.

Yep, B., Gale, N. V., & Zheng, Y. (2020). Comparing hydroponic and aquaponic rootzones on the growth of two drug-type Cannabis sativa L. cultivars during the flowering stage. *Industrial Crops and Products*, *157*, 112881. doi:10.1016/j.indcrop.2020.112881

Chapter 11
Role of Aquaponics and Hydroponics in Modern Agriculture

Muhammad Zakwan Hadeed
University of Veterinary and Animal Sciences, Lahore, Pakistan

Muhammad Muhammad
Hebei Agricultural University, Baoding, China

Muhammad Safdar
https://orcid.org/0009-0006-1779-6967
University of Agriculture, Faisalabad, Pakistan

Aamir Raza
https://orcid.org/0009-0001-1867-2660
University of Agriculture, Faisalabad, Pakistan

Sobia Shahzad
Islamia University Bahawalpur, Bahawalnagar, Pakistan

Abdul Malik
University of Agriculture, Faisalabad, Pakistan

Mubashir Ahmed
University of Agriculture, Faisalabad, Pakistan

Abdul Raheem
Muhammad Nawaz Shareef University of Agriculture, Pakistan

ABSTRACT

The chapter explores aquaponics and hydroponics, two innovative and environmentally friendly farming techniques. It provides a comprehensive overview of the concepts, their history, and scientific principles. The chapter also explores the structure and elements of these systems, emphasizing their ecological and financial benefits. It also discusses plant growth dynamics in hydroponic systems, focusing on nutrient management and growth variables. The chapter also covers fish species selection, health, and nutrition in aquaponics, integrated pest management, and technological advancements in system optimization. It provides an impartial evaluation of the sustainability and environmental consequences of these systems, showcasing case studies and practical applications.

DOI: 10.4018/979-8-3693-3807-0.ch011

Copyright © 2024, IGI Global. Copying or distributing in print or electronic forms without written permission of IGI Global is prohibited.

INTRODUCTION

Aquaponics and Hydroponics

Hydroponics, the practice of growing plants in water that has a high concentration of nutrients, has been in existence for more than a hundred years (Sharma , *et al.*, 2018). The system employs several closed-loop mechanisms in which the roots of plants are suspended in a fluid that contains vital minerals and dissolved oxygen. By circumventing the constraints of soil absorption, this method of directly delivering nutrients results in accelerated growth rates and increased yields in comparison to conventional agriculture.

Aquaponics enhances hydroponics by integrating it with aquaculture, which involves the cultivation of aquatic creatures such as fish (Rakocy, 2012). The nutrient-rich effluent from the fish tank serves as a fertilizer for the plants cultivated in the hydroponic system in this symbiotic arrangement. Beneficial microorganisms transform ammonia derived from fish waste into nitrates that can be readily absorbed by plants, creating a self-sustaining environment that saves water and reduces waste. Aquaponic systems have the capacity to use up to 90% less water compared to conventional agriculture, making them very suitable for dry regions or countries experiencing water constraint (Obirikorang , *et al.*, 2021).

Aquaponics and hydroponics are modern agricultural methodologies that provide a departure from conventional soil-based farming methods. Aquaponics is a self-contained system that combines aquaculture with hydroponics, creating a mutually beneficial connection between aquatic creatures and plants. This system utilizes fish excrement as a source of nutrients for plant development, while the plants in turn cleanse the water for aquatic life.

Historical Background and Evolution

Aquaponics and hydroponics are agricultural technologies that have transitioned from old approaches to modern solutions. The origins of these technologies may be traced back to the Aztec Chinampas circa 1000 AD, when fish played a vital role as fertilizers, producing a mutually beneficial system akin to contemporary aquaponics (Henkel, 2015). The Hanging Gardens of Babylon, built around 600 BC, serve as early illustrations of hydroponics, demonstrating the practice of growing plants using water supplied via complex networks. The Floating Rice Gardens in Southeast Asia showcased a sophisticated comprehension of nutrient fluxes in aquatic ecosystems by combining the development of rice paddies with fish. The use of "water of dung" for plant nourishment, arising from modernization and technological progress, anticipated the principles of wastewater usage in present-day aquaponics.

In the 1970s, the New Alchemy Institute was founded with a specific emphasis on combining aquaculture with hydroponics. This laid the groundwork for the development of contemporary aquaponic systems. The efficiency and potential of aquaponics and hydroponics have been further boosted by technological developments in automation, sensor technology, and vertical farming. The historical evolution of various agricultural technologies demonstrates humanity's ongoing quest for sustainable and efficient practices. The development of aquaponics and hydroponics, from historical experimentation to contemporary scientific methods, demonstrates our continuous commitment to enhancing food production while reducing environmental harm. Analyzing this historical backdrop offers a more profound comprehension of humanity's persistent endeavor to develop a more sustainable and productive relationship with the earth.

PRINCIPLES

Hydroponics

Hydroponics, a revolutionary technique, involves growing plants without soil by manipulating their basic requirements for survival. It provides a structure for achieving the most favorable conditions for plant development in this soil-free environment. The fundamental tenets of hydroponics include the provision of nutrients, the provision of root support and oxygenation, the maintenance of regulated conditions, and the implementation of effective water management as shown in figure 1. Efficient nutrient delivery is vital in hydroponic systems, since it directly supplies necessary minerals to the plant's roots via a nutrient solution, hence optimizing absorption and promoting maximum growth potential. Root support and oxygenation are crucial in hydroponic systems, which use various substrates like perlite, rockwool, or recycled materials like coconut coir to anchor roots and facilitate oxygen passage.

Controlled settings, such as greenhouses, provide meticulous control over light, temperature, and humidity, therefore reducing plant stress and maximizing growth potential. Some crops, such as lettuce, may be harvested with a much higher speed of up to 50% when using hydroponic systems as opposed to traditional soil-based approaches (Wang , *et al.*, 2023). Effective water management is crucial in hydroponics, since closed-loop systems that recirculate the nutrient solution greatly decrease water use in comparison to traditional agriculture. Effective control of microbial populations inside the system may facilitate the decomposition of nutrients and safeguard plants against detrimental diseases.

Hydroponic systems have the capacity to reduce water use compared to conventional agriculture, making them well-suited for locations facing water scarcity. Utilizing hydroponics, vertical farming may greatly enhance the food production capacity in metropolitan regions that have limited land resources. Hydroponics offers a method for achieving sustainable and effective food production by controlling the plant's surroundings and accurately supplying necessary supplies.

Aquaponics

Aquaponics is an environmentally-friendly agricultural technique that integrates aquaculture with hydroponics to create a mutually beneficial and efficient system (Pantanella, 2018). The system utilizes a closed-loop recirculating water system, resulting in significantly reduced water use, making it more water-efficient compared to conventional soil-based agriculture as shown in figure 2 below. The fundamental principle of aquaponics is in the symbiotic interdependence between fish and plants, where fish waste acts as a valuable food supply for plants, while plants effectively cleanse and purify the water. Biological filtration, also known as the nitrogen cycle, is a vital process in aquaponics. During this process, beneficial bacteria transform ammonia, which is present in fish waste, into nitrites and nitrates. These compounds are critical nutrients that promote plant development. This mechanism guarantees an equilibrium and enduring ecology. Aquaponics also integrates systems for pH and water quality regulation, guaranteeing the ideal pH levels for the well-being of both fish and plants.

The efficacy of aquaponics in promoting plant development, resulting in greater basil harvests in aquaponic systems as compared to conventional hydroponics. In addition, aquaponics enhances resource efficiency by using fish waste as a fertilizer supply for plants, therefore reducing the environmental consequences linked to conventional farming. Aquaponics tackles issues like water shortage and nutrient contamination by fostering a symbiotic relationship between aquatic and plant organisms. The fundamentals of aquaponics include a mutually beneficial interaction between fish and plants, hydroponic farming, the recycling of nutrients, the management of water, the nutrition of plants, the chemistry of water, and the ecology of microorganisms.

Plant nutrition is crucial for achieving optimum development, and aquaponic plants require the same vital nutrients as plants cultivated on soil. The chemical composition of water has a vital role in determining the availability of nutrients and the activity of microorganisms. Additionally, water quality and the health of plants are influenced by factors such as temperature and the amount of dissolved oxygen present. An assortment of advantageous bacteria is crucial for the processes of nitrification, waste breakdown, and disease prevention. It is important to provide favorable circumstances for these microorganisms in order to maintain the general well-being of the aquaponic system. Aquaponics systems may attain agricultural yields that are 2-3 times greater than those of conventional agriculture for certain crops. Additionally, it mitigates soil erosion, fertilizer runoff, and pesticide use, so promoting a more sustainable food production system.

SYSTEM DESIGN AND COMPONENTS

Hydroponics

The design of a hydroponics system is a crucial factor that directly impacts the effectiveness and achievement of soilless growing. For successful implementation and optimization of hydroponic systems, it is crucial to have a thorough understanding of the related components as discussed below (Conn , *et al.*, 2013,Yanes , *et al.*, 2020).

Designs of the System

The Nutrient Film Technique (NFT) is a widely used method for cultivating leafy greens and herbs. In NFT systems, a constant flow of a thin film of nutrient solution is circulated over the roots of the plants. Channels or troughs serve as containers for the plants, enabling their roots to hang and directly receive nutrients from the flowing fluid. This method guarantees optimal root contact and effective absorption of nutrients.

Deep Water Culture (DWC) is a suitable method for cultivating bigger plants such as tomatoes and peppers. In DWC systems, the plants are suspended in floating rafts or baskets inside a deep reservoir filled with nutritional solution. The roots penetrate the fluid, consistently immersed in an abundant supply of nutrients and oxygen. Close observation of water temperature and levels of dissolved oxygen is essential for the proper functioning of this system.

The Ebb and Flow system, also known as Flood and Drain, is a flexible method that cyclically inundates the cultivation substrate with a nutrient solution, followed by letting it to recede back into a reservoir. The alternating rise and fall of water levels generates a cycle of aeration, which supplies the roots with new oxygen as the water is drained. It is versatile and may be used with many types of growth soil and crops.

Aeroponics is an advanced method of gardening that involves suspending plants in mist-filled chambers. The roots of the plants are sprayed with a nutrient solution at regular intervals, allowing them to grow without being rooted in soil. This technique completely removes the need for a substrate, so optimizing the flow of air and perhaps expediting the growth process. Nevertheless, it requires careful regulation of nutrition administration and the frequency of misting.

Vertical farming utilizes hydroponics to optimize space by using vertical systems. Utilizing stacking grow beds, hanging towers, or multi-tiered racks optimizes land use, making it well-suited for urban agriculture and locations with limited land availability.

Components

The system comprises a reservoir, pump, timer, Nutrient Dosing System, optional Growing Medium, Lighting, and Environmental Control Systems. The reservoir functions as a container for nutrient solution and serves as the central hub, its dimensions determined by the number of plants and the design. The pump enables the movement of the solution, ensuring even distribution and oxygenation. The Nutrient Dosing System dispenses concentrated nutrients to ensure optimal nutrient levels are maintained. The pH and EC meter measure the acidity and electrical conductivity of the solution, enabling adjustments to ensure optimal nutrient availability. Perlite, rockwool, and coconut coir are types of growing mediums that offer structural support to plant roots. Lighting enhances or replaces natural sunlight, ensuring the most favorable conditions for photosynthesis. Environmental Control Systems employ sensors to regulate temperature, humidity, and ventilation to achieve optimal plant growth.

Appropriate System Selection

The decision-making process of an appropriate system design and components is contingent upon several aspects, such as the selection of plants, the allocated money, space limitations, and technical expertise. Various crops possess distinct requirements and root structures, which impact the appropriateness of specific systems. The expenses are contingent upon the intricacy of the system, level of automation, and the components used.

The success of hydroponic systems is characterized by the tight coordination of components and thorough design. Every component, from the dynamic core of the reservoir to the regulated dispersion of mist in aeroponics, serves a crucial function in establishing a harmonious and supportive setting for the cultivation of plants without soil. Gaining knowledge about the system designs and fundamental components enables individuals to effectively traverse the dynamic field of hydroponics and harness its capacity for sustainable and efficient food cultivation.

Aquaponics

Aquaponics is a complex combination of aquaculture and hydroponics that depends on a well-planned system to enable the mutually beneficial interaction between fish and plants. The interplay among components is crucial for establishing a sustainable and efficient environment. Exploring the intricacies of system design uncovers the captivating interaction between technology and biology, highlighting the cleverness behind this sustainable food production technique.

Components

The aquaponic system comprises a fish tank, grow beds, pumps, piping, biofilter, sump tank, and monitoring and control devices. The fish tank accommodates many freshwater species such as tilapia, catfish, and koi, and its dimensions and structure may vary depending on its size. Grow beds provide both structural reinforcement and a spacious surface area for efficient nutrient absorption, while also allowing for customizable water flow and system intricacy. Pumps and pipes enable the movement of water containing nutrients from the fish tank to plant beds and filters.

The biofilter houses microorganisms that facilitate the conversion of ammonia, derived from fish waste, into nitrates that are used by plants. Additional designs include media filters, moving bed filters, and plant-based systems. The sump tank functions as a storage unit for accumulated water and houses submersible pumps, enabling the force of gravity to return the water to the fish tank. Advanced systems may include sensors and controllers to monitor water quality parameters like pH, temperature, and nutrient levels. Automated adjustments may be made to guarantee optimal conditions for aquatic life and plants.

System Explanation

Aquaponic systems use a synergistic blend of technology and biology to achieve sustainable food production. The systems consist of many components such as horizontal or vertical grow beds, bell-siphon, media filters, and plant-based filters. These systems need a substantial space for efficient functioning and necessitate periodic maintenance. Biofilters, such as moving bed filters, media filters, and plant-based filters, provide a substantial surface area for bacterial colonization, but they need a constant flow of water.

Surveillance and management systems are crucial for ensuring the optimal operation of the system. By using sensors, it is possible to monitor various parameters like pH, temperature, dissolved oxygen, and nutrient concentrations. Controllers may then make automated adjustments to pumps, lights, and ventilation systems according on the data collected. Aquaponic systems may be customized to accommodate diverse specifications and circumstances. Typical variations of aquaponics include recirculating aquaponics, which employs a closed-loop system with continuous water circulation, non-recirculating aquaponics, which involves periodic water changes, and integrated aquaponics, which combines hydroponics with a small component.

Understanding the system architecture and components of aquaponics unveils the intricate interaction between technology and biology, facilitating the feasibility of this sustainable method of food production. Every element, ranging from the fish tank sanctuary to the grow beds, has a vital function in fostering the mutually beneficial connection between plants and fish. Through a thorough analysis of these factors, we can fully comprehend the capacity of aquaponics to revolutionize the practice of food growing and provide a more ecologically sustainable future.

BENEFITS OF AQUAPONICS AND HYDROPONICS

Environmental Benefits

Water Conservation

Aquaponics and hydroponics represent a fundamental change in how we use water, making a substantial contribution to worldwide water conservation initiatives. Traditional agriculture is well recognized for its significant water requirements, with estimates suggesting that it may use more than 2,000 liters of water to produce a single kilogram of tomatoes (Haq , *et al.*, 2021). On the other hand, aquaponic and hydroponic systems exhibit exceptional water efficiency by using recirculating systems that limit water waste. Aquaponic systems exhibit a water use reduction of up to 90% compared to traditional soil-based agriculture (Goddek , *et al.*, 2015). Hydroponic systems, on average, use 70-90% less water in comparison to conventional agricultural techniques (Elmulthum , *et al.*, 2023).

Reduced Land Use

Aquaponic and hydroponic systems provide a compact and regulated structure, enabling the practice of vertical farming and optimal usage of space. Conventional farming requires extensive areas of cultivable land, leading to deforestation and the destruction of natural habitats. Hydroponic towers and aquaponic systems, on the other hand, enable the cultivation of crops in urban areas and the conversion of non-arable ground for agricultural purposes. Vertical hydroponic systems have the ability to enhance crop productivity per unit area, making them well-suited for urban agriculture. Aquaponics allows for the efficient combination of fish farming with crop production, resulting in a mutually beneficial relationship that maximizes the use of space (Pantanella, 2012).

Lower Chemical Usage

Aquaponic and hydroponic systems mitigate the need for synthetic fertilizers and pesticides, hence reducing chemical dependency. Conventional agricultural methods often include the use of chemical inputs, resulting in soil deterioration and water contamination. Hydroponics involves growing plants in water solutions that are rich in nutrients, which removes the need for fertilizers found in soil. On the other hand, aquaponics uses the waste produced by fish to provide nourishment for plants. Hydroponic systems intrinsically decrease the need for chemical fertilizers, since nutrient solutions are customized to meet the specific needs of plants (Velazquez-Gonzalez , *et al.*, 2022). Aquaponics obviates the need of synthetic fertilizers by depending on the nutrient conversion process enhanced by beneficial microorganisms .

Economic Benefits

Increased Yield

One of the main economic benefits of aquaponics and hydroponics is their ability to provide greater crop yields in comparison to conventional soil-based agriculture. The regulated conditions and exact distribution of nutrients in hydroponic systems, along with the mutually beneficial connection in aquaponics, result in ideal circumstances for plant development. Hydroponic systems have been shown to accelerate plant development by 30-50%, resulting in higher crop yields (Yep , *et al.*, 2020). Aquaponic systems often demonstrate superior productivity in comparison to conventional farming, which may be due to the presence of nutrient-dense water originating from the fish tanks (Colt , *et al.*, 2022).

Lower Transportation Costs

Aquaponics and hydroponics, especially when used in urban or local environments, have the benefit of being close to markets. This minimizes the need for elaborate transportation and distribution networks, leading to cost reductions and a reduced carbon impact. Localized hydroponic and aquaponic farms in close proximity to metropolitan areas may substantially decrease transit distances, leading to reduced carbon emissions and transportation expenses.

Market Potential

Aquaponics and hydroponics have a promising economic feasibility that translates into significant market potential. Aquaponics and hydroponics are well-positioned in the market because to the rising consumer demand for fresh, locally produced products and the growing interest in sustainable agricultural techniques. Worldwide hydroponics industry is expected to see significant expansion because to its rising popularity in both commercial and residential environments. Aquaponics has become more popular in specialized markets, thanks to its sustainable and efficient production techniques. There is also possibility for further growth and development. To summarize, aquaponics and hydroponics provide economic advantages that go beyond higher crop yields. These advantages include cost-efficient local production, potential for market growth, and the generation of job prospects. As these systems progress and win more widespread approval, their favorable economic influence is set to become more substantial.

PLANT GROWTH AND NUTRITION IN HYDROPONIC SYSTEMS

Nutrient Solutions and Their Management

Hydroponic systems depend on nutrient solutions, which are a mixture of vital components that directly provide plants with the nutrients they need. The solutions consist of 16 vital elements, comprising macronutrients such as nitrogen (N), phosphorus (P), potassium (K), calcium (Ca), magnesium (Mg), and sulfur (S), as well as micronutrients like iron (Fe), manganese (Mn), zinc (Zn), copper (Cu), boron (B), molybdenum (Mo), chlorine (Cl), and nickel (Ni) (Trejo-Téllez and Gómez-Merino, 2012). Each element has a crucial function in plant physiology, including nitrogen, phosphorus, potassium, calcium, magnesium, sulfur, iron, manganese, zinc, copper, boron, molybdenum, chlorine, and nickel.

The absorption of nutrients is influenced by several parameters including pH, electrical conductivity (EC), temperature, oxygen levels, and light exposure. The ideal pH range for most plants is between 5.5 and 6.5. Any variations from this range might impede the absorption of nutrients and lead to shortages. Consistent monitoring, fine-tuning, substitution, and thorough cleaning of the system are crucial for preserving the ideal nutritional solution. Other factors to take into account include the kind of plant, the stage of development it is in, and the quality of water. Based on research conducted by the International

Society of Soilless Culture, hydroponically cultivated crops have the potential to provide a 20-50% higher yield compared to conventional soil-based approaches.

Effective fertilizer management may enhance water use efficiency by 70-80% in comparison to traditional agricultural practices. Advanced hydroponic systems have the capability to automate the monitoring and modification of nutrient solutions, resulting in decreased labor costs and assuring constant plant development. Ultimately, hydroponic cultivators have the ability to establish an optimal setting that enhances both crop output and quality, while simultaneously limiting the use of resources. Effective fertilizer management may significantly enhance water usage efficiency by 70-80% in comparison to traditional agricultural practices.

Factors Affecting Plant Growth

Hydroponic systems use light, temperature, humidity, air movement, and carbon dioxide (CO2) to maximize plant development. The intensity of light plays a vital role in the process of photosynthesis, while the length of light exposure controls the growth of plants in terms of both vegetative and reproductive development. The shape and yield of plants are influenced by the quality of light, with blue and red wavelengths playing a crucial role in photosynthesis and blooming.

Temperature is crucial for achieving optimum development, since the majority of plants flourish within a narrow range of 65-75°F (18-24°C). Optimal nutrient absorption and root growth are promoted by maintaining lower root temperatures. Keeping the relative humidity at a modest level of 50-70% helps to maintain a balance between transpiration and avoid excessive water loss. Proper air circulation minimizes the accumulation of still air and decreases moisture levels, hence lowering the likelihood of sickness and facilitating the interchange of gases.

Physical aspects include the choice of a suitable growth medium, such as inert media like rock wool, perlite, or coconut coir, as well as providing enough support for climbing plants or heavy fruits. Water quality should be pristine and devoid of any pathogens, while the system design should guarantee optimal water circulation and frequent examination to avert blockages and equipment malfunction.

Additional elements include plant diversity, insect infestations and diseases, as well as the practice of monitoring and analyzing data. Well-regulated hydroponic systems have the potential to enhance productivity by 20-50% in comparison to conventional soil-based agriculture. By using LED lights that have been fine-tuned to emit certain light spectra, it is possible to improve both the speed and quality of plant development. Additionally, ensuring that the temperature of the root zone remains at an ideal level may accelerate the absorption of nutrients and result in a yield increase of up to 15%. Through comprehending and controlling these pivotal elements, hydroponic cultivators may establish an optimal setting for their crops, hence enhancing efficiency and reducing resource use.

FISH PRODUCTION IN AQUAPONIC SYSTEMS

A new way to grow food is aquaponics, which combines fish farming and plant growth in a way that works well for both. In this complicated environment, plants eat the nutrients that fish waste, and the plants clean the water for the fish. To get the most out of aquaponics, you need to understand how fish generation works in this mutual dance.

Selection of Suitable Fish Species

Picking the right fish species is important for aquaponic output because it helps with food cycle, keeping the system stable, and making money. Temperature tolerance, pH tolerance, ammonia tolerance, nutrient conversion efficiency, feed conversion ratio (FCR), system compatibility, market demand, and revenue are some of the most important things to think about when choosing fish.

Tilapia are often used in aquaponics because they are hardy, eat a variety of foods, convert nutrients quickly, and are omnivore. Channel catfish grow faster, but they like water that is cooler. Rainbow trout, on the other hand, do better in colder areas. Yields of 1-2 kg per square meter of water area per year are possible with tilapia farming, which is about the same as with standard aquaculture methods.

Disease resistance is another thing to think about. This helps the system stay healthy and reduces the need for action. Also, fish that are nice to look at, like koi carp, can make your aquaponic system more appealing, especially if it's in a public or teaching space. You can make the best choice for your aquaponic system in terms of fish health, plant growth, and overall success by carefully considering these factors and doing study on specific fish species. Remember that picking the right fish is the first and most important thing you need to do to make an aquaponic food production system that works and lasts.

Fish Health and Nutrition Management

For an aquaponic system to work well, it is very important to keep the fish healthy and fed properly. A healthy fish population not only gives plants the nutrients they need to grow, but it also helps keep the environment in balance and the water quality stable. Let's look at the science side of controlling these important parts:

Water Quality and Fish Health

- **pH:** Most fish do best in a range from 6.5-7.8. Changes outside of this range can be stressful for fish, which can hurt their growth and defense systems. Check the pH often and change it as needed by adding natural bases or acidifiers.
- **Temperature:** The best temperature for different types of fish is different. Tropical fish like water that is 75 to 82°F (24 to 28°C), while colder fish do better in water that is 60 to 70°F (16 to 21°C). Keep the temperature ranges stable so as not to stress the fish.
- **Dissolved Oxygen (DO):** Fish need dissolved oxygen levels of 5 to 8 mg/L in order to breathe. Not getting enough DO can cause breathing, tiredness, and eventually death. Check the DO level often and use air pumps if needed.
- Ammonia and nitrite are byproducts of fish waste that are harmful to fish even in small amounts. If the biofilter works, it can change them into nitrate, which is less dangerous. Keep a close eye on the amounts of ammonia and nitrite and fix the problem if they get too high. You can do this by changing some of the water or making changes to the biofiltration.
- **Pathogens and Parasites:** Using good cleanliness, putting in place quarantine rules, and choosing fish breeds that don't get diseases help lower the chance of outbreaks. If problems happen, they need to be found quickly and treated with natural or organic means.

Nutrition Management

- **Diet:** Pick full, high-quality fish food made for aquaponics systems, taking into account the nutritional needs of the fish type and stage of life.
- **Feeding Rate:** Don't feed them too much, because that can damage the water quality and cause nutrition problems. Feed the fish at the right rate for their size, temperature, and amount of movement.
- **Feed Type:** Try different types of feed, like pellets, flakes, or real food, to make sure the animal likes them and gets all the nutrients they need.
- **Feeding:** To keep things consistent and avoid overfeeding, use automatic feeds or set feeding plans.
- **Monitoring:** Watch how the fish behave and what they eat to see if they might be having stomach problems or not getting enough nutrients. Check the vitamin content of fish waste to see how well the feed is being used and make changes to the diet if needed.

Considerations

Aquaponic systems can significantly reduce fish deaths by up to 50% compared to traditional farming methods. Biodiverse fish groups can enhance resilience and improve water quality through biofiltration. Fish fed hydroponic meals efficiently use nutrients, reducing waste and requiring fewer chemicals for plants. Stress management is crucial for fish health and immune system function. Monitoring fish growth, feeding rates, and water quality trends helps identify trends and make necessary adjustments. Continuous learning ensures long-term success in aquaponic systems. By controlling water quality, providing the right food, and prioritizing fish health, a healthy population of fish is essential for the longevity and success of an aquaponic system. A healthy fish population is the foundation of an effective and long-lasting aquaponic system.

INTEGRATED PEST MANAGEMENT IN SOILLESS AGRICULTURE

Soilless agriculture, including hydroponics and aquaponics, has many benefits compared to conventional soil-based techniques. Nevertheless, pest and disease problems may still occur in these regulated ecosystems. Integrated Pest Management (IPM) is a scientifically grounded approach that aims to avoid pest issues and reduce the need for strong chemical pesticides. Integrated Pest Management (IPM) plays a vital role in soilless agriculture because it addresses the specific difficulties associated with this method, including restricted biodiversity, demanding management practices, and environmental considerations (Barzman , *et al.*, 2015).

The main principles of Integrated Pest Management (IPM) in soilless agriculture are implementing preventive measures, maintaining cleanliness and hygiene, conducting regular monitoring, using biological and physical controls, and optimizing environmental conditions. Preventative techniques include the selection of pest-resistant types, frequent cleaning and sterilization of equipment, tanks, and growth systems, vigilant monitoring for early indications of pest activity, introduction of beneficial insects or predators, application of bacteria or fungus, and use of insect barriers and traps. Physical control is manipulating variables such as temperature, humidity, and ventilation to hinder the growth of pests and provide an inhospitable environment for their survival.

Other factors to take into account include the maintenance of records, ongoing education, and available local support. Recording pest sightings, techniques for managing them, and the efficacy of these tactics aids in shaping future Integrated Pest Management (IPM) approaches. Remaining informed about novel pest management methodologies and developing risks guarantees the ability to adjust and achieve efficient pest control. The use of indigenous beneficial organisms and management strategies adds to the long-term sustainability and environmental compatibility.

Ultimately, implementing an Integrated Pest Management (IPM) strategy in soilless agriculture may effectively address pest-related obstacles while simultaneously enhancing plant well-being, ensuring food safety, and upholding environmental stewardship. To ensure successful crop growth and a sustainable future for soilless agriculture, it is crucial to prioritize prevention and implement preventive measures together with a diverse management plan.

Challenges of Pest and Disease Management

The control of pests and diseases in soilless systems, such as aquaponics and hydroponics, has distinct issues. The aforementioned components consist of a closed-loop system, regulated environmental conditions, and water serving as a transmission medium. The self-contained structure of these systems restricts the occurrence of natural predation and the spread of pest populations, potentially resulting in a fast increase in outbreaks if not controlled. Moreover, the absence of soil-dwelling predators and decreased competition for resources in these environments might amplify the prevalence of certain pest species.

In the controlled environment, it is crucial to maintain continual surveillance in order to detect early indications of pest activity, since outbreaks have the potential to rapidly propagate within the limited area. Proficiency in technical knowledge is essential for taking proactive measures in pest management. However, the availability of intervention methods is restricted owing to worries about water pollution and fish health, which adds further difficulties.

Notable difficulties in aquaponics include fungal infections, proliferation of algae, bacterial pathogens, insect pests, nutritional imbalances, and root illnesses. Pythium and root rot, fungal diseases, may be worsened in aquaponics because to the warm and humid conditions, as well as the presence of organic materials from fish waste. Aphids, thrips, whiteflies, and fungus gnats are prevalent pests in hydroponics. Additionally, imbalances in nitrogen levels may compromise plant health, rendering them more vulnerable to infestations and diseases.

By using proactive integrated pest management (IPM) tactics in soilless agriculture, it is possible to achieve a reduction in pesticide when compared to traditional approaches. Successful pest and disease management in soilless systems necessitates a comprehensive strategy that encompasses timely identification, proactive intervention, and the use of a variety of biological, physical, and cultural control techniques.

Strategies for Sustainable Pest Control

Implementing sustainable pest management in aquaponics and hydroponics entails taking preemptive actions to mitigate pest issues prior to their occurrence. This encompasses the process of choosing pest-resistant cultivars, optimizing the environmental conditions for growth, ensuring cleanliness and hygiene, and closely monitoring for first indications of pest presence. Biological control includes the use of indigenous predators like as ladybugs, lacewings, and predatory mites, together with naturally existing microbes like Bacillus thuringiensis and Trichoderma species.

Physical control approaches include several techniques, including the use of insect barriers, water flow management, light manipulation, cultural control practices, crop rotation, intercropping, and nutrient management. These methods are quite easy and have little influence on the environment. Utilizing local resources may help reduce environmental impact and enhance system resilience. Keeping records provides valuable information for developing future Integrated Pest Management (IPM) plans and enhances the efficiency of pest control measures in the long run.

Empirical evidence demonstrates that the use of Integrated Pest Management (IPM) techniques in soilless agriculture may result in a remarkable reduction in pesticide usage, reaching up to 80% when compared to traditional approaches. The introduction of beneficial insects, like as ladybugs, may result in a 50% decrease in aphid populations in hydroponic systems. Additionally, the use of certain light spectra can interfere with pest egg-laying and lower pest pressure by up to 30%. Ultimately, it is crucial to use a wide range of sustainable pest management methods in order to establish a flourishing and resilient ecosystem in soilless agriculture. To establish a flourishing and resilient ecosystem in an aquaponic or hydroponic system, it is crucial to prioritize prevention and use proactive measures, biological control, and a holistic approach.

TECHNOLOGICAL ADVANCEMENTS AND AUTOMATION

Technology's Role in Aquaponic and Hydroponic Systems Optimization

The progress in technology is transforming the enhancement of aquaponic and hydroponic systems, providing more accuracy, automation, and insights based on data. These technological developments result in improved efficiency, production, and sustainability. Precision monitoring and control include the utilization of sensors and data recorders to oversee critical characteristics such as pH, temperature, dissolved oxygen, nutrient concentrations, and water flow. Environmental control systems use sensor data to provide a steady growing environment, while nutrient dosing systems administer vital nutrients according to plant requirements and water analysis. Advanced automation and robotics include the use of automated feeding systems, robotic plant manipulation, and automated water management systems.

Machine learning algorithms, cloud-based platforms, and decision support tools are used to provide data analysis and decision assistance. Machine learning algorithms process past data and current sensor readings to enable predictive maintenance, proactive control of pests and diseases, and improvement of system settings. Cloud-based solutions provide remote monitoring and control of systems in real-time, hence reducing geographical constraints. AI-driven systems assess data and provide suggestions for modifying system parameters, nutrient concentrations, and pest management tactics.

The advantages of these technologies include heightened efficacy and output, diminished labor expenses, augmented sustainability, and enhanced food safety and quality. Research indicates that the use of automation in hydroponic systems may enhance crop production by a maximum of 20% while concurrently decreasing water consumption by 30%. Utilizing data-driven analysis from sensor networks may result in a 15% decrease in expenses related to fertilizers, while employing artificial intelligence-based models for predicting pests and diseases can avert up to 50% of possible crop losses.

Ultimately, technology is swiftly revolutionizing the terrain of soilless agriculture, providing a potent set of tools for enhancing aquaponic and hydroponic systems. However, comprehending the fundamental scientific ideas and customizing these innovations to particular systems are essential for attaining best outcomes.

Automation and Monitoring Technologies

The field of soilless agriculture, including hydroponics and aquaponics, is undergoing a significant transformation as a result of technical breakthroughs and more automation. These instruments facilitate environmental monitoring and control, nutrient management, and automation. Sensors and data collecting systems oversee crucial elements such as pH, temperature, conductivity, dissolved oxygen, and humidity, guaranteeing prompt modifications. Automated control systems regulate the distribution of nutrients, control the temperature, and manage water flow in order to sustain ideal circumstances for plant development. Remote monitoring and management enable farmers to remotely observe the functioning of their systems and get notifications about possible problems.

Nutrient management and automation include the use of automated dosing systems, nutrient level monitors, and closed-loop nutrient recycling systems. These systems enhance resource efficiency, mitigate wastage, and facilitate nutrient recycling while lowering water use. Robotics and automation in soilless agriculture include many technologies like as automated sowing and transplanting, autonomous harvesting robots, and AI-driven plant monitoring and analysis.

Research indicates that the use of automation in hydroponics might result in a significant decrease in water consumption, up to 90%, in comparison to conventional soil-based farming methods. Utilizing precision nutrient dosing devices may enhance crop yields by up to 20% while also reducing fertilizer waste. Algorithms using artificial intelligence may effectively detect and identify pests and illnesses, resulting in a reduction of crop losses by up to 30%. To summarize, technical progress and automation have the capacity to greatly transform soilless agriculture. Through the use of sensors, robotics, and AI, farmers may enhance their operations by achieving higher levels of accuracy, productivity, and environmental consciousness. This allows for the optimization of resource utilization, waste reduction, and increased output.

SUSTAINABILITY AND ENVIRONMENTAL IMPACT

Assessment of the Sustainability

Aquaponic and hydroponic systems are environmentally-friendly techniques of producing food that provide notable benefits in terms of efficient resource use, regulated conditions, and the possibility for enhanced output. Hydroponics may achieve a substantial reduction in water use compared to conventional agriculture. This is mostly due to the implementation of closed-loop systems and efficient delivery techniques, resulting in a potential decrease of up to 90% in water usage. Aquaponics further repurposes fish waste into fertilizers for plants, therefore significantly reducing the water impact.

Both systems have a smaller land footprint, making them well-suited for urban settings or regions with limited land resources. Additionally, they maintain a regulated environment, therefore eliminating the need for chemical fertilizers and pesticides. This practice fosters environmental conservation and diminishes the likelihood of contamination hazards. Vertical farming allows for a more efficient use of land compared to traditional agriculture, since it may be conducted in a vertical manner.

Nevertheless, they also possess vulnerabilities, such as excessive energy consumption, reliance on technology, waste disposal, and economic feasibility. Hydroponic systems significantly depend on artificial lighting and temperature control, resulting in a substantial energy consumption, especially when non-renewable energy sources are used. In addition, aquaponics produces fish waste that requires appropriate management and disposal to prevent water contamination and maintain nutritional equilibrium.

Evaluating the sustainability of aquaponic and hydroponic systems is a multifaceted undertaking. Although they have benefits in terms of resource efficiency, regulated settings, and potential for higher production, it is important to carefully assess their reliance on technology, energy usage, and waste management difficulties. Maximizing the beneficial environmental effect and long-term sustainability of these systems requires optimizing system design, leveraging renewable energy sources, and applying sustainable waste management methods.

Environmental Impact and Carbon Footprint Analysis

Soilless agriculture, including hydroponics and aquaponics, is an innovative method of food production that provides substantial yields, optimal resource use, and little land requirements. It has various beneficial effects, including as water preservation, reduced land use, recycling of organic waste, enhanced crop production and food safety, and less reliance on chemicals. Nevertheless, there are certain adverse consequences associated with it, including the consumption of energy caused by the operation of pumps, climate control systems, and automation technologies, the generation of electronic waste, the contamination of nutrients and water, the restriction of crop diversity, and the higher initial investment costs for small-scale farmers.

To evaluate the carbon footprint of soilless agriculture, a comprehensive analysis must be conducted using a lifecycle assessment methodology. This involves examining every step of the process, including system building, operation, product transportation, and disposal. Key determinants of the carbon footprint include the use of sustainable energy sources such as solar or wind power, the selection of energy-efficient machinery and sustainable materials, the mitigation of emissions related to product transportation, and the implementation of responsible waste management practices. Research indicates that soilless systems might potentially reduce environmental effect by up to 80% in comparison to traditional agriculture, taking into account factors such as water use, soil erosion, and pesticide usage.

By using renewable energy sources and utilizing efficient equipment, the carbon footprint of soilless agriculture may be further reduced via energy consumption optimization. Life cycle evaluations demonstrate that the carbon footprint of soilless agriculture may be equivalent to or even lower than that of conventional agriculture when it is responsibly maintained. Soilless agriculture may become a really sustainable and ecologically conscious method of food production by embracing renewable energy sources, emphasizing local production, and practicing appropriate waste management.

CHALLENGES AND FUTURE DIRECTIONS

Current Limitations and Challenges

Soilless agriculture, while it has the potential for sustainable food production, encounters several limits and problems that impede its widespread implementation. The challenges include substantial upfront capital, specialized knowledge, continuous system fine-tuning and upkeep, reliance on energy resources, restricted variety of cultivated crops, control of pests and diseases, and addressing nutritional imbalances and inadequacies.

From a technological standpoint, soilless systems need specialized equipment, sensors, and automation technologies, resulting in greater upfront expenses in comparison to conventional farming. Proficiency in technical skills is necessary for the management of intricate systems that include precise nutrition control, temperature regulation, and automation. Continuous attention and skill are necessary for system optimization and maintenance, which may be challenging for large-scale operations or resource-limited environments.

From a biological standpoint, soilless systems may encounter difficulties when it comes to cultivating some crops, especially root vegetables and others that depend on intricate soil interactions. Efficient pest and disease control methods are essential for soilless systems, since they have unique difficulties arising from the enclosed habitat and the scarcity of natural predator populations. Plant health may be rapidly impacted by imbalances and deficits in nutrients, requiring ongoing monitoring and modification of nutrient solutions.

From an economic standpoint, market acceptability and consumer perception exhibit significant variation, but long-term profitability may be substantial as a result of enhanced yields and resource efficiency. Nevertheless, the initial expenses associated with investment and the intricacy of operations might initially impede the financial feasibility, particularly for farmers operating on a small scale. Access to inexpensive equipment, training programs, and information resources pertaining to soilless agriculture is essential for the acceptance and development of this practice, especially in poor nations.

Evidence suggests that the initial capital required to establish a hydroponic system might be 2-5 times more than that of conventional soil-based agriculture. Additionally, up to 20% of crop losses in hydroponic systems can be linked to problems related to pests and diseases. Insufficient knowledge among consumers and limited acceptability in the market might result in higher prices for soilless-produced products when compared to traditionally grown alternatives.

Notwithstanding these difficulties, soilless agriculture continues to be a hopeful remedy for tackling global food security issues and advancing sustainable food production. By surmounting these obstacles and promoting cooperative endeavors, we may establish the groundwork for a future in which soilless agriculture flourishes, guaranteeing the sustenance of food, the optimization of resources, and the preservation of the environment for future generations.

Future Trends and Potential Areas

Soilless agriculture, including aquaponics and hydroponics, is undergoing fast development and has significant promise for future food cultivation. One emerging trend is precision agriculture, which utilizes artificial intelligence, robots, and data analytics to optimize fertilizer delivery, light spectrum, and climatic conditions tailored to the particular requirements of each plant. This approach may increase yields by up to 30% compared to conventional approaches. Vertical and rooftop soilless systems are becoming popular in urban farming since they provide fresh food while using less land and reducing transportation emissions.

The integration of circular economy entails the use of waste heat from buildings or renewable energy sources for the operation of systems, as well as the utilization of plant biomass from processing plants for the production of biofuels or compost. The advanced study of aquaponics is on the integration of various fish species, the optimization of biofiltration systems, and the use of alternate water sources such as saltwater or treated wastewater.

Possible research areas encompass comprehending the intricate interconnections among plant roots, beneficial microorganisms, and nutrient absorption, formulating novel probiotic supplements, investigating sustainable growth substrates, establishing closed-loop water management systems, employing advanced sensors and AI algorithms for instantaneous monitoring, and breeding new plant varieties tailored for soilless environments.

The projected value of the worldwide soilless agriculture market is anticipated to reach USD 12.7 billion by 2027. Furthermore, it is estimated that expenditures in research and development would have a yearly growth rate of 15% in the near future. Enhancing resource efficiency and achieving greater closure in soilless systems may result in a 50% reduction in their environmental effect compared to existing methods. The potential of aquaponics and hydroponics in addressing pressing global issues such as food hunger, resource scarcity, and climate change is very promising.

CONCLUSION

Aquaponics and hydroponics are cutting-edge, adaptable, and effective methods of food production that tackle urgent agricultural challenges such as water shortages, land constraints, and environmental degradation. They decrease water use, need less acreage, and diminish dependence on artificial fertilizers and pesticides. Aquaponics and hydroponics facilitate diversification by allowing the growing of a broader range of crops, especially in difficult regions or metropolitan areas. The future prospects for aquaponics and hydroponics include the use of AI, robots, and sophisticated sensors, which will result in highly tailored growth conditions and improved plant well-being, productivity, and resource efficiency. Soilless systems provide a great opportunity for sustainable and localized food production in urban environments, which helps promote healthy cities and decrease transportation emissions.

Advancements in the study of microbe-plant interactions, development of new growth medium, and implementation of closed-loop water management will enhance efficiency and broaden the potential of these systems. Genomics and plant breeding may promote productivity and profitability by creating cultivars specifically adapted to soilless conditions, with increased disease resistance, greater nutrient absorption, and higher yields. Nevertheless, the key obstacles to widespread adoption still revolve around surmounting the upfront financial expenses, ensuring proper handling of electronic trash, and facilitating the diffusion of information on a larger scale. By adopting these systems, promoting research and innovation, and implementing them responsibly, we can unleash a future that is both sustainable and food-secure for future generations.

REFERENCES

Barzman, M., Bàrberi, P., Birch, A. N. E., Boonekamp, P., Dachbrodt-Saaydeh, S., Graf, B., Hommel, B., Jensen, J. E., Kiss, J., Kudsk, P., Lamichhane, J. R., Messéan, A., Moonen, A.-C., Ratnadass, A., Ricci, P., Sarah, J.-L., & Sattin, M. (2015). Eight principles of integrated pest management. *Agronomy for Sustainable Development*, *35*(4), 1199–1215. doi:10.1007/s13593-015-0327-9

Colt, J., Schuur, A. M., Weaver, D., & Semmens, K. (2022). Engineering design of aquaponics systems. *Reviews in Fisheries Science & Aquaculture*, *30*(1), 33–80. doi:10.1080/23308249.2021.1886240

Conn, S. J., Hocking, B., Dayod, M., Xu, B., Athman, A., Henderson, S., Aukett, L., Conn, V., Shearer, M. K., Fuentes, S., Tyerman, S. D., & Gilliham, M. (2013). Protocol: Optimising hydroponic growth systems for nutritional and physiological analysis of Arabidopsis thaliana and other plants. *Plant Methods*, *9*(1), 1–11. doi:10.1186/1746-4811-9-4 PMID:23379342

Elmulthum, N. A., Zeineldin, F. I., Al-Khateeb, S. A., Al-Barrak, K. M., Mohammed, T. A., Sattar, M. N., & Mohmand, A. S. (2023). Water Use Efficiency and Economic Evaluation of the Hydroponic versus Conventional Cultivation Systems for Green Fodder Production in Saudi Arabia. *Sustainability (Basel)*, *15*(1), 822. doi:10.3390/su15010822

Goddek, S., Delaide, B., Mankasingh, U., Ragnarsdottir, K. V., Jijakli, H., & Thorarinsdottir, R. (2015). Challenges of sustainable and commercial aquaponics. *Sustainability (Basel)*, *7*(4), 4199–4224. doi:10.3390/su7044199

Haq, S., Boz, I., & Shahbaz, P. (2021). Adoption of climate-smart agriculture practices and differentiated nutritional outcome among rural households: A case of Punjab province, Pakistan. *Food Security*, *13*(4), 913–931. doi:10.1007/s12571-021-01161-z

Obirikorang, K. A., Sekey, W., Gyampoh, B. A., Ashiagbor, G., & Asante, W. (2021). Aquaponics for improved food security in Africa: A review. *Frontiers in Sustainable Food Systems*, *5*, 705549. doi:10.3389/fsufs.2021.705549

Pantanella, E. (2012). Nutrition and quality of aquaponic systems. *Sustainable Agriculture.*

Pantanella, E. (2018). Aquaponics production, practices and opportunities. *Sustainable Aquaculture*, 191-248.

Rakocy, J. E. (2012). Aquaponics—integrating fish and plant culture. *Aquaculture production systems.*

Sharma, N., Acharya, S., Kumar, K., Singh, N., & Chaurasia, O. P. (2018). Hydroponics as an advanced technique for vegetable production: An overview. *Journal of Soil and Water Conservation*, *17*(4), 364–371. doi:10.5958/2455-7145.2018.00056.5

Trejo-Téllez, L. I. (2012). Nutrient solutions for hydroponic systems. *Hydroponics-a standard methodology for plant biological researches.*

Velazquez-Gonzalez, R. S., Garcia-Garcia, A. L., Ventura-Zapata, E., Barceinas-Sanchez, J. D. O., & Sosa-Savedra, J. C. (2022). A review on hydroponics and the technologies associated for medium-and small-scale operations. *Agriculture*, *12*(5), 646. doi:10.3390/agriculture12050646

Wang, L., Ning, S., Zheng, W., Guo, J., Li, Y., Li, Y., Chen, X., Ben-Gal, A., & Wei, X. (2023). Performance analysis of two typical greenhouse lettuce production systems: commercial hydroponic production and traditional soil cultivation. *Frontiers in Plant Science*.

Yep, B., Gale, N. V., & Zheng, Y. (2020). Comparing hydroponic and aquaponic rootzones on the growth of two drug-type Cannabis sativa L. cultivars during the flowering stage. *Industrial Crops and Products*, *157*, 112881. doi:10.1016/j.indcrop.2020.112881

Chapter 12
Role of Remote Sensing and GIS for Sustainable Crop Production Under Climate Change

Muhammad Zakwan Hadeed
University of Veterinary and Animal Sciences, Lahore, Pakistan

Muhammad Safdar
https://orcid.org/0009-0006-1779-6967
University of Agriculture, Faisalabad, Pakistan

Aitizaz Ali
Unitar International University, Malaysia

Muhammad Adnan Shahid
https://orcid.org/0000-0001-6403-0290
University of Agriculture, Faisalabad, Pakistan

Ting Tin Tin
https://orcid.org/0000-0001-7634-1686
INTI International University, Malaysia

Abdul Malik
University of Agriculture, Faisalabad, Pakistan

Aqsa Arif
University of Agriculture, Faisalabad, Pakistan

Ramsha Rashid
University of Agriculture, Faisalabad, Pakistan

Aamir Raza
https://orcid.org/0009-0001-1867-2660
University of Agriculture, Faisalabad, Pakistan

Hafsa Muzammal
University of Agriculture, Faisalabad, Pakistan

ABSTRACT

Geospatial techniques are being increasingly used in agriculture to optimize production, manage resources, and mitigate climate change impacts. These techniques include GPS, GIS, and remote sensing, which map and monitor crop growth and yield variability, allowing for more efficient application of inputs like fertilizers, water, and pesticides. They also create models that simulate crop growth under different climate scenarios, allowing farmers to plan for future conditions and adapt their practices accordingly. Geospatial data is used to create detailed maps of soil properties, helping farmers make

DOI: 10.4018/979-8-3693-3807-0.ch012

Copyright © 2024, IGI Global. Copying or distributing in print or electronic forms without written permission of IGI Global is prohibited.

better decisions about crop selection, planting, and nutrient management. These techniques also help monitor and manage water resources, particularly in areas prone to drought or flooding, by mapping groundwater reserves, analyzing precipitation patterns, and predicting crop water demand. Geospatial data can evaluate climate hazards, guide crop selection and management, and analyze land use patterns, aiding farmers and policymakers in decision-making.

INTRODUCTION

Definition of Geospatial Techniques and Climate Change

Geospatial approaches encompass the utilization of advanced technology and instruments for the purpose of acquiring, examining, and administering geographic data. The methodologies encompass the utilization of geographic information systems (GIS), remote sensing, & global positioning systems (GPS) for the purpose of gathering and scrutinizing data pertaining to the Earth's surface. Geospatial approaches play a crucial role in comprehending the effects of climate change on global ecosystems, water resources, as well as meteorological patterns (Kocaman et al., 2019).

Climate change is a global phenomenon causing significant changes in Earth's temperature, precipitation, and weather patterns due to human activities like fossil fuel combustion, deforestation, and industrial operations. It impacts the environment, human health, and economy. Geospatial tools are crucial for studying climate change effects, monitoring and analyzing changes in land cover, vegetation, and water resources. The application of geospatial methodologies in climate change studies has led to significant progress in understanding the Earth's consequences (Lu et al., 2021).

Importance of Geospatial Techniques in Agriculture

Geospatial techniques are increasingly important in agriculture due to their ability to provide accurate data on soil conditions, weather patterns, and crop development. These techniques use satellite images, GPS, and GIS to acquire, analyze, and manage agricultural data, making them crucial in various aspects of the agricultural sector (Ge & Thomasson, 2017).

Geospatial tools are being utilized in precision agriculture to allocate resources, improve efficiency, and reduce costs. They enable farmers to monitor crops, identify pests, diseases, and nutrient deficits, and make timely decisions for planting, irrigation, and harvesting. Geospatial methods also help in identifying crop choices resilient to changing weather patterns and assessing regions susceptible to flooding or drought, playing a crucial role in climate change adaptation (Gebbers et al., 2016).

This chapter explores the importance of geospatial techniques in the agricultural sector, particularly in the context of climate change. It covers various techniques like remote sensing, GPS, GIS, and UAVs, and their applications in agriculture such as precision farming, crop modeling, soil mapping, water management, climate risk assessment, land use planning, and yield prediction. Geospatial techniques offer operational effectiveness, reduced costs, and increased crop productivity, but also present financial challenges and limited specialized knowledge. The chapter also discusses case studies and future trends in geospatial technology, offering a summary of the benefits and challenges of employing these techniques in agriculture.

TYPES OF GEOSPATIAL TECHNIQUES USED IN AGRICULTURE

Geospatial techniques are crucial in modern agriculture for optimizing farming practices, enhancing productivity, and enabling informed decision-making. This discourse presents a list of effective geospatial strategies and provides scholarly references to support their practical implementation in the agricultural sector. Overview of the main types of Geospatial Techniques used in agriculture under changing climate.

Remote Sensing

Satellite imagery and remote sensing are geospatial techniques used in agriculture to monitor crop health, moisture levels, and nutrient concentrations. Satellite imagery provides data about an entity or region without physical interaction, while remote sensing uses sensors to gather data across multiple wavelengths of light. These techniques help farmers determine chlorophyll content in crops, identify nutrient deficiencies, and make informed fertilization and irrigation decisions (Zhang et al., 2019).

Hyperspectral remote sensing uses light wavelengths to identify botanical compositions and terrestrial substrates, enabling early disease detection, plant species identification, and crop health monitoring. Synthetic Aperture Radar (SAR) generates high-resolution images, enabling farmers to monitor crops in adverse weather conditions and assess parameters like crop height, moisture content, and biomass. LiDAR uses laser technology to accurately determine spatial separation between sensors and terrestrial surfaces, generating detailed three-dimensional representations of topography and flora (Adamchuk & Hummel, 2016).

Global Positioning Systems (GPS)

The use of satellites and receivers accurately determines Earth's surface coordinates, aiding in monitoring agricultural equipment, field demarcation, and soil attributes. Global Positioning System (GPS) technology is used to delineate field borders and accurately pinpoint the placement of agricultural sensors, ensuring accurate tracking of agricultural operations. GPS technology provides precise direction for agricultural machines, ensuring the precise positioning of seeds, fertilizers, and pesticides. This approach reduces input wastage, improves operational efficiency, and minimizes environmental impact, thereby enhancing the overall efficiency of agricultural operations.

Geographic Information Systems (GIS)

The framework is designed to collect, store, modify, analyze, and display various geospatially referenced data, aiding in soil analysis, field layout planning, and crop yield monitoring, and utilizing Geographic Information Systems (GIS) for agricultural production analysis across seasons and locations. GIS is a tool that integrates various data layers, including soil composition, weather patterns, and historical yield data, to predict crop output, aiding agricultural practitioners in strategic production processes and resource allocation (Srivastava, 2020).

Unmanned Aerial Vehicles (UAVs)

The study explores the use of unmanned aerial vehicles (UAVs) equipped with multispectral sensors for crop health monitoring and nutrient deficiency detection, aiming to acquire high-resolution imagery of agricultural fields and land characteristics, thereby enhancing crop vitality and evaluating crop impairment. UAVs, equipped with multispectral or thermal cameras, are used for crop health evaluation, capturing high-resolution footage of crops, enabling comprehensive information extraction on plant health, stress levels, and disease outbreaks (Limpisathian, 2011).

APPLICATIONS OF GEOSPATIAL TECHNIQUES IN AGRICULTURE

Precision Farming

Precision farming, also known as smart farming or precision agriculture, is an agricultural method that uses advanced technology like sensors, drones, satellite imagery, GPS, and data analytics to monitor soil conditions, crop growth, and environmental factors, enhancing agricultural productivity. Precision farming aids farmers in making informed decisions, reducing resource inefficiencies, and improving crop yield. Real-time field data allows farmers to monitor and manage crops effectively, enabling targeted adjustments based on specific requirements (Khan et al., 2020).

Precision farming techniques, utilizing advanced technologies and data analysis, are crucial in the agricultural sector to optimize resource allocation, boost productivity, and mitigate climate variability. GIS and remote sensing tools help assess soil attributes, water accessibility, and crop health, enabling tailored management strategies and stress-reducing methods (Liang, 2020).

Sensor-based technology, such as soil moisture sensors and weather stations, collect real-time data to inform agricultural decisions, improving efficiency and adaptability to climate conditions. Decision support systems offer farmers tailored guidance, optimizing inputs, reducing costs and mitigating environmental impacts. Implementing precision farming techniques enhances agricultural productivity and climate change response (Khan, 2021). The multiple components of precision farming, most of which are greatly enhanced by GIS.

Here are some key points and practices associated with precision farming (Dabas et al., 2020):

- **Remote Sensing and Imaging:** Technological advancements like sensors, satellites, and drones collect vital data on nutrient content, soil water level, disease prevalence, and crop health, aiding agricultural practitioners in resource allocation.
- **GPS and Mapping:** GPS technology aids farmers in precise mapping and monitoring of fields, enabling accurate crop growth and crop distribution, and enabling tailored management practices based on soil attributes and properties.
- **Variable Rate Technology:** Virtual Reality Technology (VRT) optimizes resource allocation and utilization in agriculture by modifying input applications like climate data, irrigation, and fertilizers, minimizing waste and enhancing efficiency.
- **Robotics and Automated Machinery:** Automated farming systems, integrated with mobile devices, perform tasks like seed sowing, fertilizer application, crop watering, and harvesting, reducing labor and improving operational efficiency (Arslan & Colkesen, 2021).

- **Decision Support Systems and Data Analytics:** Advanced analytical approaches analyze large data sets in precision agriculture, aiding farmers in data interpretation, pattern identification, and informed decision-making on crop management strategies (Pathan et al., 2021).

Advantages of Efficient Precision Farming

Precision farming techniques can enhance crop growth and agricultural productivity by judicious application of inputs like fertilizers, herbicides, and water. This practice reduces the environmental footprint and financial burden. It optimizes inputs, reduces wastage, and promotes environmental sustainability. Precision farming also enhances profitability by allowing farmers to make informed decisions based on data. Real-time data and advanced analytical tools facilitate decision-making, enhancing risk assessment, crop management, and resource allocation. Sustainable agriculture is crucial in today's global agriculture, offering opportunities to improve profitability, productivity, and sustainability while mitigating environmental impact (Qureshi et al., 2020).

Crop Modeling

Crop modeling, also known as farm modeling or crop simulation, is a method used to predict and simulate the growth and development of crops under different environmental conditions. It involves the use of computational and mathematical models to represent the physiological processes and interactions that occur within plants, as well as their response to external factors such as climate, soil conditions, and management practices (Boote et al., 2019). Crop modeling is crucial for agricultural productivity, especially in the context of climate change. These models replicate crop growth and development across various environmental conditions, helping farmers predict potential effects. They integrate climatic data, soil information, and crop physiology to aid in informed decision-making, resource optimization, and adaptation methods. Process-based models, which focus on understanding and representing physiological processes like photosynthesis, respiration, and nutrient absorption, are used to forecast agricultural productivity. Assessing climate change's impact on crop yields, water requirements, and nutrient dynamics can help design appropriate adaptation measures (Asseng et al., 2019).

Crop models aim to simulate growth, yield, and behavior over time, allowing farmers, researchers, and policy makers to understand and predict the effects of different inputs. plant work. These models are often based on scientific knowledge of plant physiology, ecology, and agronomy, and they combine empirical data with experimental results to make accurate predictions.

Here are some key parts and applications of crop plant production (Janssen et al., 2017):

- **Growth and development of crops:** crop growth models simulate the process of plant growth and development, from budding to maturity. They consider factors such as leaf area expansion, total biomass, flowering, fruiting, and senescence. By following these steps, culture models provide information about the timing and duration of critical growth stages.
- **Environmental factors:** Plant models include environmental data, such as climate, temperature, solar radiation, rainfall, and soil characteristics. These factors affect plant growth, photosynthesis, transpiration, nutrition, and seed production. By incorporating environmental data, crop types can determine how crops respond to different climates and soils.

- **Management practices:** Plant models can evaluate the effects of different management practices on plant performance. These practices include planting dates, irrigation schedules, nutrient management, pest control, and crop rotation. By designing different management scenarios, farmers and researchers can improve crop production strategies and identify sustainable practices.
- **Decision Support Systems:** Crop models are often used as part of Decision Support Systems (DSS) to help farmers and agricultural professionals make informed decisions. By combining environmental data with crop types, DSS software can provide recommendations for planting, watering, fertilizing, and pest management based on predicted crop responses.

Advantages of Crop Modeling

Crop models are essential tools for farmers to predict crop growth, development, and yield in various conditions, enabling them to make informed decisions and implement measures to enhance production and manage risks. Utilizing different crop varieties can maximize resource allocation, such as water, fertilizers, and pesticides, and inform decisions on irrigation scheduling, fertilizer rates, and pest control tactics (Jones et al., 2017).

Crop models are vital for researchers and agricultural scientists, facilitating the investigation of plant physiology, responses, and the interaction between plants and their environment. Agriculture models can contribute to the development of novel cultivars and agricultural methods. They also play a significant role in shaping agricultural policy formulation and long-term planning processes in politics. Crop modeling helps farmers adjust to environmental fluctuations, optimize resource exploitation, and improve production and sustainability. Overall, crop modeling is crucial for understanding crop dynamics, enhancing management practices, and expediting decision-making in agriculture (McBratney et al., 2003).

Soil Mapping

Soil mapping, also known as land survey or land description, is a method of systematically analyzing and documenting the features and characteristics of land in an area. This includes collecting data on various soil characteristics, such as nutrients, organic matter, texture, EC, drainage, and compaction, among others. Soil maps provide valuable information about the spatial distribution and variability of soils, helping land management decisions and agricultural practices (Bouma, 2016).

Soil mapping techniques are crucial in agricultural output, especially in the context of climate change. These techniques gather and analyze soil information to understand its characteristics, productivity, and heterogeneity. These techniques enable farmers to optimize crop management practices, adjust to climatic conditions, and uphold sustainable agricultural production principles. Soil sampling and laboratory analysis are essential tools for farmers, providing insights into soil properties and potential limitations. Proximal soil sensing uses sensors to assess soil parameters in real-time, enabling detailed maps of soil properties. This information helps farmers make precise management choices for specific locations (Minasny, 2017).

Remote sensing techniques like satellite imaging and drones enable soil mapping, providing farmers with detailed information on soil attributes. This data aids in executing tailored management strategies and aids in precision agriculture techniques like variable rate fertilizers and site-specific irrigation. Digital soil mapping (DSM) uses statistical and geospatial approaches to generate comprehensive maps, enhancing productivity and sustainability in the face of climate change (McKenzie et al., 2019).

Here are the main steps and procedures involved in Soil Mapping (McKenzie et al., 2019):

- **Soil Survey:** Soil survey is the first step in soil classification. This involves collecting representative soil samples from different locations in the study area. Samples are often taken at different depths and locations to capture changes in soil characteristics. Sampling methods may include targeted sampling, grid sampling, and transect sampling based on known soil variations.
- **Laboratory analysis:** The collected soil samples are sent to the laboratory for analysis. Various laboratory tests are carried out to determine soil characteristics such as electrical conductivity (EC), cation exchange capacity (CEC), texture, Organic matter content, pH, nutrient level, and other factors.
- These tests provide quantitative data that form the basis of the soil mapping.
- **Geospatial data collection:** Land maps include the collection of geospatial data, such as elevation and coordinates, using techniques such as GIS (Geographic Information System) and GPS (Global Positing System). Georeferencing of soil samples and field observations enables accurate mapping and spatial analysis.
- **Data Integration and Interpretation:** Soil data from laboratory surveys, field observations, and geospatial data are combined to determine the relationship and pattern between soil properties and topographic features. This definition helps to understand the factors that affect the landscape and changes in the study area (Hartemink, 2008).
- **Packs and Monitoring:** Data that refers to defining and creating a map and represents the distribution of food. The series, such as split contours, and system interpolation (e.g., Geostatistical, work to produce a map. These maps can be displayed in a variety of formats, including digital maps, paper maps, or online interactive maps.

Advantages of Soil Mapping

Soil maps are essential for land management decisions, including land use planning, natural resources management, and agricultural techniques. They help farmers determine the appropriate application of fertilizer, irrigation systems, and crop selection based on soil characteristics. They also aid in precision agriculture by strategically allocating inputs like water, herbicides, and fertilizers to areas where they are most needed, enhancing crop development and optimizing resource utilization. Soil maps also aid in environmental planning by identifying areas susceptible to harm, degradation, or pollution, enabling the implementation of erosion control strategies and conservation plans (Hengl & Heuvelink, 2013):

Soil conservation is facilitated by soil maps, enabling the adoption of practices like conservation tillage, cover cropping, terracing, contour plowing, strip cropping, and rainwater collection. Soil maps also serve as valuable tools in scientific research and education, allowing researchers to study soil relationships, construct predictive models, and aid in agricultural education initiatives. Overall, soil classification is crucial for understanding land changes, informed land management choices, and sustainable agricultural practices (Viscarra Rossel & Behrens, 2010).

Water Management

Water management refers to the planning, control, and conservation of water resources to meet various needs including industries, agriculture, environment, and housing. It involves the efficient use, distribution, and treatment of water to ensure water sustainability, reduce waste and reduce water-related hazards. Effective water management practices are key to addressing the improvement of water quality and promoting long-term water security and water scarcity (Singh & Frevert, 2018).

Water management is crucial in agricultural production, especially in the face of climate change. Farmers must employ efficient strategies to maximize irrigation efficiency, preserve water resources, and promote long-term sustainability. Remote sensing technology, precision irrigation methods, rainwater collection and conservation, crop selection and rotation, and integrated water resource management (IWRM) are essential tools for effective water management. Remote sensing allows farmers to monitor evapotranspiration rates, identify crop water stress, and detect changes in surface water bodies, enabling the optimization of irrigation scheduling and distribution. Precision irrigation methods, such as drip irrigation, sprinklers, and micro-irrigation systems, help farmers distribute water directly to the root zone, reducing losses caused by evaporation and runoff (Pereira, 2019).

Rainwater harvesting strategies, such as ponds, reservoirs, and rainwater harvesting systems, help address water scarcity caused by climate change and reduce reliance on groundwater supplies. Crop rotation strategies, which involve the systematic rotation of deep-rooted and shallow-rooted crops, can improve water management methods and adapt to shifting climatic circumstances. Integrated water resource management (IWRM) promotes holistic water resource management within a specific watershed or basin, synchronizing planning efforts, enhancing water utilization efficiency, and involving stakeholders for sustainable water management. Effective water management strategies can alleviate the impacts of limited water availability, promote adaptation to climate change, and foster sustainable agricultural practices (Sharma, 2021).

Here are some key features and strategies of water management (Pereira et al., 2012):

- **Water Resources Assessment:** Water management begins with an assessment of available water resources, including natural water (rainfall), surface water (rivers, ponds, lakes), underground aquifers, and drainage systems. Evaluating the quantity and quality of water sources helps determine their suitability for various purposes and informs decisions on water allocation.
- **Water conservation and efficiency:** Water conservation aims to reduce water consumption and waste through efficient use of water. This includes measures such as water-saving technology, efficient irrigation systems, leak detection, water recycling, and rainwater harvesting. Improving water use improves water quality and reduces water loss.
- **Irrigation Management:** Irrigation is the largest use of water in agriculture. Good irrigation management includes practices such as drip irrigation, efficient irrigation, recycling of wastewater, ocean, or seawater, and scheduling irrigation based on soil moisture levels and crop water needs. Proper irrigation management can reduce water use and increase crop production.
- **Water infrastructure and distribution systems:** Well-planned and maintained water infrastructure, including pipelines, canals, reservoirs, and dams, is necessary to store, transport and distribute water efficiently. An efficient water distribution system allows you to get the water resources in the right way, reduce water loss and make the release of water controlled for different uses (Allen et al., 1998).
- **Water Quality Management:** Water quality management is important for both environmental health and human consumption. This includes monitoring and controlling pollutants, treating and recycling wastewater and industrial waste, and implementing best practices to prevent pollution

and protect water sources, and also to increase the yield of crops. Good water management ensures safe and clean water for various uses.

- **Water governance and policy:** Strong water governance systems, policies, and regulations play an important role in water management. These include regulations to control pollution, water distribution systems, water performance standards, water pricing systems, and water rights. Effective governance ensures that water resources are used responsibly.
- **Integrated water management:** integrated water management takes a holistic approach to water resources, taking into account the connection between water systems, the environment, and human activities. This includes the planning and implementation of water projects in areas such as environmental protection, industry, agriculture, and urban development. An integrated approach helps balance competing water needs and promotes sustainable water management practices.

Advantages of Effective Water Management

Effective water management practices are crucial for a sustainable water supply, ensuring consistent and reliable water supply for various purposes like agriculture, industry, and residential use. These practices minimize water waste and enhance water utilization, reducing environmental footprint, cost savings, and water quality. They also protect biodiversity, water resources, and wetlands, preserving habitat and natural balance. Climate change considerations are integrated into water management systems, enhancing resilience to extreme weather events, droughts, and floods (Sharma & Minhas, 2010):

Adaptive strategies, such as integrating water management practices, efficient irrigation systems, and water conservation techniques, can mitigate climate-related risks and ensure sustained water resource availability during crises. Economic benefits of water management include reduced costs associated with water treatment and supply, promotion of water-dependent industries like tourism, aquaculture, and recreational activities, and enhanced agricultural output (Molden et al., 2010).

Climate Risk Assessment

Climate risk assessment is the process of analyzing and quantifying the adverse effects of climate change on various aspects, including natural processes, human activities, and the social and economic sectors. This includes assessing the vulnerability of various systems to climate risks and assessing their ability to adapt and withstand these risks. Climate risk assessment plays a critical role in managing and understanding the potential impacts of climate change, informing decision-making processes, and developing appropriate adaptation strategies (IPCC, 2014). Climate risk assessment approaches are crucial in the agricultural sector, especially in the context of climate change. These methodologies help farmers and policymakers assess the consequences of climatic variability and change on agricultural systems, identify vulnerable regions, and formulate suitable adaptation plans. Climate modeling and scenarios involve mathematical models that predict future climate conditions, providing valuable insights into probable future climates. Impact assessment models, which incorporate climate scenarios, agricultural data, crop models, and socio-economic aspects, assess the effects on crop yields, water availability, soil quality, and other variables.

Risk mapping and geographical analysis use Geographic Information System (GIS) tools to visually represent regions susceptible to adverse consequences from climate change effects. These methods can identify areas susceptible to climate-related hazards such as drought, flood, heat stress, and other dangers by incorporating climatic data, soil characteristics, topography, and crop-specific information. Spatial analysis approaches help identify hotspots and evaluate the vulnerability of agricultural systems, leading to the development of targeted adaptation plans and resource allocation (Porter, 2014).

Participatory methodologies involve farmers and local stakeholders in climate risk assessment processes, involving their viewpoints and knowledge into decision-making procedures. This approach enhances stakeholder ownership, knowledge exchange, and community resilience in the context of climate change impacts. Climate risk assessment methods serve as a fundamental framework for understanding and effectively addressing climate change hazards in agricultural production, helping farmers and policymakers identify vulnerable areas, determine key areas for adaptation activities, and allocate resources efficiently (Wollenberg, 2016).

Here are the main steps in a climate risk assessment (UNFCCC, 2012):

- **Risk Identification:** The first step in climate risk assessment is to identify the potential climate-related risks in a region or region. These risks may include extreme weather events such as changes in precipitation patterns, sea level rise, heat waves, droughts, floods, and heavy rains. Understanding the possible risks is key to assessing their impact on different areas and systems.
- **Exposure Assessment:** Exposure assessment involves determining the extent to which a system or sector is exposed to climate-related hazards. This includes assessing the spatial or temporal distribution and intensity of hazards and identifying resources, infrastructure, and populations at risk. Exposure assessment helps identify areas and systems most vulnerable to climate impacts.
- **Vulnerability Assessment:** Vulnerability assessment search to understand the sensitivity and adaptive capacity of a system or sector to climate-related risks. This includes assessing system characteristics, sustainability, and resilience, including infrastructure, social, human welfare, and environmental condition. Vulnerability assessment helps identify areas of high vulnerability and potential climate risk areas.
- **Impact assessment:** Impact assessment involves evaluating the consequences and impacts of climate-related risks on different regions and systems. This includes the assessment of the indirect effects of climate change on the environment, agriculture, economic activity, water resources, and human health. The impact assessment provides insight into the magnitude and nature of potential risks (Ford et al., 2013).
- **Risk assessment:** Risk assessment combines information gathered from an exposure assessment, vulnerability assessment, and impact assessment to quantify and rate or range climate risks. This includes assessing the likelihood of a hazard occurring, assessing the extent of potential impacts, and determining the likelihood and severity of the hazard. Risk assessment helps to prioritize actions and allocate resources for climate change adaptation.
- **Planning:** Based on the findings of the risk assessment, planning involves developing plans and actions to reduce the negative effects of climate change. This includes building resilience, policy implementation, identifying adaptation strategies, and incorporating climate change considerations into decision-making processes. Adaptation strategies aim to reduce vulnerability, build adaptive capacity, and promote sustainable development in the face of climate risks.

Advantages of Climate Risk Assessment Tools

Climate risk assessment is a vital tool for decision-making across various levels, including policy makers, government agencies, enterprises, and communities. It helps identify areas requiring attention, optimize resource allocation, and provide valuable insights for climate change policy formulation. Risk assessment also contributes to resilience by understanding and quantifying climate-related hazards, enabling proactive measures to mitigate climate change's adverse impacts. It aids in sector planning and management, forming strategies in public health, urban planning, infrastructure, water resources, and agriculture. Incorporating climatic hazards into long-term development plans is a key objective of sector planning. Overall, climate risk assessment is essential for understanding, measuring, and addressing climate change risks (Carter et al., 2007).

Land Use Planning and Conservation

Planning and conservation of land to use is an important process that seeks to guide the use of natural resources efficiently and effectively while protecting the environment and protecting environmental standards. These processes include determining and evaluating the use of appropriate land, taking into account factors such as social needs, sustainability, environmental integrity, and economic viability for a long time (FAO, 2017).

Land use planning and conservation strategies are crucial in addressing climate change challenges in agricultural production. Precision agriculture, a strategic approach that uses advanced technologies like GPS, remote sensing, and data analytics, enhances resource utilization and productivity by accurately assessing soil variability and monitoring crop development. This helps farmers make informed decisions about irrigation, fertilization, and pest management, reducing resource wastage and mitigating environmental impacts. Conservation agriculture prioritizes soil health, erosion mitigation, and biodiversity protection, reducing soil disturbance, improving water infiltration, and promoting crop rotation diversification (Scharf, 2016).

Agroforestry, a land management practice that integrates trees within agricultural systems, enhances soil fertility, manages microclimate, and diversifies income sources. It also contributes to carbon sequestration, greenhouse gas emissions mitigation, and biodiversity enhancement. Land conservation and restoration initiatives aim to preserve ecological systems, including natural habitats, wetlands, and riparian zones. These practices improve soil health, water retention capacity, and carbon sequestration potential. Protected areas and buffer zones around agricultural lands enhance resilience against extreme weather events and maintain ecological equilibrium. Implementing these strategies helps farmers reduce climate-related risks, adapt to changing climatic conditions, and contribute to the sustainability and resilience of agricultural systems (Kumar, 2018).

The main components and objectives of land conservation planning are (Liu, 2016):

- **Land Resources Assessment:** Land use planning involves assessing the physical and ecological characteristics of land, such as topography, soil quality, water resources, biodiversity, and habitat, to identify suitable areas for different uses and areas requiring protection. It involves delineating and allocating areas for specific purposes, such as conservation, recreation, agriculture, industrial, commercial, and residential, ensuring proper planning and resource allocation.

- **Environmental protection:** Land use planning involves preserving environmental areas and conserving natural resources through the designation of protected areas, nature reserves, green belts, and conservation areas, aiming to maintain wetlands, conserve water, prevent habitat fragmentation, and balance the environment.
- **Sustainable development:** Land use planning promotes sustainable development by supporting economic growth, environmental protection, and social well-being. It considers factors like social equity, climate change mitigation, renewable energy, and resource production, ensuring present needs are met without compromising future generations' ability to meet their own needs.
- **Community Involvement:** Land use planning requires community participation, public discussion, and diverse perspectives to consider community needs and aspirations, fostering ownership and sustainable practices, and promoting a sense of ownership.
- **Monitoring and Management of Change:** Land use planning is a continuous process that needs regular monitoring and coordination to assess the effectiveness of management plans and identify necessary adjustments, while flexible management allows for adaptation to new information and challenges.

Advantages of Effective Land Use Planning and Conservation

Land use planning is a vital process that optimizes land resources, reduces risks, promotes sustainable urban development, and preserves natural resources for future generations. It also plays a significant role in environmental conservation, preserving habitats, biodiversity, and ecosystems by establishing protected areas and mechanisms (Bennett & Dixon, 2017).

Land use planning improves urban quality of life by ensuring essential services, protecting natural areas, developing transportation systems, and creating sustainable cities. It also reduces disaster risk by relocating development activities. Land use planning is essential to combat climate change by promoting sustainable management systems, promoting renewable energy, reducing carbon emissions through afforestation and reforestation, and enhancing community and ecosystem adaptability. It considers environmental, social, and economic factors for a sustainable landscape (Malinga & Totolo, 2017).

Yield Prediction

Yield forecasting, also known as crop planning, is a crucial process for predicting the production of a crop, aiding in agricultural planning, resource allocation, market forecasting, and monitoring food security, using various methods and factors (Hoffmann et al., 2016):

Yield prediction methodologies utilize statistical models, machine learning algorithms, and remote sensing data to estimate agricultural yields and assess climate patterns. Accurate projections aid farmers, policymakers, and academics in making informed decisions, optimizing resource allocation, and developing effective adaptation strategies (Ruiz-Ramos, 2019).

Statistical models are mathematical representations used to analyze data on historical correlations between climate variables and agricultural yields, incorporating parameters like temperature, precipitation, solar radiation, and crop management practices, but may have limitations (Lobell, 2019).

Machine learning algorithms like artificial neural networks, random forests, and support vector machines are increasingly used in yield prediction, capturing nonlinear associations between climate variables and crop yields, enabling more precise prognostications (White, 2011).

Crop growth models use climate, soil conditions, and management practices to predict agricultural productivity. Remote sensing techniques and satellite data enhance precision and promptness of yield projections by monitoring vegetation indices and evaluating conditions like drought stress (Thenkabail, 2019).

Advantages of Yield Prediction in Agriculture

- **Analysis of historical data:** Forecasting and crop modeling are methods used to predict future agricultural yields and crop production, using historical data and mathematical models to simulate plant growth and development processes, thereby understanding the impact of changing variables.
- **Remote Sensing and Satellite Imagery:** Remote sensing tools like satellite imagery and aerial surveys offer valuable insights into plant health, weed indices, and vegetation landscape disparities. These data can help analyze crop conditions, identify stressors, and develop strategic plans. Combining remote sensing with machine learning can predict agricultural yields over large areas, using metrics like height, leaf area index, biomass, and development stage (Rezaei et al., 2019).
- **Expert knowledge and crop advice:** Crop forecasting relies on specialist knowledge and local experience, with information from agricultural extension agencies, farmers, and experienced farmers. Factors like climate change, disease outbreaks, and market conditions influence yield estimation. Models face uncertainties due to data availability and complexity. Using diverse methodologies and data sources enhances yield prognostications. Technology advancements improve forecasting precision and timeliness, enabling informed decision-making in agriculture (Wang et al., 2019).

BENEFITS OF USING GEOSPATIAL TECHNIQUES IN AGRICULTURE

Benefits of Using Geospatial Techniques in Agriculture, Such as Increased Efficiency, Reduced Costs, and Improved Yields

Geospatial techniques, which include the use of geographic information systems (GIS), remote sensing, and global positioning systems (GPS), offer many benefits when applied to agriculture. Here are some of the key benefits of using geospatial systems in agriculture (Gebbers & Adamchuk, 2010):

- **Efficient farming:** Geospatial systems enhance farming precision by allowing farmers to plan farm management across multiple locations. By combining geospatial data with topography, climate, soil, and vegetation indices, farmers can optimize fertilizer, water, and resource application, improving crop quality, increasing efficiency, and reducing waste.
- **Site-Specific Management:** Geospatial systems offer precise information on soil properties, enabling farmers to implement site-specific management strategies. This includes adjusting seeding rates, selecting suitable crops, and applying soil amendments, thereby maximizing plant growth and yield potential (Huang et al., 2010).
- **Crop health monitoring:** Visual monitoring data, like satellite imagery and aerial surveys, can help farmers identify potential stressors in real-time, such as low plant vigor, nutrient deficiencies,

pest and disease outbreaks, or water stress. Early detection can improve yields, reduce crop losses, target pest control, and improve nutrient supplementation or water reform.

- **Crop classification and monitoring:** Geospatial systems are utilized for crop monitoring and mapping, with GPS seed monitors in harvesters recording seed data and creating a seed map. This information helps farmers identify high and low productivity areas, guiding future management decisions like irrigation changes or crop density adjustments.
- **Land and resource management:** Geospatial systems aid in efficient land and resource management in agriculture, enabling analysis of soil quality, land use planning, and crop rotation. Integrating geospatial data with climate models helps farmers assess risks, anticipate impacts, and make informed decisions, promoting sustainable resource management and improved land use efficiency (Chakraborty et al., 2012).
- **Data-Driven Decisions:** Geospatial systems offer farmers accurate, up-to-date spatial information, enabling data-driven decisions. By combining data with historical records, weather forecasting, and agronomic models, farmers can assess risks, optimize resource allocation, and plan operations more efficiently. This leads to informed decision-making, cost reduction, and improved agricultural productivity. Geospatial systems support farmers' operations, reduce environmental impact, and address changing agricultural challenges (Mishra et al., 2018).

Challenges of Using Geospatial Techniques in Agriculture

Geospatial systems offer significant advantages in agriculture, but their implementation presents several challenges, including the need for effective data collection and analysis (Kaur, 2020):

- **Availability and quality of data:** Reliable geospatial data is crucial for effective implementation of measures, but access can be challenging, especially in developing or limited resource regions. Collecting, processing, and organizing data requires technical expertise, financial resources, and time.
- **Cost and technology requirements:** Geospatial technologies often require specialized hardware and software, which can be costly for small farmers or regions with limited resources. The initial investment in equipment like GPS receivers, drones, or satellite imagery can hinder adoption (Rai et al., 2018).
- **Data collection and analysis:** The integration of diverse datasets, including satellite imagery, weather data, and geospatial information, can be challenging due to differences in methods, resolutions, and processing. This requires advanced computational tools, GIS expertise, and can be particularly challenging for small farmers or those without technical assistance (Swain et al., 2020).
- **Spatial and Temporal Variation:** Agricultural landscapes show significant variations in soil, climate, and crop characteristics, but accurately achieving these changes is challenging due to geospatial systems' large-scale operation, uncertainty in data at the field level, and careful consideration of spatial patterns.
- **Interpretation and Decision Making:** Geospatial data offers valuable information but interpreting it into actionable decisions can be challenging. Farmers and managers must understand its importance, regional conditions, and management objectives, and integrate it with local knowledge.

- **Infrastructure and Connectivity:** Geospatial systems rely on digital infrastructure, but limited access to reliable internet connections in rural or remote areas can hinder data sharing, real-time monitoring, and efficient use of geospatial tools.
- **Adoption and awareness:** The adoption of geospatial systems in agriculture necessitates awareness, training, and demonstration of practical benefits to farmers and stakeholders. Many farmers may be unfamiliar with these technologies or perceive them as complex and inaccessible. Overcoming barriers requires capacity building and knowledge transfer (Weldon et al., 2018).

To overcome challenges in agriculture, resources, capacity building, data-sharing, and collaboration among stakeholders are crucial. Technological advancements, cost reduction, and knowledge expansion can enhance geospatial systems in agriculture, leading to better decision-making, increased productivity, and sustainable farming methods. This is especially important as costs are reduced and knowledge expands (Geng & Rasmussen, 2021).

CASE STUDIES AND FUTURE OF GEOSPATIAL TECHNIQUES IN AGRICULTURE

Case Studies of Successful Applications of Geospatial Techniques and Their Impact on Crop Agricultural Production

The application of geospatial techniques has had a profound impact on contemporary agriculture, as it has effectively improved precision and sustainability within the industry. An exemplary case of success is the implementation of precision agriculture methodologies by John Deere, a prominent manufacturer of agricultural machinery. The use of GPS technology into their machines facilitated the exact planting, management, and harvesting of crops for farmers. This methodology has resulted in significant advantages, such as enhanced agricultural productivity, diminished utilization of resources, and decreased operational expenditures. (John Deere, 2021).

This paper provides a comprehensive analysis of the geospatial techniques employed in the agricultural sector and evaluates their influence on crop output, operational efficiency, and cost reduction. Geospatial approaches utilized in the field of agriculture involve a range of technologies, including the Global Positioning System (GPS), Geographic Information Systems (GIS), & remote sensing. The Global Positioning System (GPS) facilitates the acquisition of exact location information for agricultural machinery, so enabling meticulous planting and harvesting practices, minimizing redundancy, and yielding time and resource savings. Geographic Information Systems (GIS) enable farmers to do spatial data analysis, encompassing factors such as soil quality and weather patterns, in order to make well-informed determinations regarding crop management. The utilization of satellites and unmanned aerial vehicles (UAVs) in remote sensing facilitates the surveillance of crop vitality and the timely identification of illnesses or nutrient insufficiencies. These strategies collectively aim to optimize the allocation of resources, mitigate environmental effect, and deliver greater agricultural production, enhanced efficiency, and cost savings. (Thompson et al., 2017).

The present study examines the emerging trends in the utilization of Geospatial Techniques in the agricultural sector. The agricultural sector is currently experiencing rising trends that have the potential to significantly transform farming methods. An observable phenomenon is the convergence of artificial intelligence (AI) & machine learning techniques with geographical data. The integration of these components facilitates the application of predictive analytics, so assisting farmers in proactively anticipating and mitigating potential obstacles, including disease outbreaks and weather-related adversities. Furthermore, there is a growing adoption of blockchain technology in the agricultural sector, which provides the advantage of establishing transparent supply chains and ensuring the traceability of food products throughout the entire production process, from the origin at the farm to the final consumption at the table. The Internet of Things (IoT) facilitates the interconnection of sensors and equipment within agricultural settings, enabling the acquisition of real-time data and augmenting the practice of precision agriculture. (Chen et al., 2020).

The potential influence of innovations in geospatial technology on agriculture. Advancements in geospatial technology exhibit significant potential for the agricultural sector. One example of a technology that can offer comprehensive insights into crop health as well as soil conditions is hyperspectral imaging. This advanced imaging technique enables the collection of detailed data, facilitating accurate nutrient delivery and effective disease management strategies. The emergence of autonomous farming represents a noteworthy advancement, as it involves the utilization of drones and robotic machines that feature geospatial technology to carry out various agricultural activities such as planting, weeding, and harvesting. These technological advancements possess the capacity to substantially enhance efficiency, diminish labor expenses, and enhance total production inside the agricultural sector. (Lu and Xie, 2020).

Future Challenges and Opportunities for the Use of Geospatial Techniques in Agriculture

Geospatial approaches have demonstrated significant potential; yet their extensive implementation is hindered by various constraints. It is imperative to address concerns pertaining to data privacy and security in order to foster confidence among farmers when it comes to exchanging sensitive information. There is a pressing need to address the digital divide that exists in rural areas, characterized by restricted availability to high-speed internet & geospatial technology. Furthermore, it is imperative to give thorough thought to the regulatory concerns pertaining to data ownership and usage rights. Conversely, there are several prospects for the utilization and advancement of geospatial technologies in developing regions, which can contribute to the improvement of climate resilience by means of enhanced prediction and adaption tactics, as well as the mitigation of global food security concerns. The establishment of partnerships among governmental entities, academic institutions, and private enterprises is of utmost importance in ensuring the long-term viability of geospatial techniques in the agricultural sector. It is imperative to address and overcome the obstacles associated with these techniques in order to fully exploit their capabilities. (Büyüksalih et al., 2021).

CONCLUSION

Geospatial techniques have significantly improved agriculture efficiency, reduced costs, and yields. However, challenges like reliable data, specialized hardware, and spatial variations persist. Technological advancements, cost reduction, and knowledge expansion can enhance these systems, leading to better decision-making, increased productivity, and sustainable farming methods. Emerging trends in geospatial technology include artificial intelligence, machine learning, blockchain, and the Internet of

Things. These techniques are crucial for understanding climate change's effects on global ecosystems, water resources, and meteorological patterns. They provide accurate data on soil conditions, weather patterns, and crop development, enhancing crop growth and productivity.

Crop modeling, also known as farm modeling or crop simulation, is crucial for agricultural productivity, especially in the context of climate change. Soil mapping is vital for optimizing crop management practices and adjusting to climatic conditions. Climate risk assessment helps identify vulnerable regions and formulate suitable adaptation plans. Land use planning and conservation are essential processes in agriculture, guided by precision agriculture, conservation agriculture, agroforestry, land conservation, and yield forecasting.

REFERENCES

Tropogo. (n.d.). *Application of drones in acriculture in India.* Toropogo. https://tropogo.com/blogs/application-of-drones-in-agriculture-in-india

Adamchuk, V. I., & Hummel, J. W. (2016). Remote sensing and imaging sensors for site-specific management in precision agriculture. *Journal of Applied Remote Sensing*, *10*(1), 015009.

Allen, R. G., Pereira, L. S., Raes, D., & Smith, M. (1998). Crop evapotranspiration: Guidelines for computing crop water requirements. *FAO Irrigation and Drainage Paper*, 56.

Arslan, O., & Colkesen, I. (2021). Precision Farming Applications and Future Perspectives: A Review. *Computers and Electronics in Agriculture*, *190*, 106335.

Aruna, M. S., & Pradesh, M. (2012). *A Review: The application of Remote Sensing*. GIS and GPS in Precision Agriculture.

Asseng, S., Martre, P., Maiorano, A., Rötter, R. P., O'Leary, G. J., Fitzgerald, G. J., Girousse, C., Motzo, R., Giunta, F., Babar, M. A., Reynolds, M. P., Kheir, A. M. S., Thorburn, P. J., Waha, K., Ruane, A. C., Aggarwal, P. K., Ahmed, M., Balkovič, J., Basso, B., & Ewert, F. (2019). Climate change impact and adaptation for wheat protein. *Global Change Biology*, *25*(1), 155–173. doi:10.1111/gcb.14481 PMID:30549200

Bennett, G., & Dixon, J. (Eds.). (2017). *Farming Systems and Food Security in Africa: Priorities for Science and Policy Under Global Change*. Routledge.

Boote, K. J., Jones, J. W., White, J. W., Asseng, S., & Lizaso, J. I. (2019). Putting mechanisms into crop production models: Challenges and opportunities. *Agronomy Journal*, *111*(2), 535–555.

Bouma, J. (2016). Soil science contributions to precision agriculture. In *Soil Mapping and Process Modeling for Sustainable Land Use Management* (pp. 1–19). CRC Press.

Büyüksalih, G. (2021). Challenges and Opportunities of Geospatial Technologies in Agriculture. In Handbook of Research on Smart Technology Integration in the Agricultural Sector (pp. 1-16). IGI Global.

Carter, T. R., Jones, R. N., & Lu, X. (2007). Assessing Climate Change Impacts on Agriculture: The Value of Climate Analogues. In Ecosystems and Human Well-Being: Current State and Trends (pp. 253-263). Island Press.

Chakraborty, D., Ghosh, S., & Joshi, P. K. (2012). Application of geospatial technology in precision agriculture: A review. *Journal of Agricultural Physics*, *12*(1), 1–15.

Chen, Z. (2020). Geospatial big data-driven smart agriculture: A review. *Computers and Electronics in Agriculture*, *176*, 105612. doi:10.1016/j.compag.2020.105612

Dabas, M., Pandey, S. K., & Singh, S. (2020). Precision Farming Technologies for Sustainable Agriculture: A Review. *IOP Conference Series. Earth and Environmental Science*, *431*(1), 012057.

FAO. (2017). *Land-use planning: Making sustainable land-use decisions at the local level*. Food and Agriculture Organization of the United Nations.

Ford, J. D., Berrang-Ford, L., Lesnikowski, A., Barrera, M., & Heymann, J. (2013). How to track adaptation to climate change: A typology of approaches for national-level application. *Ecology and Society*, *18*(3), 40. doi:10.5751/ES-05732-180340

Ge, Y., & Thomasson, J. A. (2017). *Geospatial techniques for managing agriculture and natural resources*. CRC Press.

Gebbers, R., & Adamchuk, V. I. (2010). Precision agriculture and food security. *Science*, *327*(5967), 828–831. doi:10.1126/science.1183899 PMID:20150492

Geng, X., & Rasmussen, J. (2021). Overcoming barriers to precision agriculture adoption: A systematic review. *Agricultural Systems*, *188*, 103015.

Hartemink, A. E. (2008). Geostatistics and soil mapping. *Geoderma*, *146*(1-2), 90–99.

Hengl, T., & Heuvelink, G. B. M. (2013). Soil mapping and soil monitoring: A review. *International Journal of Applied Earth Observation and Geoinformation*, *23*, 1–14.

Hoffmann, H., Zhao, G., Asseng, S., Bindi, M., Biernath, C., Constantin, J., Coucheney, E., Dechow, R., Doro, L., Eckersten, H., Gaiser, T., Grosz, B., Heinlein, F., Kassie, B. T., Kersebaum, K.-C., Klein, C., Kuhnert, M., Lewan, E., Moriondo, M., & Ewert, F. (2016). Impact of spatial soil and climate input data aggregation on regional yield simulations. *PLoS One*, *11*(8), e0151782. doi:10.1371/journal.pone.0151782 PMID:27055028

Holzworth, D. P., Huth, N. I., deVoil, P. G., Zurcher, E. J., Herrmann, N. I., McLean, G., & Snow, V. O. (2014). APSIM - Evolution towards a new generation of agricultural systems simulation. *Environmental Modelling & Software*, *62*, 327–350. doi:10.1016/j.envsoft.2014.07.009

Huang, J., Yang, X., Rozelle, S., Li, N., & Zhang, J. (2010). Irrigation, agricultural performance and poverty reduction in China. *Food Policy*, *35*(6), 455–464.

IPCC. (2014). *Climate Change 2014: Impacts, Adaptation, and Vulnerability. Contribution of Working Group II to the Fifth Assessment Report of the Intergovernmental Panel on Climate Change*. Cambridge University Press.

Janssen, S., Porter, C. H., Moore, A. D., Athanasiadis, I. N., Foster, I., Jones, J. W., & Antle, J. M. (2017). Towards a new generation of agricultural system data, models and knowledge products: Design and improvement. *Agricultural Systems*, *155*, 255–268. doi:10.1016/j.agsy.2016.09.017 PMID:28701817

John Deere. (2021). *Precision Agriculture: The Future of Farming*. John Deere. https://www.deere.com/en/technology-products/precision-agriculture/

Jones, J. W., Antle, J. M., Basso, B., Boote, K. J., Conant, R. T., Foster, I., & Porter, C. H. (2017). Toward a new generation of agricultural system data, models, and knowledge products: State of agricultural systems science. *Agricultural Systems*, *155*, 269–288. doi:10.1016/j.agsy.2016.09.021 PMID:28701818

Kaur, H. (2020). Challenges and opportunities in the use of remote sensing and geospatial technology for sustainable agriculture. *Journal of Spatial Science*, *65*(2), 241–263.

Khan, M. S. (2021). Role of remote sensing in sustainable water resources management in agriculture: A review. *Agricultural Water Management*, *248*, 106770.

Khan, Z. H., Bhat, S. A., Khare, D., & Wani, S. H. (2020). Precision Farming in Agriculture: A Review. *Journal of Pure & Applied Microbiology*, *14*(3), 1925–1937.

Khosla, R., & Reich, R. M. (2016). The role of geospatial techniques in precision agriculture. In M. Gebbers & A. Adamchuk (Eds.), *Precision agriculture basics* (pp. 17–29). Springer.

Kocaman, S., Demircioglu, M. B., & Yilmaz, E. (2019). The use of geospatial techniques in climate change research. *Journal of Environmental Science and Engineering B*, *8*(6), 173–185. doi:10.17265/2162-5263/2019.06.002

Kumar, B. M. (2018). Agroforestry: A pathway to sustainable land use and climate resilient farming. *Current Opinion in Environmental Science & Health*, *5*, 106–113.

Liang, S. (2020). Applications of remote sensing in precision agriculture: A review. *The Science of the Total Environment*, 142063.

Limpisathian, P. (2011). *Geographic Information System in Agriculture and Precision Farming*. Academic Press.

Liu, B. (2016). Statistical models for yield prediction in agriculture: Opportunities and challenges. *Journal of Agricultural Biological & Environmental Statistics*, *21*(3), 361–380.

Lobell, D. B. (2014). Remote sensing and geospatial technologies for agriculture. In G. Backhaus & R. J. Hussey (Eds.), *The Routledge handbook of environment and communication* (pp. 250–258). Routledge.

Lobell, D. B. (2019). The use of satellite data for crop yield gap analysis. *Field Crops Research*, *237*, 104–110.

Lu, B., & Xie, W. (2020). Geospatial technology applications in autonomous farming: A review. *Computers and Electronics in Agriculture*, *175*, 105565.

Lu, X., Zhang, X., Li, J., & Zhang, H. (2021). Application of geospatial techniques in climate change research: A review. *Remote Sensing*, *13*(3), 541. doi:10.3390/rs13030541

Malinga, R., & Totolo, O. (2017). The role of land use planning in achieving sustainable land management in rural Botswana. *Journal of Environmental Planning and Management*, *60*(10), 1861–1881.

McBratney, A. B., Mendonça Santos, M. L., & Minasny, B. (2003). On digital soil mapping. *Geoderma, 117*(1-2), 3–52. doi:10.1016/S0016-7061(03)00223-4

McKenzie, N. J., Grundy, M. J., Webster, R., & Minasny, B. (2019). *Digital Soil Mapping Encyclopedia of Soil Science* (3rd ed.).

Minasny, B. (2017). Soil sensing for precision agriculture. In *Precision Agriculture* (pp. 223–241). Springer.

Mishra, A. K., Raghuwanshi, N. S., & Singh, R. (2018). Geospatial techniques in agriculture: An overview. *Current Science, 114*(3), 575–586.

Mkhabela, M. (2020). An overview of precision farming technologies used in the agricultural industry. *Agrotechnology, 9*(2), 1–8.

Molden, D. J., Oweis, T., Steduto, P., Bindraban, P. S., Hanjra, M. A., & Kijne, J. W. (Eds.). (2010). Improving water productivity in agriculture: Editor's overview. *Agricultural Water Management, 97*(4), 517–523. doi:10.1016/j.agwat.2009.03.023

Pathan, S., Sood, A., & Ahmad, T. (2021). A Review on Precision Agriculture: Techniques, Challenges, and Opportunities. *Computers and Electronics in Agriculture, 185*, 106131.

Pereira, L. S. (2019). Improving irrigation water use efficiency in agriculture: A review. *Agronomy for Sustainable Development, 39*(1), 1–22. PMID:30881486

Pereira, L. S., Cordery, I., & Iacovides, I. (2012). Coping with water scarcity: An action framework for agriculture and food security. *Agriculture, Ecosystems & Environment, 144*, 54–65.

Porter, J. R. (2014). Food security and food production systems. In *Climate Change 2014: Impacts, Adaptation, and Vulnerability. Part A: Global and Sectoral Aspects. Contribution of Working Group II to the Fifth Assessment Report of the Intergovernmental Panel on Climate Change*. Cambridge University Press.

Qureshi, M. F., Hussain, M., Shabbir, R. N., & Baig, I. A. (2020). Precision Agriculture: A Review. *Advances in Horticultural Science, 34*(3), 393–410.

Rai, R., Sharma, A., & Sinha, A. (2018). Challenges in the adoption of geospatial technology for agricultural decision-making: A review. *International Journal of Remote Sensing & Geoscience, 7*(4), 2482–2489.

Rezaei, E. E., Siebert, S., Ewert, F., Webber, H., Gaiser, T., Naumann, G., & Zhao, G. (2019). Assessing yield gaps at the country level using an adapted FAO EcoCrop model. *European Journal of Agronomy, 103*, 79–88.

Ruiz-Ramos, M. (2019). Adaptation to climate change through the choice of cropping system and sowing date in northern Spain. *European Journal of Agronomy, 103*, 97–109.

Scharf, P. C. (2016). Precision nutrient management. In *Precision Agriculture* (pp. 399–432). Springer.

Sharma, B. R., & Minhas, P. S. (2010). Water management in agriculture: A global perspective. *Journal of Applied and Natural Science, 2*(1), 88–94.

Sharma, D. (2021). Rainwater harvesting: A boon for sustainable agriculture. *Journal of Water Supply: Research & Technology - Aqua, 70*.

Singh, V. P., & Frevert, D. K. (2018). *Handbook of Applied Hydrology* (2nd ed.). McGraw-Hill Education.

Srivastava, A. (2020). A GIS-based approach for crop yield estimation using multiple regression analysis. *Geocarto International, 35*(3), 225–239.

Swain, D., Roy, P., & Panda, S. S. (2020). Challenges in the adoption of geospatial technology for precision agriculture in India: A review. *International Journal of Scientific Research in Science, Engineering and Technology*, *6*(4), 373–381.

Tetteh Kwasi Nuer, A., Agbeko, D., Worlali, S., Mwangi Thiga, M., Ndogo Ndungu, S., Wangari Mutiga, M., & Babcock, L. H. (2018). *Why Invest in ICTs for agriculture?* CTA.

Thenkabail, P. S. (2019). Advancements in remote sensing of agriculture: Context description, existing operational monitoring systems, and major information needs. *Remote Sensing*, *11*(18), 2107.

Thompson, J. A. (2017). Geospatial applications in precision agriculture: An overview. *Applied Geomatics*, *9*(1), 67–74.

UNFCCC. (2012). *Technical Guidelines for the National Adaptation Plan Process*. United Nations Framework Convention on Climate Change.

Viscarra Rossel, R. A., & Behrens, T. (2010). Using data mining to model and interpret soil diffuse reflectance spectra. *Geoderma*, *158*(1-2), 46–54. doi:10.1016/j.geoderma.2009.12.025

Wang, Z., Ma, D., Guo, W., & Zhang, Y. (2019). Crop yield estimation based on machine learning: A review. *Sensors (Basel)*, *19*(19), 4302. PMID:31590234

Weldon, L., McBride, M. F., & Earley, R. (2018). Challenges to implementing precision agriculture techniques: A case study of farmer attitudes in Ireland. *Land Use Policy*, *70*, 439–446.

White, J. W. (2011). Field-based approaches for developing maize germplasm adapted to drought and heat stress. *Crop Science*, *51*(3), 1–15.

Wollenberg, E. (2016). Climate change mitigation and adaptation in the land sector: From complementarity to synergy. *Environmental Research Letters*, *11*(3), 033001.

Zhang, Y., Liu, X., Liang, S., & Yang, X. (2019). Applications of remote sensing in precision agriculture: A review. *Journal of Agricultural Science*, *11*(2), 42–54.

Chapter 13
Smart Application in Climate Change Policy for India

Mitrajit Biswas
O.P. Jindal Global University, India

ABSTRACT

The world has always been divided based on access to technology. Since the dawn of humanity, the idea of evolving and utilizing the best of resources gave birth to the advent of technological evolution. Amidst all of this, our human lives have been transformed by technology in the form of personal belongings, like a phone, which is now smart enough to do almost any function that you can imagine. It states that technology has definitely made our lives easier and engaging for the most part; however, one must not forget that there are challenges of the technological prowess as well. Technology is now taking the next step already where data is the fuel and based on that, technology is stepping up to play the role in the world of government policies and governance initiative. This chapter wants to highlight the importance of technology and governance of urban resources especially with the onset of the more intense discussions on climate change, global warming, etc.

INTRODUCTION

Technology is the biggest game changer in the world of today. The way we live in today's world is driven and articulated by technology. However, technology is dynamic and never stops at a particular junction. The entre human civilization has been a story of the growth of technology and the evolution of the human race along with it. The world has always been divided based on access to technology. Since the dawn of humanity, the idea for evolving and utilizing the best of resources gave birth to the advent of technological evolution. Amidst all of this our human lives have been transformed by technology in the form of personal belongings be it a phone which is now smart enough to do all possible functions that you can imagine. It states that technology has definitely made our lives easier and engaging for the most part, however one must not forget that there are challenges of the technological prowess as well. Technology is now taking the next step already where data is the fuel and based on that, technology is stepping up to play the role in the world of government policies and governance initiative. This article

DOI: 10.4018/979-8-3693-3807-0.ch013

Copyright © 2024, IGI Global. Copying or distributing in print or electronic forms without written permission of IGI Global is prohibited.

wants to highlight the importance of technology and governance of urban resources especially with the onset of the more intense discussions on climate change, global warming etc. It has got a direct connection with the high emission levels.

Energy has been a critical component for the growth and advancement of any human civilization. As we have evolved over the ages, the role of making use of data for the betterment of human lives has been at questions, Food, Health, Transportation are already the important areas that have been discussed. However, when it comes to the question of harnessing the best possible way for individual output of energy, governance is still negligible in most of the developing world. This is where the role of technology becomes important. So, the paper would try to connect Food, Health, and Transportation and Energy optimization as the four pillars and how can technology play a role in the overall process of the governance is going to be dealt here. The idea is to build for the understanding that how does governance play a part in the overall human society and what has India been doing in these regards. In the world's most populated country over the last few years it is well known that Aadhar card, UPI have been some of the biggest game changers related to citizen data management as well as financial technology. However, the onset of technology in the world of food production/agriculture, Health care services, Transportation and last but not the least Urban Energy Management will play a very important role and have been already for the governance and policy decisions in India. The change is slow but now the role of technology in these areas of governance is coming through.

There has been already a changing paradigm of technology and governance as it is already evident in the works of **Almgren & Skobelev, D. (2020).** The paper speaks of the "fourth technological paradigm (wave) (1930–1985) was characterized by power engineering, machinery manufacturing, and new synthetic materials and communications equipment production, and contributed towards high-volume manufacturing of consumer commodities, weaponry, motor passenger vehicles and trucks, field engines, planes, and to the growing importance of computers and software products. Characteristic features of the fourth wave are still seen in all, even very advanced economies. Industrial sectors of the fourth wave are those consuming large amounts of natural resources (including energy).

The fifth technological paradigm is based on computer science, microelectronics, biotechnology, new types of energy sources and energy generation, genetic engineering, materials, satellite communications and space exploration. It is also a period of migration from singular 'stand-alone' companies to an intertwined electronic net of small, medium-sized and large enterprises, interacting closely in the fields of technology, product quality control, and innovation planning. A distinctive feature of the fifth wave is the enhanced role of microelectronic components. The advantage of the fifth paradigm lies in the individualization of production and consumption, an increase in production flexibility, and a strong attention to resource efficiency.

The origins of the sixth technological paradigm can be traced back to around 2010. Biotechnology and nanotechnology, genetic engineering, membrane and quantum technologies, photonics, micromechanics, and thermonuclear energy are becoming more and more conventional solutions. Experts expect that the synthesis of these areas will eventually lead to quantum computing and artificial intelligence, and provide access to a fundamentally new development level of governmental, societal, and economic systems. Experts predict that the sixth technological paradigm will enter the maturity phase after 2040. It is expected that a new scientific, technical and technological revolution based on achievements in the aforementioned basic technological areas will take place during 2020–2025. There are reasons for making such estimations: in 2010, the most economically developed countries had 20% of their productive power in fourth technological paradigm, 60% in the fifth, and about 5% in the sixth. Currently we are observing the structural reorganization of the world economy. We can try to predict the birth of a new technological paradigm out of IT and communication technologies and bioengineering with a certain measure of nano-technological solutions in well-developed economies of the 'first world', which will ultimately lead to a beneficial 'long wave' of growth. A drop in oil prices is a characteristic sign of the end of the 'delivery' period; and a new technological paradigm will grow exponentially in no small part thanks to a 'diffusion' of innovative, resource-efficient technologies and overall production energy intensity reduction."

The Evolution of Technology and Governance in a Coordinated Relationship

However, before the article moves towards that part of the discussion, it has to be remined that India has already utilized technology in the form of data capturing for its 1.4 billion plus citizens in the form of Aadhar Card. The system is like a social security card for USA had been completed all over India. Barring the fact of the geographical, demographic diversity as well as other social factors, this article would like to cite the example of this gigantic exercise as one of the classic examples of how data is captured and utilized for governance related purposes. This has been already implemented in the way of the government's policy related implementation in India. Capturing the data of the common people for their welfare related schemes has been a game changer for millions in India. This is what has proved that technology and governance have now become an important part of our lives. In terms of the overall contribution of Aadhar which has got a motto of "Right for every common citizen" has indeed turned out to be true. From wire transfer, government deposit schemes have become easier with the bank accounts being linked to the Aadhar card numbers. One of the important steps in India was to stop the seepage of leaks when it came to money transfer and the corruption of the touts. This is where it leads to an intersection and as well as tipping point of the correlation between technology and policy governance.

Davis et al. (2012) speaks in their paper that "governance can be affected through a wide variety of mechanisms, including military action, transfers of funds, promulgation of legal instruments, publication of scientific reports, advertising campaigns. Different technologies of governance involve generation and allocation of different kinds of resources, including both material resources such as money or personnel, and intangible resources such as status and information. Different technologies also exert different kinds of influence over the governed. So, for instance, financial auditing as a technology of corporate governance may be influenced especially strongly by a combination of legal regulation and detailed self-regulation, while environmental auditing is shaped by pressures from a more diffuse set of actors articulating less detailed norms. The ways in which such governance operates are often immensely intricate, creating substantial empirical and analytical challenges in efforts to understand the roles of indicators as a technology of such governance."

As **Canedo et al. (2020)** puts forth that with "well-defined governance processes, organizations can gain a strategic advantage over others as they systematically evaluate and improve their processes and services, leading the organization to perform better and, consequently, to be more competitive. There is a link between Information and Communication Technologies (ICT) and improved governance, providing a competitive advantage for organizations and citizens." To make matters clear if someone is confused as to how the role of the technology can be seen in the form of Aadhar card then one must understand the workings of the system. It is based on a biometric system that captures all the data and is hard to duplicate although fake Aadhar ID cards can be found. Nevertheless, with privacy being an issue of genuine concern, government of India has also been able to effectively run campaigns related to polio, tuberculosis, water related schemes as well as other welfare programs. Duplication of efforts have reduced making redundancy less of an issue and although census in India is yet to be done since 2011, the government has got data on record for most of the major campaigns that it wants to undertake. This is where the idea of the technology in the form of Big Data and government policy preferences make a handshake. The formalization of the economy has also happened in the form of the fin tech sector making inroads in India. This can be seen in the form of the utilization of UPI scanners around the country including unorganized sector sellers/vendors having an UPI system. The formalization of Indian banking and finance system has started to take steps through the advent of this technology which has played a significant role in giving financial self-sufficiency and literacy. These are the prime examples of the role of technology in governance function within India holistically with ever expanding presence.

Role of India in the Global South in Varied Sectors of Technology and Governance

The first sector that has been immensely helped in terms of the policy and governance is the food sector. Speaking of India and its immense population especially that is still marginalized, one of the important applications for the government had been to provide food. The amount of data generated, its utilization for the food distribution program had been immensely successful for a country like India. The way of the food distribution has been going on in India since independence. However, the growth of the technological impact is important to be considered whether the policy interventions are actually making a difference or not. During the era of covid related challenges which had been significant, the idea for the role of data capturing cannot be emphasized enough. To capture an actual data on how the management of the data had been done in terms of the information generated and then acting upon it may be significantly difficult. However, the preliminary understanding of the fact that data can be crucial for taking major policy decisions cannot be overlooked. The earlier system of falsified ration cards and as well as information forgery is still out there today. However, the level and the impact of the technological impact have been generated enough when it comes to the food distribution for a large section of the marginalized people. This cannot be overlooked as it gave the direction and objective to the governance aspect of India for a very big population that is backed by technology.

Next comes the sector for health, where India has had a huge challenge thanks to its huge population. The advent of the Covid era created huge trouble around the world. It was also the time when technology came in handy for the crafting of health-related major steps. Related to this the first and the foremost issue at hand was to deliver vaccines. The role of technology not only for the development of vaccines was important but also played a major role in the advent of the data related governance to whomsoever the vaccines should be forwarded. It played a major role in the tracking of the vaccination program as well as the number of vaccines that was in supply and being acquired. Leakage of the vaccines was lower compared to the huge number of vaccines that was being circulated across India. This was a game changer in the role of the government and its distribution of the vaccines which amounted to a billion plus. Indeed, this was the biggest discovery of technology for the government related to tracking of vaccination related information in the real time. The onset of websites such as cowin and others helped people to book appointments, availability of vaccine for the ones who had access to the online related information. This is how one can actually see the role of technology in moments that actually threw challenges. It is therefore important to reflect that another step that is being taken by India is now medical records.

India is running the world's biggest health program in terms of the government insurance. Every one of the beneficiaries has also been provided with a card. Similarly on paper, even the state level governments have put in their own health insurance schemes. Health sector is important when it comes to data management. Most importantly, health related data is private including the list for maintaining ailments, diseases or illness and how much of the government related health insurance coverage. In an immensely populated country with so much of data at stake the role of technology and governance is definitely the one when it comes to data management and policy intervention is important for a sensitive area like health. The role of vaccination, health related insurance as well as keeping the records of the population has its own risks as well. The idea of technology cannot be overlooked as the developing countries are trying to put in their social policies in perspective. This is where the idea of the data being used for policy making decisions comes in. India has been at the forefront of data related policy changes. This is where most of the technology in terms of the customized policies from private companies to general government ideas based on the greater needs of the population. Heath sector has been one of that amazing stories of evolving into an example of how much data can play a part in the sector related to governance of a major related policy area as health.

In other areas such a s transportation and traffic related areas, especially for car registration, traffic tickets as well as toll tax playing technology has already started to play the role. This can be seen already in the form of the fast tags being used for toll tax payment. Technology has come in handy to make things easier for the collection of taxes. Easier movement of vehicles as well as collection of data related to traffic violations makes it easier for the government to track vehicles as well as other policy related changes that can be initiated. This can be already seen in the way when Delhi introduced its even-odd policy which did not work the way it was supposed to. However, there was a genuine area where the policy interventions can be done more effectively through the data collected and how it can play a role for technological impact. It can never be emphasized that technology is not just limited to understanding of computers, mobile phones, washing machines, water purifiers etc. They don't make sense until and unless there is information or data. It is the fuel based on what technology can give an edge and make for the next step in policy related evolution. The best of the policy related implementation and governance is definitely enriched when there is data to back it up. Every citizen is a data storage mechanism which the governments can and have to work on in the 21st Century which is a digital book.

The idea for the smart cities and urban governance is dependent on technology and governance. As **Canedo et al. (2020)** suggests that "While digital technologies can improve urban problem solving, the frequent alignment between the smart city framework and a more general ideology of technology remains problematic, influencing how we think about, govern, and participate in the city. This has produced a "smart mentality", whereby "cities are made responsible for the achievement of smartness —i.e. adherence to the specific model of a technologically advanced, green and economically attractive city, while 'diverse' cities, those following different development paths, are implicitly reframed as smart-deviant". This is where the policy initiative of the government of India especially in terms of the smart cities initiative can be seen as the juxtaposition of technology and governance coming in a part of the urban management. This can be seen in the smart waste management cities to transform cities like Indore and even in other places such as New Town, Kolkata and Bengaluru for example.

The idea for the growth of the technology in India has been something that has actually helped in the disbursement of a lot of policy initiatives. In terms of the paper looking at the role of technology and governance, the idea of urban management especially in terms of the climate related policies is something to look at. The growing importance of technology is of paramount importance here to control and look at the amount of emission. In a country like India where carbon emissions are one of the prime concerns, the ideal scenario would be to utilize control mechanism for industries, individuals as well as other units that are involved in emission. This is where the prime concern for resources optimization as well as energy efficiency can be brought to a better level especially in the urban area. The previous discussions related to areas of health, technology, transportation all play a role in the control of the emissions related efficiency which can play a major role in governance related to the creation of smart urban cities. The future planning of Indian smart cities is taking baby steps to build an eco-system and an urban environment that is built on the idea of an energy efficient city. It involves building an environmentally friendly system that can actually build on technology for an energy efficient system. Most of the modern and advanced urban societies are already utilizing the power of energy efficiency which is an important step for collaboration technology and governance.

Indian cities have an immense challenge coming up with the rise of global temperatures, unplanned urbanization and uncontrolled emission. To make things better for urban governance especially in matters of pollution reduction and emission control it would take immense use of technology. It can be already seen in the way how pollution purifiers have been placed. It has been put forth based on the data that has been collected based on the emissions related data. It is important for the urban governance especially from the local policy making decision levels to introduce technology. The governance aspect will come in based on a proper understanding of the importance of data. Technologies in the form of controlling air conditioners, electric cars, other amenities in the form of escalators, lifts are all part of one urban society. Collectively the rise and the growth of the urban societies in India have been not planned and it has been hasty. This is where the role of the public policy institutions needs to expand beyond the cities into smaller towns and provide directions for the way that the next steps can be taken for the growth and evolution of the governance related matters which is technologically assisted. This is the focus of the Indian government since the advent of smart city planning mechanism. It has been in phase for the last decade and there are some cities which are showing tremendous potential although the optimization and unity for resources utilization is still waiting to be achieved.

The rise of the telecom industry has been one of the key areas that have propelled the information and telecom revolution that had propelled India. It is the spread and the growth of telecom expanse all over India that has allowed for the common citizens and the government to connect at a better level. Removing the elitism and clutter of English, even other languages have now come up with their expanse in the way they have been able to reach out to the masses. Mobile phones had become one of the prime instruments to reach out to the farmers, other marginalized section and of course the urban elite. However, without confusing all the three sections, if one takes a look at the farmers then for a long time the role of technology can be found in the way farmers have been given information from the government in terms of weather, soil patterns etc. It is a classic example of the way how technology and governance can play a role in the modern societies of developing economies even amongst stratified communities. The role of governance is important when it comes to impacting a larger section needs to be impactful and the role of private players also plays a very important role. The data rates becoming cheaper allows for the access to the important policies that are meaningful and can impact the larger section. Agriculture in India had been dependent on rudimentary elements for a long time since the technological advent.

The rise of the policies such as the green revolution is a result of the technological prowess that transformed India. It is one of the pioneering examples of the way how technology and governance came together to form the bridge between technology and governance. It is how the first steps for the food sufficiency in India could be seen with the advent of technology and policy related governance.

That is how, when, where, why the technological advancements can be made to count. India as a study points out that how from the global south, technology played a role with access to data and information. The critical areas which have been mentioned above such as food, health, transportation and especially related to sustainable development, key tracking of data is very important. This is how India has utilized technology with the examples of green revolution, telecom revolution etc. In terms of the governance related issues, social welfare, access to government programs play a very important role in the policy interventions. This can be seen in the way how certain important milestones have been achieved by India over a period of time. However, the biggest challenge that India presently faces is the way how government can actually balance its carbon neutral goals along with sustainable development goals while not compromising on the huge challenges of development aspirations. A country from the global south is faced with this immense challenge which is not going to be easy but nevertheless the aspiration shall always remain.

The Future of Technology and Governance in the Global Sphere

Mulligan & Bamberger, (2018) speaks of "cities are made responsible for the achievement of smartness —i.e. adherence to the specific model of a technologically advanced, green and economically attractive city, while 'diverse' cities, those following different development paths, are implicitly reframed as smart-deviant"

The paper here is not proposing theoretical framework or a model. However, the objective is to track down the ways how the role of technology has been playing a part in the evolutionary process of the governance being helped by technology. The idea of technology and governance is a consistent process that every country has to utilize for the greater benefit of the society and population. It is how the overall scale of governance has been improving and increasing in the form of technology being an interventionist force. An idea of the policy related mechanism is to allow for the technological advancements to seep down in terms of the value for the common citizens. The role of technology in India if traced down since independence can be seen in the form of space tech, agricultural and food related technology, transportation and finally energy related developments. This is what makes for the story of technology and governance. Also, the article/paper did not specifically focus on one particular area. It has tried to come to a narrative that can build on the past and can provide the direction for the future. In India, the biggest threat that is going to come up is in terms of the future challenges is also dependent on the way how the governments can work on the application of technology that can actually help the commoners. It has already been mentioned in terms of the technological and policy related interventions through UPI, banking, health related previous instances.

The privacy of data is definitely one of the major points of concern in the modern times. However, in a modern world, the ideal scenario is to balance the dichotomies of the technological downsides. Governance and policy-oriented decisions cannot be without side effects. The idea is to build on the governance and technology for providing better safety, access to resources that would make for the creation of a wonderful society. This is how the modern-day human society is functioning where the trust of the government especially at urban level is at cross heads in a new age technological aided governance. Another critical area of data related governance is control over the digital space. This is one such are which has worked like a double-edged sword. Keeping track of the internet related space is a very important part of the policy related issue as well as governance decision making. This is where the next phase of the governance in the era of internet and cyberspace is important. It includes promoting safety as well as secured environment for the people especially the young and the vulnerable from cyber threats. In a country like India where cyber frauds and scams are one of the highest in the world it is important for the governance function to take policy decisions very carefully. Control over internet, WhatsApp and other technology-oriented communication channels also play a vital role in this. It is because of this scenario that a new age of technology and governance is arising.

Hutten (2019) speaks of "Good governance leads to good management, good performance, and good investment of public money, good public behavior and good outcomes. The governors of public services organizations face a difficult task. They are the people responsible for governance—the leadership, direction, evaluation and monitor of the organizations they serve. Their responsibility is to ensure that they address the goals and objectives of these organizations and that they work in the public interest. They have to bring about positive outcomes for the users, as well as providing value for the taxpayers who fund these services. They have to balance the public interest with their accountability and compliance. There is clear evidence that many have difficulties in fulfilling these responsibilities."

Case Study Narrative Approach: Four Cities in India and Implication on Urban Planning

There is no empirical data or as mentioned above theoretical framework being suggested. Rather the paper seeks to trace the evolution of the governance how it has formed, taken shape in a country like India and evolved moving into the future dimensions. The paper speaks of how the journey of technology and governance has embraced each other in a long period of time. It seeks to contribute in terms of gauging the journey of technology and its evolution in collaborating with governance. In the scholarly understanding there have been papers that have focused attention on the world of governance and technologies especially in the cyber sphere and the information technology sector have metamorphosed. Similarly, there have been even works on health and education although a combined analysis in different sectors seems to have been missing. Especially focusing on the varied areas with emphasis on India in terms of accessing the historical evolution is where the paper looks to contribute. It has shaped itself in terms of the flow of the paper as mentioned earlier. Elements of technology and governance in all the important sectors as just mentioned above in the form of education, health, cyber sphere and energy in terms of origin, evolution and the future path had taken several turns. It matters much more when in a country like India; the world's most populous nation with its myriad challenges can actually leverage technology for policy-oriented governance that impacts millions. The paper has tried to repeatedly highlight and reflect on the nuances and the challenges which had been solved or been worked on based on the intersection of technology and governance. In terms of the sectors the examples of the sectors that have been mentioned have created new direction and carved out new opportunities. The paper hopes to open up new avenues for the future research which will carry empirical data for taking forward the scope of the paper in terms of contributing to the research gaps. This paper is going to help in setting the idea of technology and governance in the global south where India fits in perfectly. As a nation that has faced immense challenges and has overcomes them despite numerous shortcomings', the works of scholars and the excerpts from this paper brings forth the highlight technology may not be the panacea for every problem. There are many areas that technology is still confused and more importantly the ones who are supposed to put it at work, that is us is also grappling with proper usage of it. The intersection of human intelligence and technology is crucial and important in the evolution of steps that governance will take in conjunction. Data driven and intersection-based component of policy making that helps governments to govern in terms of major policy steps with the advent of artificial intelligence dawns upon as the future step of the journey between technology and governance.

Challenges of a nation like India in terms of balancing the aspirations and the environment: It is no surprise to find that Indian cities constantly top the most polluted cities in the world. Lion's share of the environmental reports in the last 10 years have come out with reports over how polluted the Indian cities are. Although generally there is confusion between pollution and climate change however, the scenario should not be as such. Climate change especially in the form of global warming has generally been attributed to pollution. Now when it comes to the understanding of how to control pollution, emission and balance sustainable development the idea would be to utilize the role of governance and technology. The question always remains how? The paper has consistently been focusing on the theme of the challenges of India. As mentioned before this paper is narrative based rather than empirical. The challenges of the world's most populous nation like India are the triple frontier which will be discussed later. Indian cities have been trying to use technology in terms of its combat battle against pollution and its ancillary effects. Delhi had tried to first bring in the odd-even system as a directive of the governance by Delhi government. However, the role of technology was lagging. The question may be asked as to how technology could play a role. First and foremost, data base of cars being registered to same address be it either ending in even or odd numbers needed to be tracked to actually make the policy meaningful. That is if a family has four members with two cars who number plates are ending in odd and even respectively, the family will be allowed to take out only one car based on the number they choose. Here the role of database management and remote taxation through mobile phone payment comes in handy. India already has the technology in place but its implementation in extended space to deal with core challenges of overcoming pollution related issue is critical. India has also been focusing on the creation of the public transportation infrastructure in the form of one of the lengthiest metros in the world. Introduction of seamless ticketing and keeping the costs still affordable in Delhi metro has been one of the many ways. Next again comes the use of technology in terms of mapping AQI (Air Quality Indicators) and putting in the purifiers of air at the sensitive areas to make for breathable air quality. The pilot projects have started on these but it highlights the challenges of pollution, poverty and population.

Kolkata: Bengal and Eastern Part of India

If we begin to look at the eastern coast of India where there are important cities such as Kolkata, Bhubaneshwar etc. the role of looking at natural conservation, urban planning and technology related governance plays a very important role. Speaking of the eastern coast of India, global warming and climate change has now made it mandatory for the eastern coast cities to face at least one cyclone every year. It not only brings in tremendous hassle for the administration but several reconstruction challenges. It has been seen that the poorer developing countries such as Haiti, Nepal already face the brunt of the natural calamities and climate change. Speaking of this phenomenon, the government of West Bengal has focused on the conservation of **sundari trees (Mangrove Forests)** (Bhattacharya and Sen, 2022) and plant them even more as it has a tremendous role in mitigating the brute force of the cyclonic depressions that form every year. The trees being planted more and more allows for the space for the eastern urban settlements to face the cyclones with a less brute force especially from Kolkata. This is an example of the indigenous knowledge now being put into practice. Also, along with the indigenous knowledge being put into practice there also comes the impact and influence of knowledge which is already found with the use of geo-spatial mapping and the location of the cyclonic depression technology is also playing a key role in the role of the governance mechanism. Speaking of the geo-spatial mapping of cyclones and other facets, India has already taken the initiative of launching a satellite for monitoring climate, air pollution on behalf of entire South Asia. It is not only a novel initiative but also signifies a major role of India to create a nation brand in the form of nation taking steps to help out its neighbourhood to better monitor the climate and the pollution mapping. Now going back to the point from where it started the idea for protecting the eastern zone cities will depend on acting based upon the indigenous knowledge and then creating the application of technology for the governance with the upcoming unwarranted challenges. Now moving into the southern zone of India, the other narrative case of Bengaluru as a city has always been the center of discussion for a different set of challenge. However, that will come in later. As of now speaking of the challenges, Kolkata still has its own issue which needs to be brough forth. Kolkata had been one of the congested cities with extreme levels of pollution for which already smart initiatives marrying it with technology has already started to come in. This comes in the form of smart buses with air purifiers (HT Correspondent, 2023), recycling shops and smart garbage (Chakraborti, 2020) collection which is already happening at the other cities of India as well. More on that later but first and foremost the importance for a country like India which has a mountain of challenges. Speaking of mountain, this paper will introspect into Delhi and its huge problems related to it emission problem and rising mountain garbage. Continuing with the city of Kolkata, the city faces immense challenges in terms of the particulate matter in the air. Kolkata as a city most importantly had been an unplanned urban settlement created during the British times. The city like from any other developing or neo colonial nation always faces the challenge with certain exceptions of settler colonies such as USA, Australia, New Zealand and Canada. Now going back to Kolkata, to overcome the challenges of growing urban settlement, satellite cities to reduce the impact of emission and improve the quality of life are the first policy prescriptions. This is where, satellite enlistments such as Saltlake city and New Town, around the city of Kolkata have come up. These are planned settlements and provide for a technology enabled urban life with planning and scope for investment, expansion and betterment. The best of the examples regarding this can be in the form of New Town, Kolkata that have lower levels of pollution. Also backed with the urban planning, waste management and the connection with the initial argument of the paper which brings to the fore energy conservation, transportation and data management on urban quality of life this new development can be cited as new examples.

Bengaluru: The Silicon Valley of India

This will bring us to move away from the eastern part of the country to the southern part of the country which is Bangalore or Bengaluru. In that city there has been different set of challenges although with a common theme. Rising urban settlement, innumerable construction done hastily and desperately has created another set of challenges for this city. The rise of the city as a tech hub has changed the garden city of India created once a military barrack settlement into a modern urban settlement. However, the level of the urban issues related to quality of life and the dealing with the boom in the urban population has still been left wanting (Mohan and Tiwari, 2023). Bangalore as a case narrative brings to the fore the issue of rapid urbanization, urban emission and the issue of dealing with quality of life, sustainable development and emission. All of this can be looked into and resolved through the advocacy and focus on public transportation. It is one area where the aspiration status of Indian citizens to own private cars aping the earlier materialism of the western society as well as lack of government initiatives to invest in public transportation especially mass urban transit even 5 years back did hurt Bangalore. It has been only seen in the last 5 years that seems to be a rush in improving the public infrastructure of Bangalore. That itself has got issues with the urban settlement coming in and taking in urban spaces. Bangalore following the global steps and even from the ones in Indian cities finally is trying to come with a coordinated response that involves the same kind of policy prescription much like the one discussed about Kolkata. Everything said and done, Bangalore in the way it has dealt with urban problems seems to not have yet fully understood the way to deal with its problems. The city already faces increasing pressure with the increased complex challenge of urban living. Unplanned urbanization has been a strict challenge which can only be solved with data governance, corruption free administration and technology management of the components of urban planning. However, it is easier said than done and it is not just about writing it on a piece of a paper. The idea should be to redistribute population, cut down on private transportation and then get the investments for the public infrastructure projects. On paper the investments are happening for the public transport that allows for the ease and movement of traffic. Bangalore has been one of those cities that have faced the brunt of booming urbanization without planning much ahead of it. The infrastructure and the bridging of the technology for urban governance is surely something that India needs to take care of for the future generation. Bengaluru has been a classic case where the urban mismanagement of the urban resources had led to the problems. It is ironical that despite being dubbed as the "Silicon Valley of India", the urban planning needs to be better and precise. Bangalore although does not fall into the category of cities with extreme pollution although it has its own challenges related to traffic congestion and urban amenities. The role of technology and governance epitomizes how the idea of a good quality of life depends on developing urban amenities which can also bring down the impact of global warming, climate change accelerated by urban traffic, emissions and population pressure. In an urban setting to create a sustainable development mechanism the point already mentioned especially for cities in India would be focused on the upcoming challenges and pressures. The renovation of water bodies, incentives on using public transportation, discouraging private transportation as well as improving the urban infrastructure (Prasher, 2023). It is the way forward for the development of urban infrastructure as well as the growth of the cities in a manner that can lead India to move towards its goal of carbon net neutrality by 2070. The challenges are extreme in the context of rapid urbanization and the extreme challenges that are coming in already. Bengaluru can take in a bit of leaf from pushing in smart solutions that are simple but achievable. Traffic systems, energy monitoring and urban planning have already been spoken about in terms of the coordinated efforts of technology with governance. Next comes the question of simple steps which can be made into policy prescriptions.

Mumbai: The Maximum City of India From the West Coast

All said and done this narrative can be taken forward into the western coast of India with focus on Mumbai, one of the most congested and widely affected city of India and the world by burgeoning population and creaking public infrastructure including residential problems. Mumbai represents the challenges of urban development in the present times without slowing or slacking away. The city in the western coast of India has tremendous problems to solve especially as the city is vulnerable to urban population pressure, rising sea levels as well as an infrastructure nightmare with residential problems. Mumbai has already started to work on it in a three-pronged manner. The first comes in the form of reclaiming land from the sea to build transportation projects and also related to infrastructure. This has an important implication to reduce pressure on transportation which has always been an issue because of unplanned roads, lack of open space and increasing private cars in the roads. This is where the second facet comes in which is to push for aggressive expansion in the public infrastructure especially transportation. Mumbai has always faced the brunt of excessive pressure on local train network with people coming in from suburbs where the population pressure is extreme. This is where the expansion of rapid mass transit plays a major role and has been focused upon. Next is the issue of urban quality of life in Mumbai which is severe and extreme in many ways. Mumbai has the unenviable recognition of hosting the biggest slum of Asia. This is where the development of satellite townships and cities play a major role in reducing urban settlement and reduces squatting. Navi Mumbai is the example of that project with better space management, improved public amenities and lower urban emission. Connecting the two urban settlements is the first step for Mumbai. Similar project is being done in Kolkata, Delhi and even in Bengaluru and other cities which has not been mentioned here. India wants to project itself as a super power and a proud emerging nation, however for that to happen barring the political and socio-economic challenges the first thing that needs to be focused upon is giving citizens a good quality of life. Standard of living if measured in terms of the material possession can be found in slums of Dharavi (Asia's biggest slum). However, if one visits their squat like rooms standing on the edge of the city near the Airport, they are full of settlements that are comparable to other countries of the developing world battling urban poverty and the new set of urban challenges that comes along with it. Mumbai and Indian government need to and have to certain extent been working on creating a city that can be better or provide access to a better prospect of life for the next generation. The city of Mumbai which is also known as the maximum city like any other city of India faces the issue of citizens who may not feel as responsible or may lack the knowledge or intent (TNN, 2023). The cities of a developing nation face the issue of active participation of citizens or lacking the moral fiber as well as the education to take part. Mumbai may well be a part of that category which can be also said to be other cities in India but possibly this issue in Mumbai is very much pronounced. Discipline, active citizenship participation and the role of technology and governance had come in India during the time of covid19. This is how the story of every Indian city can move ahead although there are also deep-rooted corruption and other issues as well. In an overly populated country such as India where the jobs or the source of jobs is located around metropolitan cities the impetus on urban resettlement has already been emphasized enough. The allocation of resources is critical for the challenges of the 21st century. India is a big country with numerous challenges and the paper here brings in the narrative of the four major cities across the nation which has common themes of urban population pressure, lack of public transportation facilities and urban infrastructure. Mumbai like the other two mentioned above has been focusing on this but amidst all of this the paper does not want to lose track of the role of technology and the governance. Mumbai as discussed before has faced so many challenges and yet has come back stronger. The new projects such as the coastal road project about to open later this year is a modern-day example of innovative infrastructure development that will cut down on mobility time. The concept of ease of mobility in terms of traffic data records and incentives as well as punishment can also play a critical role to ease out pressure on city life. As for urban redevelopment the entire Dharavi slum is now to undergo a reconstruction in terms of the space allocation and being redesigned based on providing basic amenities (Lal, 2023). These are small and critical steps in the making of the new experience of living in heavily congested cities of the global south. This brings us to last but not the least, Delhi in terms of the urban challenges.

Delhi and its Capital City Conundrums

Delhi is the older city which represents the age-old history but is also a reminder of the stagnant, over populated, troublesome urban life. New Delhi was the new version of the city that came up as an urban settlement to deal with the crises and the problems of the city. Delhi consistently ranks as the most polluted city in the world related to which the steps have been taken but has yet to be proved insufficient. Increase in the number of electric buses, Gas operated buses to cut down on fossil fuel emission have still proved to be difficult to have cut down. This has certain commonalities with the urban density, private transportation and the factors that have been mentioned earlier for the other cities. However, Delhi, amongst all of the Indian cities has the lengthiest amount of urban transit covered. The real problem however lies in the industrial emissions as well as the agriculture related emissions that come in from the neighbouring states surrounding Delhi. Speaking of thus matter let us try to reconnect back to the theme of role of technology and governance especially related to the problems of global warming, climate change and urban management. Delhi can be the prime example of this initiative and how can the government of India play a role in this. Delhi government has been working with the neighbouring states to cut down on the emission and make for proper policy changes (Express News Service, 2023). Technology can play a key role in the ways mentioned earlier through geo spatial imaging, heatmapping and bringing in policy to install dehumidifiers, air purifiers. Although these are huge money-based programs but it is doable. In fact, the World Bank has already provided certain loans to implement certain infrastructure based on these lines. Other critical role for technology implementation would be based on the carbon credits and green incentive program in the industrial areas around Delhi which has already polluted Delhi River, Yamuna and is a contributor to one of the most toxic air of the world. Lack of initiative on a bigger scale to deal with such problems has already created a lot of problems. The one with the issue of urban waste management is another headache for Delhi with rising mountains of garbage seen on the outskirt of Delhi. This can be the prime area where the role of technology and evidence-based governance can come in handy. It is because in an urban setting such as Delhi which has multiple challenges, the issues started with lack of planning and investment. In terms of waste management in the same country a city like Indore although less populated and congested the startups are working on waste segregation. Delhi is one of those cities which is literally on borrowed time one must say. The government of Delhi in its bureaucratic tussle with the central government has missed out on a lot of policy implementations that would have been vital. Role of technology and governance is key to the solving of problems but mere suggestion on paper would not work as the focus on citizen centric governance is still lacking in implementation. Delhi being the capital of India has faced socio-economic and political issues apart from global warming and climate change that has been a real issue. A multipronged approach is needed to settle the issues of urban challenges of India. Delhi government has tried to increase investment on public transportation and infrastructure development which has still fallen short. Not to mention the issues of the agricultural waste emission and industrial belt emission this has been mentioned. This kind of pressure on particular metropolitan cities is even there in the global context but the population pressure of China, India and Asia in general makes it even more difficult task. In the modern times per capita emission of India is still lower than USA and other western countries from Europe. Delhi has consistently been failing in its commitment towards creating a city that can live up to permissible standards for living especially being the capital of India and government headquarters for administration. It has proved to be a poster image of a failed city and although planned cities around the national capital region of India centered on Delhi such as Noida and Chandigarh have come up but the issue of making a sustainable city of Delhi had been lacking (Kumar, 2022).

A Hope for a Sustainable Global Future With India Leading the Way From Asia to the World

Therefore it can be said that several challenges will be coming up in the near future. India as a study highlights the challenges of urban planning. It would be imperative for the entire world to come together and especially Asian countries need to take a stronger initiative for the same. Although even other continents in Africa especially the Sahel region has faced severe drought and Amazon rainforest in Brazil have faced destruction due to forest fires still. The continent that holds 2/3rd of humanity needs to play a key role and amidst all of this the most populated country India and its neighbour the second most populated country China has such an important role. Smart urban planning, river ecosystem management such as the **Nawami Gange project (Rejuvenation of Ganga)** (News Desk, 2023) listed by U.N. as one of the top 10 projects related to sustainable development in the world can be a beacon of change. The challenges are going to be intense and the government alone cannot play a role, the citizen's needs to be more proactive as the burden of the world going ahead will come up to the three most populated nations in the form of China, India and USA. Therefore, India under its current ambitious program related to environmental safeguard and sustainable development has come with an initiative such **as L.I.F.E. (Lifestyle for Environment)** (Langa, 2022) and **International Solar Alliance (I.S.A.)**. (Gupta, 2023). Both of these initiatives along with government push to reduce fossil fuel usage for automobiles and replacing it with bio-fuel such as Ethanol made from sugarcane waste. There is also a push for electric vehicle adoption under **F.A.M.E. (Faster Adoption and Manufacturing of Electric and Hybrid Vehicles**) (Mohan, 2023) scheme. Although electric vehicles are yet to be proven conclusively to cut down on pollution levels yet it is a step by step towards the ambitious goal to cut down the emission to net zero by 2070. For a nation like India it is a big deal and even if one looks at the present emission figures it is the lowest amongst bigger nations on per capita and one of the only nations to be in line with the target set for **Sustainable Development Goals for 2030**. There is still a long way to but India even in its ancient time had the knowledge and notion for using resources that are sustainable and biodegradable such cooking gas from cow dung named more aesthetically bio waste gas. India holds the mantle for the present world and contemporary human civilization to be moving ahead. Clean tech and agro-start-ups coming in from the smaller cities of India are trying to lend their hand for the solving of the future challenges. This is how the bigger picture of India moving ahead in the 21st century is trying to balance between the aspirational needs of a rapidly developing nation vis-à-vis the challenges for maintaining sustainable development. It is this catch22 situation that India with its diverse landmass and ethnic varieties apart from geographical realities. Gandhi once said "***We have enough for every man's needs but not for one man's greed***". This Gandhian philosophy needs to be followed not only by India but by all the nations however most importantly India for the reasons mentioned above. Keeping it short and to finally conclude this is a reflective narrative structured paper and the future directions for the paper can retrospect on these points mentioned and move ahead with empirical data to prove and disprove to an extent on the narrative of the paper through empirical findings.

REFERENCES

Almgren, R., & Skobelev, D. (2020). Evolution of technology and technology governance. *Journal of Open Innovation*, *6*(2), 22. doi:10.3390/joitmc6020022

Barile, S., Orecchini, F., Saviano, M., & Farioli, F. (2018). People, technology, and governance for sustainability: The contribution of systems and cyber-systemic thinking. *Sustainability Science*, *13*(5), 1197–1208. doi:10.1007/s11625-018-0621-y

Bhattacharya, S., & Sen, S. (2022) *In West Bengal, ambitious efforts to plant mangroves yield limited results*. Scroll.in. https://scroll.in/article/1032297/in-west-bengal-ambitious-efforts-to-plant-mangroves-yield-limited-results

Butcher, J., & Beridze, I. (2019). What is the state of artificial intelligence governance globally? *RUSI Journal*, *164*(5-6), 88–96. doi:10.1080/03071847.2019.1694260

Chakraborti, S. (2020, October 16). New Town gets one-stop waste-to-wealth store: Kolkata news - times of India. *The Times of India*. https://timesofindia.indiatimes.com/city/kolkata/new-town-gets-one-stop-waste-to-wealth-store/articleshow/78689888.cms (Accessed: 10 June 2023).

Correspondent, H. T. (2023, June 6). *West Bengal govt launches buses with air purifiers in Kolkata to beat pollution*. Hindustan Times. https://www.hindustantimes.com/cities/kolkata-news/west-bengal-govt-launches-buses-with-air-purifiers-in-kolkata-to-beat-pollution-101686042102914.html

Davis, K. E., Kingsbury, B., & Merry, S. E. (2012). Indicators as a technology of global governance. *Law & Society Review*, *46*(1), 71–104. doi:10.1111/j.1540-5893.2012.00473.x

Dias Canedo, E., Morais do Vale, A. P., Patrão, R. L., Camargo de Souza, L., Machado Gravina, R., & Eloy dos Reis, V., & T. de Sousa Jr, R. (2020). Information and Communication Technology (ICT) Governance Processes: A Case Study. *Information (Basel)*, *11*(10), 462. doi:10.3390/info11100462

Express News Service. (2023, May 16). *Experts brainstorm on strategies to improve air quality in Delhi*. The Indian Express. https://www.newindianexpress.com/cities/delhi/2023/May/15/experts-brainstorm-on-strategies-to-improve-air-quality-in-delhi-2575552.html

Finger, M., & Pécoud, G. (2003). From e-Government to e-Governance? Towards a model of e-Governance. *Electronic. Journal of E-Government*, *1*(1), 52–62.

Gupta, U. (2023, June 15). India, France discuss ISA priorities for accelerating global energy transition. *PV Magazine*. https://www.pv-magazine-india.com/2023/06/15/india-france-discuss-isa-priorities-for-accelerating-global-energy-transition/

Hütten, M. (2019). The soft spot of hard code: Blockchain technology, network governance and pitfalls of technological utopianism. *Global Networks*, *19*(3), 329–348. doi:10.1111/glob.12217

Juiz, C., Guerrero, C., & Lera, I. (2014). *Implementing good governance principles for the public sector in information technology governance frameworks*. Open Journal of Accounting. doi:10.4236/ojacct.2014.31003

Khalil, S., & Belitski, M. (2020). Dynamic capabilities for firm performance under the information technology governance framework. *European Business Review, 32*(2), 129–157. doi:10.1108/EBR-05-2018-0102

Kumar, M. (2022). State Pollution Control Boards in India neither have enough staff nor expertise. *Scroll.in*. https://scroll.in/article/1036752/state-pollution-control-boards-in-india-neither-have-enough-staff-nor-expertise

Lal, N. (2023, January 17). *Mumbai slum residents stand up against Adani's redevelopment plan*. Nikkei Asia. https://asia.nikkei.com/Spotlight/Asia-Insight/Mumbai-slum-residents-stand-up-against-Adani-s-redevelopment-plan

Langa, M. (2022, October 20). PM Modi launches Mission LiFE in the presence of U.N. Secretary General. *The Hindu*. https://www.thehindu.com/news/national/pm-modi-launch-mission-life-presence-u-n-secretary-general-antonio-guterres/article66035847.ece

León, L. F. A., & Rosen, J. (2020). Technology as ideology in urban governance. *Annals of the American Association of Geographers, 110*(2), 497–506. doi:10.1080/24694452.2019.1660139

Mittal, P., & Kaur, A. (2013). E-governance: A challenge for India. *International Journal of Advanced Research in Computer Engineering and Technology, 2*(3).

Mohan, A. K., & Tiwari, A. (2023) Making sense of bengaluru's Messy Urban Development Data. *Citizen Matters, Bengaluru*. https://bengaluru.citizenmatters.in/making-sense-of-bengalurus-messy-urban-development-data-117710

Mohan, V. (2023, February 2). Govt budgets for 'green growth' but experts call it inadequate to tackle air pollution. *The Times of India*. https://timesofindia.indiatimes.com/business/budget/govt-budgets-for-green-growth-but-experts-call-it-inadequate-to-tackle-air-pollution/articleshow/97559795.cms

Mort, M., Finch, T., & May, C. (2009). Making and unmaking telepatients: Identity and governance in new health technologies. *Science, Technology & Human Values, 34*(1), 9–33. doi:10.1177/0162243907311274

Mulligan, D. K., & Bamberger, K. A. (2018). Saving governance-by-design. *California Law Review, 106*(3), 697–784.

Musso, J., Weare, C., & Hale, M. (2000). Designing web technologies for local governance reform: Good management or good democracy? *Political Communication, 17*(1), 1–19. doi:10.1080/105846000198486

News Desk. (2023, May 29). *Global Recognition to Namami Gange Programme*. The Avenue Mail. https://avenuemail.in/global-recognition-to-namami-gange-programme/

Prasher, G. (2023). Bengaluru, we have a problem: It's our Lakes. *Bangalore Mirror*. https://bangaloremirror.indiatimes.com/bangalore/civic/bengaluru-we-have-a-problem-its-our-lakes/articleshow/97289067.cms

Roco, M. C. (2008). Possibilities for global governance of converging technologies. *Journal of Nanoparticle Research, 10*(1), 11–29. doi:10.1007/s11051-007-9269-8

Sachdeva, S. (2002). e-Governance strategy in India. *White Paper on e-Governance strategy in India.*

TNN. (2023, June 2). How planning and development of Mumbai can involve citizens. *The Times of India*. https://timesofindia.indiatimes.com/city/mumbai/how-planning-and-development-of-mumbai-can-involve-citizens/articleshow/100691710.cms

Vidisha, S. (2023). Mumbai slum residents stand up against Adani's redevelopment plan. *Nikkei Asia*. https://asia.nikkei.com/Spotlight/Asia-Insight/Mumbai-slum-residents-stand-up-against-Adani-s-redevelopment-plan

West Bengal Govt launches buses with Air Purifiers in Kolkata to beat pollution. (2023) *Hindustan Times*. https://www.hindustantimes.com/cities/kolkata-news/west-bengal-govt-launches-buses-with-air-purifiers-in-kolkata-to-beat-pollution-101686042102914.html (Accessed: 11 June 2023).

Yadav, N., & Singh, V. B. (2013). E-governance: past, present and future in India. *arXiv preprint arXiv:1308.3323. Experts brainstorm on strategies to improve air quality in Delhi.* https://www.newindianexpress.com/cities/delhi/2023/may/16/experts-brainstorm-on-strategies-to-improve-air-quality-in-delhi-2575552.html

Chapter 14
Smart Environmental Monitoring and COVID-19:
AI-Based Event Detection Techniques in Morocco

Hassan Hazimze
Ibn Zohr University, Morocco

Salma Gaou
Ibn Zohr University, Morocco

Khalid Akhlil
https://orcid.org/0000-0003-3800-1693
Ibn Zohr University, Morocco

ABSTRACT

In this chapter, we explore the integration of artificial intelligence (AI) in enhancing environmental monitoring and managing the COVID-19 pandemic in Morocco. Utilizing AI-based event detection techniques, the authors extend their investigation from analyzing COVID-19 trends to applying predictive algorithms for environmental surveillance. This approach not only demonstrates AI's capacity in public health realms but also underscores its potential in environmental stewardship. Through detailed analysis, the authors illustrate how machine learning and data analytics significantly improve response strategies against pandemic threats and environmental degradation. Their findings reveal AI's critical role in developing resilient and sustainable monitoring systems, offering valuable insights for policymakers, researchers, and practitioners. This comprehensive study highlights the synergy between technological innovation and multifaceted crisis management, positioning AI as a cornerstone in advancing smart technologies for the betterment of public health and environmental sustainability in Morocco.

DOI: 10.4018/979-8-3693-3807-0.ch014

Copyright © 2024, IGI Global. Copying or distributing in print or electronic forms without written permission of IGI Global is prohibited.

INTRODUCTION

In the evolving panorama of technological advancement, artificial intelligence (AI) emerges as a beacon of innovation, reshaping the nexus between technology and societal welfare. Its application spans diverse realms, from enhancing security through fraud detection and bolstering user experience via recommendation systems, to revolutionizing public safety with facial recognition and streamlining decision-making processes. Particularly, AI's prowess in mirroring—and occasionally surpassing—human cognitive functions in specialized tasks marks the dawn of a new era in technological evolution (Ribeiro, 2022).

Central to AI's transformative potential is Machine Learning (ML) (Arık, 2021), a dynamic subfield dedicated to crafting algorithms that equip computers with the ability to learn from and adapt to data autonomously, without explicit programming. This paradigm shift towards self-learning systems unveils a spectrum of applications previously unattainable, setting the stage for machines that not only execute predefined tasks but also innovate based on evolving data landscapes.

The advent of the COVID-19 pandemic has underscored the critical need for innovative approaches to monitor, understand, and mitigate the impact of health crises. Morocco, akin to nations worldwide, confronts unique challenges in navigating the pandemic, spotlighting the imperative for inventive solutions tailored to its specific circumstances. This chapter introduces the deployment of AI and ML in crafting predictive models aimed at forecasting COVID-19 case trends in Morocco. By leveraging advanced data visualization techniques and leveraging the robust capabilities of Python for machine learning, this research endeavors to contribute significantly to the ongoing battle against the pandemic (Chew, 2021).

Structured to offer a comprehensive exploration, this chapter unfolds as follows: Section 2 synthesizes relevant literature to frame the theoretical underpinnings of our investigation. Section 3 delves into trend analysis and the formulation of a predictive paradigm. Section 4 elaborates on the sophisticated ML techniques harnessed to project COVID-19 trajectories. Section 5 encapsulates the research findings, accentuating the application of these methodologies within the Moroccan milieu. Culminating in Section 6, the discourse reflects on the broader implications and potential avenues for future inquiry, informed by the insights garnered through this study.

LITERATURE REVIEW

Overview of Scholarly Work on COVID-19

The advent of COVID-19 has prompted an unprecedented surge in research, with the scholarly community producing a robust corpus of literature to understand and combat this global health emergency. A meticulous review of the Scopus database reveals a diversification of document types, with journal articles representing the largest share, as shown in Figure 1. This pattern underscores a rapid scholarly response, albeit with a reduced presence of conference papers, likely due to the pandemic-induced cancellations of many scientific gatherings (Scopus & Eltoukhy, 2020).

Figure 1. Distribution of published document types on COVID-19
(Scopus Database)

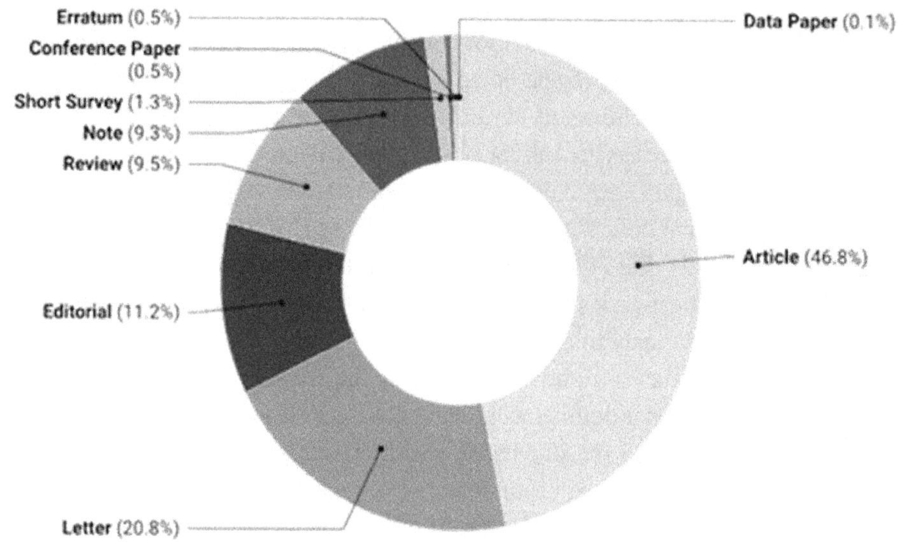

Figure 2. Domain-based classifications of COVID-19 publications
(Scopus database)

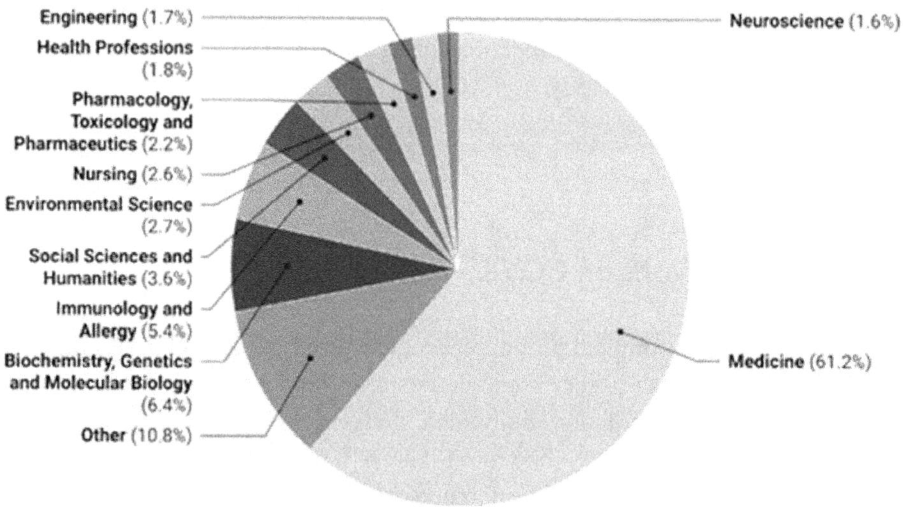

The breadth of research on COVID-19, while anchored primarily in medical science, spans across disciplines, reflecting the virus's multifaceted impact. As depicted in Figure 2, a significant portion of research is dedicated to medicine, but substantial inquiries also emerge from biochemistry, social sciences, and engineering, signaling the interdisciplinary nature of pandemic-related studies.

Research on COVID-19

This section is dedicated to a comprehensive review of the multidisciplinary research efforts launched in response to the COVID-19 pandemic, harnessing the fields of engineering and medicine. Initially, medical exploration of COVID-19 focused on identifying symptoms (Wang, 2019), characterizing the disease (Shen, 2020), and determining its incubation periods (Lauer, 2020). These foundational studies laid the groundwork for understanding the spread and impact of the virus across various populations.

Wang et al. (Wang, 2019) revealed that the elderly, due to pre-existing comorbidities, are particularly susceptible to fatal outcomes from COVID-19 (Garnier-Crussard, 2020). However, it is crucial to acknowledge that the virus does not exclusively target the elderly; younger populations are also at risk (Wei, 2020), thus highlighting the universality of the COVID-19 infection risk. Zhuang et al. (Zhuang,2020) further highlighted the phenomenon of asymptomatic carriers, adding complexity to pandemic control efforts.

The prolonged incubation period of the virus, ranging from 4 to 14 days, and the frequency of asymptomatic cases (Backer,2020) exacerbate the challenge of predicting and estimating COVID-19's impact on the population. This reality spurred interest in developing and applying innovative predictive methodologies beyond traditional statistical techniques. For instance, Remuzzi & Remuzzi (Remuzzi, 2020) and Tuite et al. (Tuite,2020) employed statistical techniques to predict the trajectory of COVID-19 cases in Italy, while similar predictions were made for Iran (Tuite & Zhuang,2020), Spain, and France (Ceylan, 2020).

Beyond these techniques, mathematical modeling and simulation have been used to forecast new COVID-19 cases in China (Batista, 2020) and Saudi Arabia (Alboaneen,2020), employing logistic growth models and the Susceptible-Infected-Recovered (SIR) model (Ivorra, 2020). Papastefanopoulos et al. (Papastefanopoulos, 2020) proposed a comparative study of time series methods, incorporating a variety of models such as ARIMA, Holt-Winters, and N-BEATS, to predict the pandemic's progression.

Research has also ventured into social media and online search trends as valuable conduits for predicting and tracking the pandemic's spread. Qin et al. (Qin, 2020) and Li et al. (Li, 2020) demonstrated the utility of these data in anticipating the emergence of new suspected COVID-19 cases. Shen et al. (Shen, 2020) and Ayyoubzadeh et al. (Ayyoubzadeh, 2020) have also reported on additional social media applications for predicting COVID-19 cases.

Nonetheless, the main limitation of statistical and modeling approaches lies in their inability to process massive volumes of data, pointing towards the necessity of leveraging advanced data analysis strategies to overcome this challenge, a topic further explored in the following section.

ANALYSIS OF TRENDS AND FORECASTING MODEL

Related Work

There have been many applications of machine learning on many different diseases and epidemics. The article by Pouriyeh et al. (Pouriyeh, 2017) discusses the application of Machine Learning techniques such as: Decision Tree, Naive Bayes, Multiplayer Perceptron, K-Nearest Neighbor, Single Conjunctive

Rule Learned, Radial Basis Function and Support Vector Machine (SVM) on heart disease from which they discovered that SVM allowed them to achieve the best results. This in turn motivated the application of machine learning techniques to the COVID-19 dataset.

Another article by Ross and Hamer (Ross, 2013) introduced the SIR mathematical model which can be used to analyze directly transmitted infectious diseases. The SIR model can be represented by a system of quadratic ordinary differential equations as follows:

$$\frac{dS}{dt} = -\beta SI \tag{1}$$

$$\frac{dI}{dt} = \beta SI - \gamma I$$

$$\frac{dR}{dt} = \gamma I$$

where $\beta > 0$ denotes the rate of disease transmission and $\gamma > 0$ denotes recovery which can also be represented in terms of duration of infection $D = 1 / \gamma$. In the above equations, S refers to susceptible cases, I refers to infectious cases, and R refers to the number of cases recovered or died.

The solution of this system of ordinary differential equations makes it possible to determine the number of susceptible, infectious and recovered cases under specific theoretical conditions.

The article by Hu et al. (Hu, 2019) have shown the use of the LSTM to analyze the trend of time series data. This chapter provided techniques such as how to represent the trend of data and the application of moving windows for time series data.

Description of The Dataset Used

The dataset for this study is a compilation of daily updated figures from the Johns Hopkins University Center for Systems Science and Engineering (CSSE) (Johns Hopkins CSSE, 2020), which meticulously tracks COVID-19 cases globally. Focused specifically on Morocco, this research amalgamates province-level data into a comprehensive national overview, capturing confirmed, recovered, and fatality figures up until March 10, 2023, when Johns Hopkins CSSE ceased its global COVID-19 data tracking. This cessation marks a significant moment in the ongoing documentation of the pandemic, shifting the responsibility of data collection and analysis to local and national health authorities as well as other global health entities.

Within this chapter, time series analysis, visualized through a series of graphs, delineates the trajectory of the pandemic in Morocco. Trends in infections, recoveries, and deaths are meticulously charted, providing a clear picture of the pandemic's evolution within the country. The dataset undergoes a crucial pre-processing step, converting figures to a logarithmic scale. This transformation is essential for the application of machine learning techniques, such as linear regression, which requires normalized data to accurately model the exponential growth patterns characteristic of infectious disease spread. This methodological preparation lays a robust foundation for our AI-driven analysis, aiming to uncover insightful correlations and forecasts pertinent to the Moroccan context.

From the analysis of Figures 3, 4, and 5, a pattern emerges: both the number of confirmed cases and deaths in Morocco show an increasing trend over the days. Simultaneously, there is a noticeable rise in the number of recovered cases, indicating the dynamic nature of the pandemic's impact. The logarithmic transformation of the data, repeated for emphasis, underscores its critical role in enabling the linear regression model to accurately learn and predict future trends based on historical data.

This dataset, despite the cessation of updates from its primary source, remains a vital tool in understanding and predicting the COVID-19 pandemic's impact in Morocco. It serves as a testament to the global effort in tracking and combating this unprecedented health crisis, providing a foundation for future research and policy-making in the face of evolving challenges.

Figure 3. Number of confirmed cases in Morocco over time

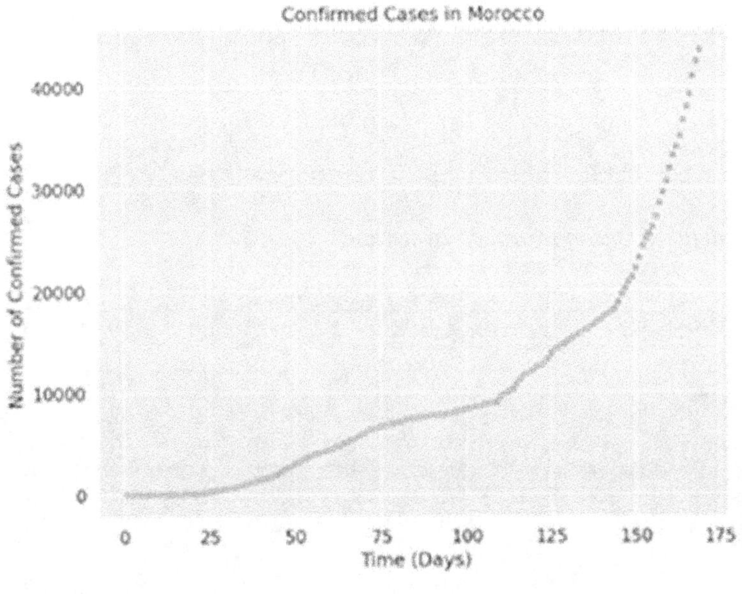

Figure 4. Number of cases recovered in Morocco over time

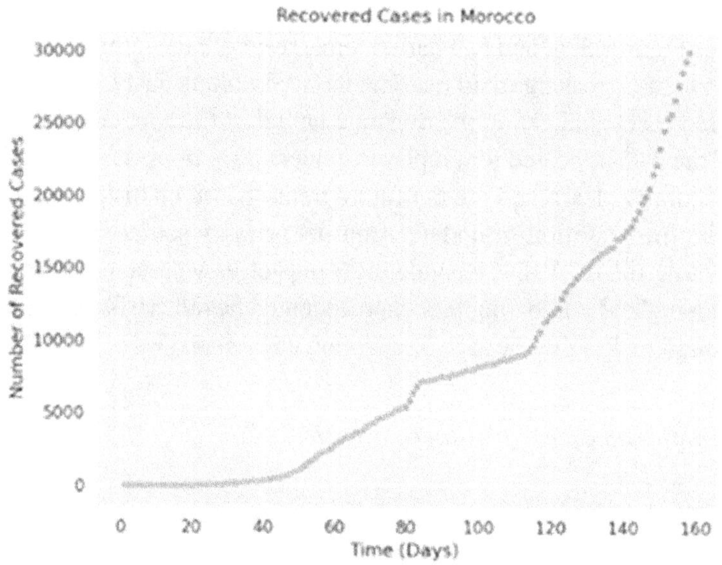

Figure 5. Number of death cases in Morocco over time

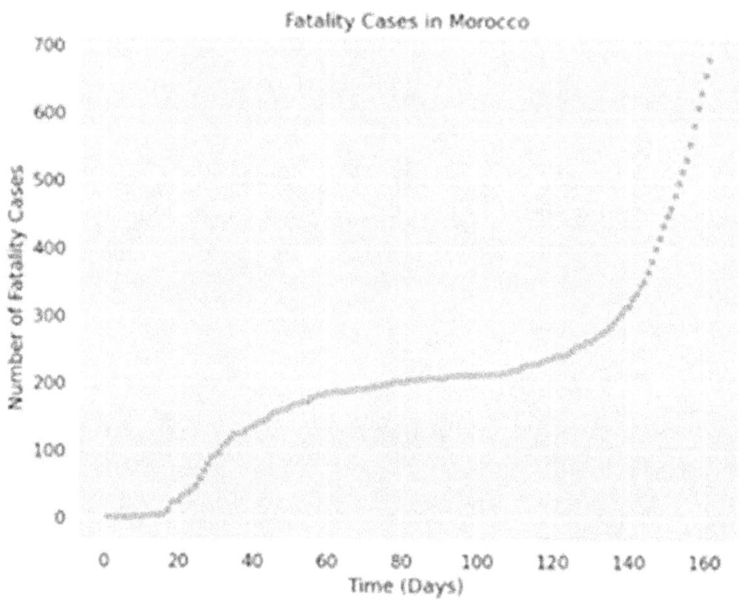

After careful feature extraction and engineering, the following entities have been added to the dataset:

- year: indicates the year the report was submitted
- month: indicates the month in which the report was submitted

- day: indicates the day of the month when the report was submitted
- day_count: indicates the count of the current day since the recordings have started
- is_weekday: indicates if the day was a day of the week
- log_cases: refers to the logarithmic conversion of the number of cases
- log_lag_i: refers to the logarithmic conversion of the offset of i days
- log_trend_i: refers to the logarithmic conversion of the observed trend compared to i days back

The offset was calculated using the following equation:

log_lag_i = log(x(t) – x(t – i)) where x (t) refers to the number of cases reported on the day being analyzed and x (t - i) to the number of cases reported t - i previous days. The trend for the current day analyzed using the following equation:

$$\log\_trend\_i = \log\left(\frac{x(t)-x(t-1)}{x(t-1)}\right)$$

where x (t) refers to the number of cases reported on the current day and x (t - 1) refers to the number of cases on a previous day. For this study, the trend and lag were calculated seven days prior. In addition to these created characteristics, demographic data of Morocco were also added, which were obtained from demographic dataset created by Prabhu (Prabhu, 2020). From the population dataset, the size, density and median age of the population were extracted and combined with the COVID-19 dataset.

This process was applied to the three datasets containing the number of confirmed cases, recovered and deaths. The final cleaned and modified dataset was then divided into 80% for the training set and 20% for the testing set.

METHODOLOGY

Sir Model

The Susceptible-Infected-Recovered (SIR) model is pivotal to our exploration of COVID-19 transmission dynamics in Morocco, offering a mathematical basis for grasping the spread of the disease. This foundational epidemiological framework categorizes the population into three separate groups: the susceptible (S), individuals at risk of infection; the infected (I), individuals currently carrying the disease; and the recovered or deceased (R), individuals who have either successfully recovered from the disease or have died.

In our study, we adapt the SIR model to reflect two hypothetical scenarios that illustrate the range of possible outcomes based on varying transmission and recovery rates. The first scenario posits a disease transmission rate (β) of 50%, aligning with the early stages of the pandemic when interactions among the population were relatively unrestricted. The recovery rate (γ), set inversely to the average incubation period of COVID-19 (approximated at 5 days), offers a baseline for gauging the disease's progression within the infected population.

The second scenario presents a reduced transmission rate of 30%, simulating the effect of public health interventions such as social distancing, mask-wearing, and lockdown measures. Maintaining the recovery rate constant allows us to isolate the impact of reduced transmission on the pandemic's trajectory. By applying these scenarios to the current Moroccan population, we estimate the potential spread and control of the virus under varied public health strategies.

The SIR model's parameters are calibrated using the latest epidemiological data, including the number of confirmed cases, recoveries, and fatalities as reported by health authorities. This approach not only grounds our model in the reality of Morocco's COVID-19 experience but also allows for the incorporation of environmental factors that may influence disease transmission rates. For instance, variations in temperature, humidity, and pollution levels could be integrated into future iterations of the model, exploring their potential impact on the virus's spread.

The analysis is conducted using the SciPy library, a powerful Python toolkit for scientific computing. Through numerical integration of the model's ordinary differential equations (Virtanen, 2020), we derive estimates for the number of susceptible, infected, and recovered individuals over time. This methodological approach provides a robust foundation for our predictions, offering insights into the efficacy of public health interventions and the potential for environmental factors to alter the pandemic's course.

In conclusion, the adapted SIR model not only highlights the dynamic nature of COVID-19's spread in Morocco but also underscores the importance of integrating environmental monitoring into our epidemiological analysis. By examining how different transmission rates can affect the disease's trajectory, we gain valuable insights into the potential for AI and machine learning to inform more targeted, effective public health strategies. Future work will delve deeper into the intersection of environmental science and epidemiology, leveraging advanced AI techniques to provide a comprehensive understanding of how to navigate and mitigate the impacts of this and future pandemics.

Adjusting the Curve

A pivotal aspect of this study involves discerning the nature of COVID-19's spread within the Moroccan context—specifically, whether the progression aligns more closely with an exponential or a logistic growth model. This determination is crucial, as it influences our understanding of the pandemic's future trajectory and informs the development of effective intervention strategies. To this end, we employed both exponential and logistic mathematical models to fit the observed case data, utilizing the following equations:

$$f(x) = a * e^{b * x + c} \tag{2}$$

$$f(x) = \frac{L}{1 + e^{-k*(x - x_0)} + b} \tag{3}$$

These equations represent two fundamentally different perspectives on the pandemic's growth. The exponential model suggests a continuous, unbounded increase in case numbers, reflecting scenarios without effective containment measures. In contrast, the logistic model anticipates a saturation point, a maximum number of cases after which the increase slows and eventually plateaus, mirroring the effect of successful public health interventions.

To ascertain the best fit for the COVID-19 data in Morocco, we utilized the SciPy library's curve-fitting function, a powerful tool that applies the least squares method to optimize the parameters of each model. The fit's quality was evaluated using two key metrics: the mean square error (MSE) and the coefficient of determination (R^2). The MSE provides a measure of the model's accuracy in predicting the observed data, with lower values indicating a closer fit. Meanwhile, R^2 offers insight into the proportion of variance in the observed data that can be explained by the model, with values closer to suggesting a high explanatory power.

This methodological approach allows for an objective comparison between the exponential and logistic models, shedding light on the underlying dynamics of COVID-19's spread in Morocco. The outcome of this analysis not only contributes to our epidemiological understanding but also underscores the potential of AI and machine learning techniques in enhancing predictive modeling. By identifying the model that best represents the pandemic's progression, we can refine our forecasts and better tailor public health measures to the Moroccan context.

In summary, the exploration of exponential versus logistic growth models forms a critical component of our study, providing valuable insights into the nature of the pandemic's spread. The findings from this analysis pave the way for further research into predictive modeling of infectious diseases, highlighting the importance of adaptable, data-driven approaches in public health planning and response.

Linear Regression Analysis

Linear regression emerges as a fundamental technique within statistical modeling and machine learning, providing a solid structure for outcome prediction through independent variables. Within the scope of our investigation into the spread of COVID-19 in Morocco, we utilized linear regression to predict forthcoming case numbers, capitalizing on its ability to delineate the association between time-series data and the pandemic's evolution.

The application of linear regression in this study involved a meticulous process of feature engineering and data preparation. Recognizing the multifaceted nature of the pandemic's spread, we expanded the dataset beyond raw case numbers to include a variety of predictive indicators. These new features were derived from both the epidemiological data and the broader population dataset, encompassing factors such as demographic characteristics, mobility trends, and environmental variables. The inclusion of these features aimed to capture the complex interplay of factors influencing COVID-19 transmission rates and outcomes.

With the enriched dataset in hand, we proceeded to split the data, allocating 80% for training the linear regression model and reserving 20% for testing its predictions. This division ensured a robust evaluation framework, allowing us to assess the model's performance against unseen data and gauge its predictive accuracy.

The effectiveness of the linear regression model was quantified using two key metrics: the root mean square error (RMSE) and the coefficient of determination (R^2). The RMSE provides a measure of the model's prediction error, offering insight into the average deviation of predicted case numbers from the actual reported figures. A lower RMSE indicates a higher accuracy of the model's predictions. Meanwhile, the R^2 value serves as an indicator of the model's explanatory power, reflecting the proportion of variance in the dependent variable that can be predicted from the independent variables. An R^2 value closer to 1 suggests that the model effectively captures the underlying dynamics of the pandemic's spread.

The training and testing of the linear regression model revealed its capability to accurately forecast the number of COVID-19 cases in Morocco. The model's performance, as evidenced by low RMSE and high R^2 values, underscores the potential of statistical modeling techniques in informing public health strategies and interventions. By providing a quantitative basis for predicting future trends, the linear regression analysis contributes valuable insights into the trajectory of the pandemic, aiding policymakers and health officials in their decision-making processes.

In conclusion, the deployment of linear regression in this study exemplifies the power of machine learning and statistical modeling in tackling the challenges posed by COVID-19. The careful selection of features, coupled with rigorous model evaluation, has enabled us to derive meaningful predictions that can support ongoing efforts to manage the pandemic's impact. Future work may explore the integration of additional variables and the application of more complex models, further enhancing our understanding and predictive capabilities in the face of this global health crisis.

RESULTS AND DISCUSSION

Sir Model

The SIR model was utilized to simulate the spread of COVID-19 in Morocco under two distinct scenarios, each varying in disease transmission and recovery rates. These scenarios offer valuable insights into how different public health strategies could influence the pandemic's trajectory.

First Scenario: High Transmission Rate

In the first scenario (Figure 6), characterized by a high transmission rate (β=0.5) and a standard recovery rate (γ=1/5), the model predicts a sharp increase in infectious cases. By day 23, the infectious cases peak at approximately 24% of the population, reflecting a rapid spread of the virus under conditions akin to minimal social distancing and limited public health interventions. Subsequently, the model shows a decline in infectious cases as recovery and fatality rates begin to outweigh new infections. The susceptible population dwindles rapidly, stabilizing at around 10% by day 35, indicating a significant portion of the population either exposed to the virus or vaccinated, thereby achieving a form of herd immunity. The cured or deceased cases rise steadily, reaching about 89% of the population by day 50, suggesting a plateau in the pandemic curve.

Second Scenario: Reduced Transmission Rate

The second scenario (Figure 7), with a reduced transmission rate (β=0.3) and an unchanged recovery rate, presents a markedly different picture. The spread of the virus is slower, with infectious cases peaking later, around day 62, at just 6% of the population. This scenario simulates the effect of rigorous public health measures, such as enhanced social distancing, widespread mask usage, and increased testing and isolation of cases. The number of infectious cases dwindles to near zero beyond day 62, while the proportion of recovered or deceased individuals grows more gradually, reaching 55% by day 110. Interestingly, a significant portion of the population, about 41% by day 95, remains susceptible, suggesting that the virus could resurge if measures are relaxed too soon.

Figure 6. The SIR model with β = 0.5 and γ = 1/5

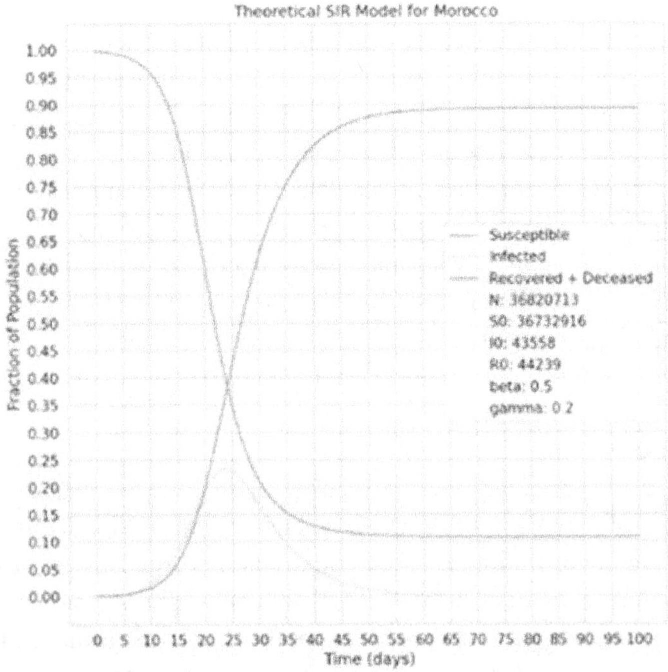

Figure 7. Le modèle SIR avec β = 0,3 et γ = 1/5

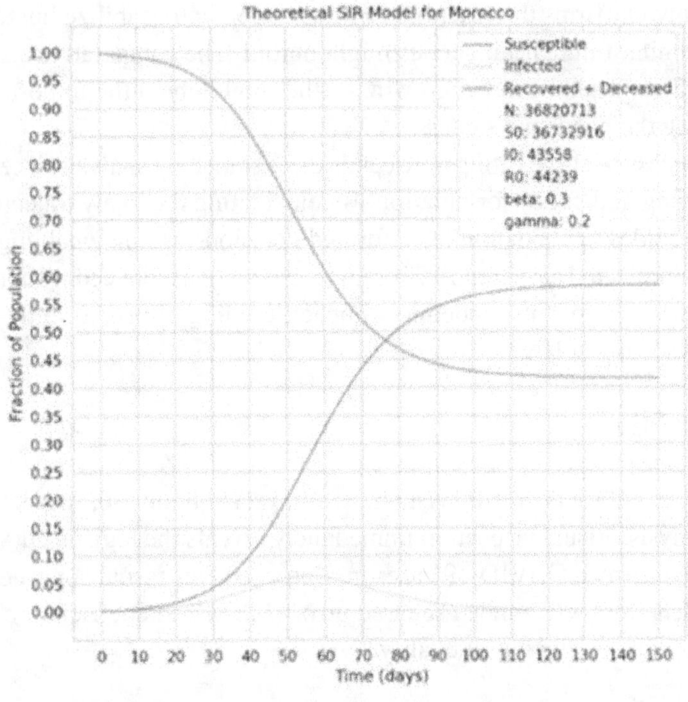

Reflections and Implications

These findings underscore the critical role of public health interventions in controlling the spread of COVID-19. The first scenario illustrates the potential for rapid virus transmission in the absence of effective containment strategies, leading to a high peak of infections but a quicker path to herd immunity. In contrast, the second scenario highlights the benefits of slowing the virus's spread, reducing peak healthcare demand, and allowing more time for vaccine development and distribution.

However, the persistence of a large susceptible population in the second scenario also warns against complacency. It indicates that while slowing the spread is crucial, it must be coupled with ongoing vigilance and preparedness for possible future waves of infection. The SIR model's projections provide a framework for policymakers and health officials to balance the immediate benefits of stringent public health measures against the long-term goal of returning to normalcy without triggering new outbreaks.

Adjusting the Curve

In our endeavor to precisely map the COVID-19 pandemic's trajectory within Morocco, the study embarked on an extensive curve-fitting exercise, employing both logistic and exponential models to encapsulate the progression of confirmed cases. This analytical approach was chosen for its ability to accommodate the nonlinear nature of infectious disease spread, offering a nuanced understanding of the pandemic's dynamics over time. By fitting these mathematical models to the real-world data, we sought to capture the essence of COVID-19's spread, from its initial exponential growth phase to the eventual plateau suggested by the logistic model.

The logistic model, in particular, provided a compelling fit to the data, resonating with the biological and epidemiological understanding that infectious diseases eventually encounter limits to growth due to factors such as population immunity and intervention measures. This model's alignment with the actual case data not only reaffirms the potential for the pandemic to stabilize but also offers a template for anticipating how similar future health crises might unfold. The comparative analysis with the exponential model, which assumes unbounded growth, further highlighted the importance of public health interventions in altering the pandemic's course.

This meticulous process of curve fitting transcends mere statistical analysis; it serves as a critical tool for public health planning and policy formulation. By elucidating the likely trajectory of COVID-19 in Morocco, the study provides a foundation for informed decision-making, enabling authorities to tailor interventions based on projected case trends. Moreover, the insights gleaned from this analysis contribute to a deeper understanding of the pandemic's impact, guiding efforts to mitigate its effects on the healthcare system and society at large.

Visualizing Curve Fits

The logistic and exponential models' fits, visualized in Figures 8 and 10, respectively, undergo a detailed comparative analysis. Visual inspection immediately reveals the logistic curve's closer alignment with the trajectory of confirmed COVID-19 cases in Morocco. This visual congruence is quantitatively supported by the models' respective mean squared error (MSE) and coefficient of determination (R^2) values, crucial metrics in assessing the fit quality.

Figure 8. Adjustment of the exponential curve of confirmed cases in Morocco

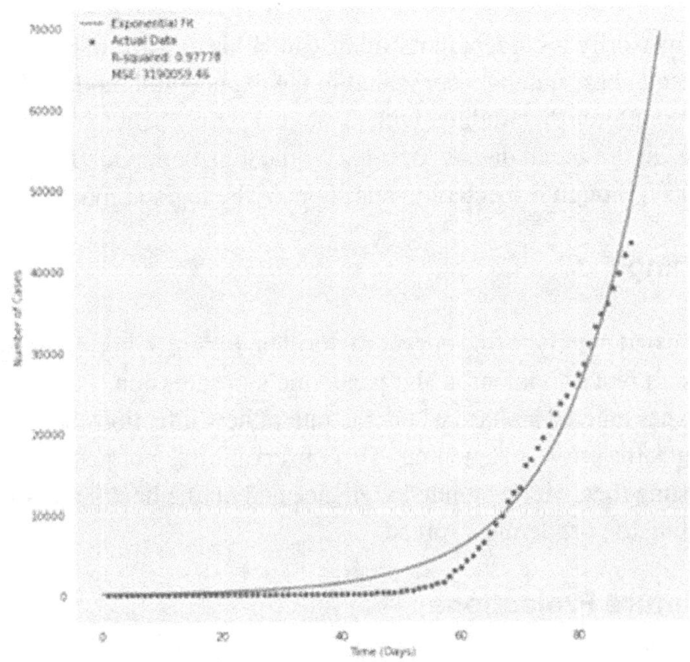

Figure 9. Adjusting the logistic curve of confirmed cases in Morocco

Quantitative Analysis

The logistic model's superiority is evident in its significantly lower MSE of 408411, indicating a closer approximation to the actual case numbers compared to the exponential model's MSE of 3190059. Furthermore, the R^2 values reinforce this finding, with the logistic curve achieving an R^2 of 0.99, suggesting that 99% of the variance in the actual data is captured by the logistic model. In contrast, the exponential model, with an R^2 of 0.97, though respectable, falls short of the logistic model's explanatory power.

Beyond Curve Fitting

The analysis extends beyond merely fitting curves to the data. Figure 9, highlighting the inflection point with a red dot, signals a pivotal moment in the pandemic's progression. Passing this inflection point indicates that Morocco has entered a phase where the rate of new infections begins to decelerate, marking the transition towards the curve's flattening. This observation is not just a statistical triumph but a beacon of hope, suggesting that, with continued vigilance and public health measures, the country is on a path towards controlling the pandemic's spread.

Implications for Future Projections

The logistic curve's fit does more than provide a snapshot of the past and present; it offers a window into the future. By accurately capturing the pandemic's current trajectory, the logistic model allows for informed projections about case numbers moving forward. This predictive capability is invaluable for policymakers and health officials, enabling strategic planning and resource allocation to mitigate the pandemic's impact effectively.

In conclusion, the meticulous adjustment of the pandemic curve through logistic and exponential models has yielded profound insights into COVID-19's behavior in Morocco. The logistic model, in particular, stands out for its accuracy and predictive utility, offering a solid foundation for future analyses. As Morocco navigates the challenging waters of this pandemic, the findings from this curve-fitting exercise illuminate the path forward, emphasizing the power of data-driven decision-making in public health.

Linear Regression

Linear regression was strategically applied to analyze COVID-19's trajectory in Morocco. This section provides a comprehensive overview of the linear regression process, from data preparation through to model evaluation and its implications for understanding the pandemic's future course.

Data Preparation and Transformation

The initial step in our linear regression analysis involved transforming the raw data of confirmed, recovered, and fatal cases into a logarithmic scale. This transformation was critical for two primary reasons: first, it normalizes the exponential growth observed in pandemic data, allowing for a more accurate linear modeling; second, it mitigates the impact of large-scale variations in case numbers, ensuring that the model remains sensitive to changes over time. Figures 10, 11, and 12 illustrate how this transformation aligns the data more closely with a linear trend, setting the stage for a more effective application of linear regression.

Figure 10. Number of confirmed cases in Morocco over time on the logarithmic scale

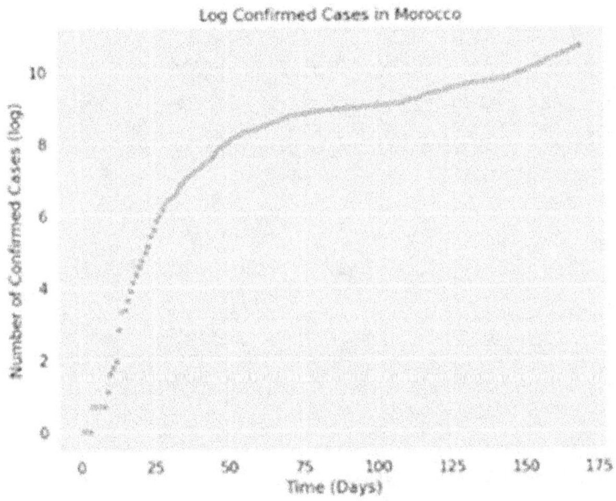

Figure 11. Number of cases recovered in Morocco overtime on the logarithmic scale

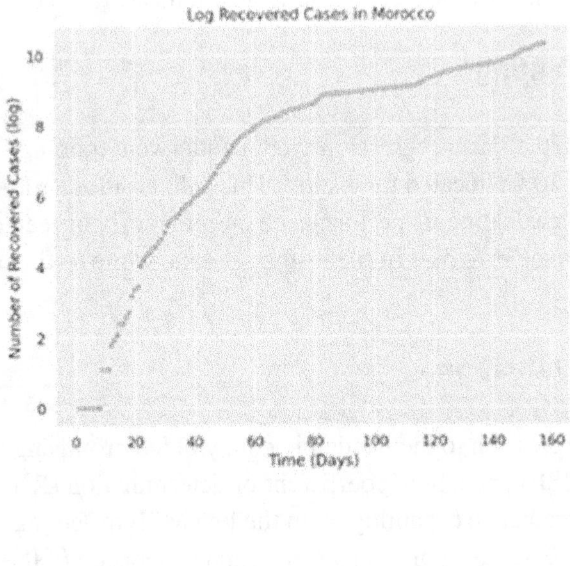

Figure 12. Number of death cases in Morocco overtime on the logarithmic scale

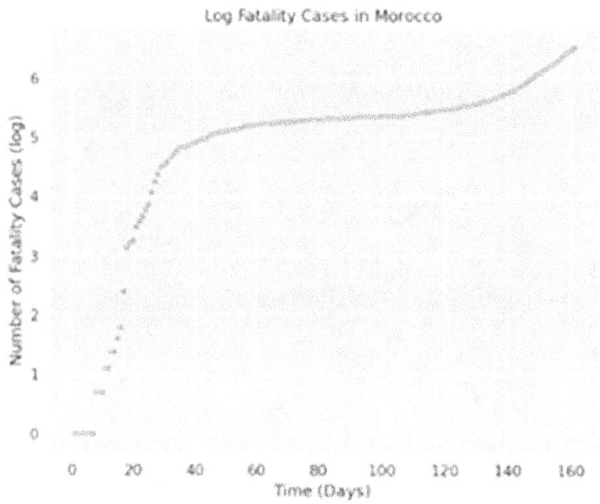

Observations From Data Visualization

Visual analysis, particularly as illustrated in the figures, reveals a noteworthy trend: a gradual decrease in the number of confirmed cases. This observation suggests that Morocco might be approaching the pandemic's plateau phase, where new cases stabilize or diminish over time—a hopeful sign that the curve may soon flatten. Such visual insights not only validate the data transformation approach but also hint at the potential effectiveness of Morocco's public health measures in controlling the virus's spread.

Model Training and Testing

Following data preparation, the linear regression model underwent training on 80% of the transformed dataset, with the remaining 20% allocated for testing. This split ensures a robust assessment of the model's predictive accuracy by evaluating its performance on previously unseen data. The training process fine-tunes the model's parameters to best fit the historical data, while testing gauges its effectiveness in forecasting future trends.

Results and Model Evaluation

The application of linear regression to the pandemic data yielded promising results, characterized by a low mean squared error (MSE) and a high coefficient of determination (R^2). These metrics are indicative of the model's strong predictive capability, with the low MSE reflecting its accuracy in forecasting case numbers and the high R^2 score signifying a significant proportion of the variance in case numbers being explained by the model. Consequently, linear regression emerges as a valuable tool for predicting the trajectory of COVID-19 cases in Morocco, offering insights that could guide public health planning and response strategies.

Implications for Pandemic Forecasting

The successful application of linear regression to COVID-19 case data underscores the potential of statistical modeling in public health. By providing a quantitative basis for forecasting the pandemic's progression, linear regression can inform proactive measures to mitigate its impact. Moreover, the model's adaptability to include additional variables offers a pathway for incorporating environmental, demographic, and social factors into future analyses, enhancing the precision and relevance of its predictions.

In summary, the detailed exploration of linear regression in this study highlights its efficacy in modeling COVID-19 case trends in Morocco. The preparatory data transformation, combined with rigorous model evaluation, lays a solid foundation for future forecasting efforts. As Morocco continues to navigate the challenges posed by the pandemic, the insights garnered from linear regression analysis will be invaluable in shaping informed, data-driven public health strategies.

COVID-19 Linear Regression Analysis

The application of linear regression to the datasets concerning confirmed cases, recovered cases, and fatalities due to COVID-19 in Morocco has yielded quantitatively robust results, as summarized in the accompanying table (Table 1). For confirmed cases, the model demonstrated remarkable precision, with a mean squared error (MSE) of 0.00025 and an R^2 score, indicative of the model's predictive accuracy, at 0.996. This high R^2 value signifies that 99.6% of the variance in the confirmed case data can be explained by the model, pointing to a strong correlation between the predictors used and the confirmed cases.

In terms of recovered cases, the linear regression model maintained high accuracy, reflected by an MSE of 0.00187 and an R^2 score of 0.990. The model's predictive capacity for recovered cases is therefore nearly as robust as for confirmed cases, with 99% of the data variance being accounted for. This suggests that the features included in the model are significantly related to the recovery rates and can be used reliably for future predictions.

The analysis of death cases through linear regression also showed a strong fit, albeit with a higher MSE of 0.00713, yet still yielding a high R^2 score of 0.979. Despite the slightly larger error, the model could explain approximately 97.9% of the variance in the death case data, underscoring the utility of linear regression in forecasting fatal outcomes of the virus.

The consistency of low MSE values across all three datasets corroborates the efficacy of linear regression in modeling the pandemic's various aspects. The high R^2 scores across the board indicate that the predictive models have captured the essential dynamics of the COVID-19 spread within Morocco. These results collectively affirm the potential of linear regression as a predictive tool, offering valuable insights for health authorities and policymakers as they strive to anticipate and manage the course of the pandemic effectively.

Table 1. Summary of results obtained after applying linear regression to each of the datasets

Data	MSE	Score R^2
Confirmed Cases	0.00025	0.996
Cases Recovered	0.00187	0.990
Death Cases	0.00713	0.979

CONCLUSION

In this chapter's conclusion, we have scrutinized the transformative role of artificial intelligence in deciphering and mitigating the impact of the COVID-19 pandemic in Morocco. The strategic application of the SIR model, coupled with linear regression and curve fitting techniques, has unraveled the complexities of the pandemic's evolution, offering invaluable foresight into its potential resolution. The alignment of the logistic curve with empirical data not only signifies a possible stabilization in case numbers but also serves as a harbinger for the strategic planning required to combat similar health emergencies in the future.

The insights garnered from this study extend far beyond the immediate realm of public health response. They touch upon the broader, more profound implications of our interaction with the environment and its subsequent impact on human health. The integration of environmental variables into the pandemic's predictive modeling process has been revelatory, underscoring the symbiotic relationship between the planet's health and our own. It has become increasingly clear that variables once considered peripheral, such as atmospheric pollution levels and meteorological patterns, exert a significant influence on the virulence and transmission pathways of pathogens.

The ramifications of this research invite a paradigm shift in how we conceive public health strategy. No longer can we afford to view health crises as isolated incidents; rather, they must be regarded within the context of global environmental stewardship. The effective management of pandemics necessitates policies that are as attentive to environmental conservation as they are to medical intervention. Our chapter advocates for a synergistic approach, whereby health and environmental governance operate in concert to fortify societal defenses against emergent health threats.

As we cast our gaze to the horizon, the prospective applications of AI in the domain of public health burgeon with promise. The exploration of advanced machine learning paradigms, including the likes of deep learning and neural networks, could potentially usher in a new epoch of predictive accuracy and diagnostic precision. The integration of real-time environmental monitoring into the analytical matrix can offer a dynamic and responsive framework for gauging the pulse of pandemic trends.

The notion of utilizing statistical bootstrapping to dissect the underlying distributions of pandemic data adds another layer of depth to our analytical arsenal. By simulating numerous iterations and sampling diverse data permutations, we stand to gain a granular understanding of the pandemic's behavior. Such methodological enhancements are anticipated to peel back the layers of complexity that shroud the nature of pandemic proliferation.

In the pursuit of advancement, the need for cross-disciplinary collaboration has become increasingly clear. Integrating insights from various scientific domains is essential for pioneering innovations in public health. By combining the expertise of epidemiologists, environmental scientists, data analysts, and AI experts, it is possible to create comprehensive models. These models are not just equipped to decode current trends but are also adept at forecasting future developments.

Thus, the narrative woven through this chapter not only addresses the contemporary challenges posed by COVID-19 but also lays a foundational framework for a future where AI-driven strategies are central to public health. By synergizing the acumen of environmental science with the predictive prowess of AI, we navigate towards a future equipped to confront pandemics with the dual swords of knowledge and technology, safeguarded by a commitment to preserving the sanctity of our planetary home.

In summation, this chapter is a testament to the resilience and ingenuity of human endeavor. It encapsulates a journey of discovery, challenges, and triumphs in the face of a global health crisis, heralding a new dawn of integrated, AI-empowered public health strategies. As we continue to combat COVID-19 and brace for the uncertainties of the future, it is the lessons drawn from such research endeavors that will illuminate the path forward, ensuring that society remains resilient, informed, and perpetually prepared to safeguard the health of all its members.

REFERENCES

Alboaneen, D., Pranggono, B., Alshammari, D., Alqahtani, N., & Alyaffer, R. (2020). Predicting the epidemiological outbreak of the coronavirus disease 2019 (COVID-19) in Saudi Arabia. *International Journal of Environmental Research and Public Health, 17*(4568), 4568. doi:10.3390/ijerph17124568 PMID:32630363

Arık, S. O., Li, X., & Menon, V. (2021). A prospective evaluation of AI-augmented epidemiology to forecast COVID-19 in the USA and Japan. *NPJ Digital Medicine, 4*(96), 146. doi:10.1038/s41746-021-00511-7 PMID:34625656

Ayyoubzadeh, S. M., Zahedi, H., Ahmadi, M., & Niakan Kalhori, R. S. (2020). Predicting COVID-19 incidence through analysis of Google Trends data in Iran: Data mining and deep learning pilot study. *JMIR Public Health and Surveillance, 6*(2, e18828), e18828. doi:10.2196/18828 PMID:32234709

Backer, J. A., Klinkenberg, D., & Wallinga, J. (2020). Incubation period of 2019 novel coronavirus (2019-nCoV) infections among travellers from Wuhan, China, 20–28 January 2020. *Eurosurveillance, 25*(2000062). doi:10.2807/1560-7917.ES.2020.25.5.2000062 PMID:32046819

BatistaM. (2020). Estimation of the final size of the COVID-19 epidemic. medRxiv. doi:10.1101/2020.02.16.20023606

Ceylan, Z. (2020). Estimation of COVID-19 prevalence in Italy, Spain, and France. *The Science of the Total Environment, 729*(138817), 138817. doi:10.1016/j.scitotenv.2020.138817 PMID:32360907

Chew, A. W. Z., Pan, Y., Wang, Y., & Zhang, L. (2021). Hybrid deep learning of social media big data for predicting the evolution of COVID-19 transmission. *Knowledge-Based Systems, 233*, 107417. doi:10.1016/j.knosys.2021.107417 PMID:34690447

Eltoukhy, A. E. E., Shaban, I. A., Chan, F. T. S., & Abdel-Aal, M. A. M. (2020). *Data analytics for predicting COVID-19 cases in top affected countries: Observations and recommendations*. Research in Public Health.

Garnier-Crussard, A., Forestier, E., Gilbert, T., & Krolak-Salmon, P. (2020). Novel coronavirus (COVID-19) epidemic: What are the risks for older patients? *Journal of the American Geriatrics Society, 68*(939–940), 939–940. doi:10.1111/jgs.16407 PMID:32162679

Hu, Y., Sun, X., Nie, X., Li, Y., & Liu, L. (2019). An enhanced LSTM for trend following of time series. *IEEE Access : Practical Innovations, Open Solutions, 7*, 34020–34030. doi:10.1109/ACCESS.2019.2896621

Ivorra, B., Ferrández, M. R., Vela-Pérez, M., & Ramos, A. M. (2020). Mathematical modeling of the spread of the coronavirus disease 2019 (COVID-19) taking into account the undetected infections. The case of China. *Communications in Nonlinear Science and Numerical Simulation*, *88*(105303), 105303. doi:10.1016/j.cnsns.2020.105303 PMID:32355435

Johns Hopkins, C. S. S. E. (2020). *2019 Novel Coronavirus COVID-19 (2019-nCoV).* Data Repository by Johns Hopkins CSSE. https://github.com/CSSEGISandData/COVID-19

Lauer, S. A., Grantz, K. H., Bi, Q., Jones, F. K., Zheng, Q., Meredith, H. R., Azman, A. S., Reich, N. G., & Lessler, J. (2020). The incubation period of coronavirus disease 2019 (COVID-19) from publicly reported confirmed cases: Estimation and application. *Annals of Internal Medicine*, *172*(577–582), 577–582. doi:10.7326/M20-0504 PMID:32150748

Li, J., Xu, Q., Cuomo, R., Purushothaman, V., & Mackey, T. (2020). Data mining and content analysis of the Chinese social media platform Weibo during the early COVID-19 outbreak: Retrospective observational infoveillance study. *JMIR Public Health and Surveillance*, *6*(2, e18700), e18700. doi:10.2196/18700 PMID:32293582

Papastefanopoulos, V., Linardatos, P., & Kotsiantis, S. (2020). COVID-19: A comparison of time series methods to forecast percentage of active cases per population. *Applied Sciences (Basel, Switzerland)*, *10*(3880), 3880. doi:10.3390/app10113880

Pouriyeh, S., Vahid, S., Sannino, G., De Pietro, G., Arabnia, H., & Gutierrez, J. (2017). A comprehensive investigation and comparison of machine learning techniques in the domain of heart disease. In *2017 IEEE Symposium on Computers and Communications (ISCC)* (pp. 204-207). IEEE. 10.1109/ISCC.2017.8024530

Prabhu, T. N. (2020). *Population by country - 2020.* Kaggle. https://www.kaggle.com/tanuprabhu/population-by-country-2020

Qin, L., Sun, Q., Wang, Y., Wu, K. F., Chen, M., Shia, B. C., & Wu, S. Y. (2020). Prediction of number of cases of 2019 novel coronavirus (COVID-19) using social media search index. *International Journal of Environmental Research and Public Health*, *17*(2365). PMID:32244425

Remuzzi, A., & Remuzzi, G. (2020). COVID-19 and Italy: What next? *Lancet*, *395*(1225–1228). PMID:32178769

Ribeiro, M. H. D. M., da Silva, R. G., Larcher, J. H. K., Mariani, V. C., & Coelho, L. D. S. (2022). Ensemble learning models coupled with urban mobility information applied to predict COVID-19 incidence cases. *Modeling, Control and Drug Development for COVID-19 Outbreak Prevention*, *24*, 821–858.

Ross, R., & Hamer, W. H. (2013). *The SIR model and the foundations of public health.* Materials Matematics.

Scopus. (2020). *Analyze Search Results, TITLE-ABS-KEY ("COVID-19" or "CoViD-19" or "Hubei Pneumonia" or "Novel Corona Virus").* Scopus. https://www.scopus.com

Shen, C., Chen, A., Luo, C., Zhang, J., Feng, B., & Liao, W. (2020). Using reports of symptoms and diagnoses on social media to predict COVID-19 case counts in mainland China: Observational infoveillance study. *Journal of Medical Internet Research*, *22*(5, e19421), e19421. doi:10.2196/19421 PMID:32452804

Shen, K., Yang, Y., Wang, T., Zhao, D., Jiang, Y., Jin, R., Zheng, Y., Xu, B., Xie, Z., Lin, L., Shang, Y., Lu, X., Shu, S., Bai, Y., Deng, J., Lu, M., Ye, L., Wang, X., Wang, Y., & Gao, L. (2020). Diagnosis, treatment, and prevention of 2019 novel coronavirus infection in children: Experts' consensus statement. *World Journal of Pediatrics*, *16*(3), 223–231. doi:10.1007/s12519-020-00343-7 PMID:32034659

Tuite, A. R., Bogoch, I. I., Sherbo, R., Watts, A., Fisman, D., & Khan, K. (2020). Estimation of coronavirus disease 2019 (COVID-19) burden and potential for international dissemination of infection from Iran. *Annals of Internal Medicine*, *172*(699–701), 699–701. doi:10.7326/M20-0696 PMID:32176272

Tuite, A. R., Ng, V., Rees, E., & Fisman, D. (2020). Estimation of COVID-19 outbreak size in Italy. *The Lancet. Infectious Diseases*, *20*(537), 537. doi:10.1016/S1473-3099(20)30227-9 PMID:32199494

Virtanen, P., Gommers, R., Oliphant, T. E., Haberland, M., Reddy, T., Cournapeau, D., Burovski, E., Peterson, P., Weckesser, W., Bright, J., van der Walt, S. J., Brett, M., Wilson, J., Millman, K. J., Mayorov, N., Nelson, A. R. J., Jones, E., Kern, R., Larson, E., & van Mulbregt, P. (2020). SciPy 1.0: Fundamental algorithms for scientific computing in Python. *Nature Methods*, *17*(3), 261–272. doi:10.1038/s41592-019-0686-2 PMID:32015543

Wang, D., Hu, B., Hu, C., Zhu, F., Liu, X., Zhang, J., Wang, B., Xiang, H., Cheng, Z., Xiong, Y., Zhao, Y., Li, Y., Wang, X., & Peng, Z. (2020). Clinical characteristics of 138 hospitalized patients with 2019 novel coronavirus-infected pneumonia in Wuhan, China. *Journal of the American Medical Association*, *323*(11), 1061–1069. doi:10.1001/jama.2020.1585 PMID:32031570

Wang, W., Tang, J., & Wei, F. (2020). Updated understanding of the outbreak of 2019 novel coronavirus (2019-nCoV) in Wuhan, China. *Journal of Medical Virology*, *92*(4), 441–447. doi:10.1002/jmv.25689 PMID:31994742

Wei, M., Yuan, J., Liu, Y., Fu, T., Yu, X., & Zhang, Z.-J. (2020). Novel coronavirus infection in hospitalized infants under 1 year of age in China. *Journal of the American Medical Association*, *323*(1313–1314), 1313. doi:10.1001/jama.2020.2131 PMID:32058570

Zhuang, Z., Zhao, S., Lin, Q., Cao, P., Lou, Y., Yang, L., & He, D. (2020). Preliminary estimation of the novel coronavirus disease (COVID-19) cases in Iran: A modelling analysis based on overseas cases and air travel data. *International Journal of Infectious Diseases*, *94*(29–31), 29–31. doi:10.1016/j.ijid.2020.03.019 PMID:32171951

Zhuang, Z., Zhao, S., Lin, Q., Cao, P., Lou, Y., Yang, L., & He, D. (2020). Preliminary estimation of the basic reproduction number of novel coronavirus (2019-nCoV) in China, from 2019 to 2020: A data-driven analysis in the early phase of the outbreak. *International Journal of Infectious Diseases*, *94*(29–31). PMID:32088335

KEY TERMS AND DEFINITIONS

AI-Based Event Detection: Method using AI to automatically identify significant incidents or anomalies in environmental or public health data.

Artificial Intelligence (AI): Technology that mimics human intelligence in machines, enabling the automation of complex tasks and enhancing data-driven decision-making.

Environmental Surveillance: The process of monitoring and analyzing environmental factors to prevent and control public health risks.

Linear Regression: Statistical technique that models the relationship between a dependent variable and one or more independent variables, used here to predict COVID-19 trends.

Machine Learning: A branch of AI that allows machines to learn from data and improve their predictions or decisions without being explicitly programmed for each task.

SIR Model (Susceptible-Infected-Recovered): Mathematical model used to predict the spread of infectious diseases by categorizing the population into three states: susceptible, infected, and recovered.

Trend Analysis: Data analysis methodology used to identify patterns or trends in data sets over time, aiding evidence-based decision-making.

Chapter 15
The Future of Water Management:
Leveraging AI for Effective Decision Support

Ahmed Elshaikh
Water Research Centre, University of Khartoum, Sudan

Jamal Mabrouki
https://orcid.org/0000-0002-3841-7755
Mohammed V University in Rabat, Morocco

Ahmed Abobaker Osman Mohamed
Water Research Centre, University of Khartoum, Sudan

ABSTRACT

This chapter delves into the incorporation of artificial intelligence (AI) technology in the realm of water management, emphasizing the potential for AI-driven decision support systems to bring about significant advancements in the monitoring and optimization of water resources. Through harnessing the power of AI algorithms and machine learning, water managers can enhance their decision-making processes, leading to more efficient resource allocation and overall improvements in water management practices. The chapter explores the transformative impact of integrating AI in water management, underscoring the opportunities it presents for promoting sustainability and effectiveness in the management of water resources. Furthermore, it highlights the promising future prospects for leveraging AI to revolutionize how water resources are monitored, analysed, and managed in a more sustainable and efficient manner.

DOI: 10.4018/979-8-3693-3807-0.ch015

Copyright © 2024, IGI Global. Copying or distributing in print or electronic forms without written permission of IGI Global is prohibited.

INTRODUCTION

Water resources management is a critical aspect of sustainable development, with increasing demands and challenges posed by population growth, climate change, and environmental degradation (Zeman et al., 2006). In this context, the integration of artificial intelligence (AI) technologies holds immense potential to revolutionize decision-making processes in the field of water resources management. Through harnessing the power of AI, water managers can access valuable insights, optimize resource allocation, and enhance overall efficiency in managing water supplies.

The envisioned structure for AI-powered decision support systems in water resources management represents a ground-breaking approach to addressing complex and dynamic challenges in the sector. Through the combination of diverse AI technologies such as machine learning, data analytics, and predictive modelling, this framework aims to provide comprehensive solutions for optimizing water management practices. Via leveraging AI-driven systems, water managers can make informed decisions based on real-time data, predictive analytics, and scenario planning to ensure sustainable use and conservation of water resources.

The advantages of integrating various AI technologies within the proposed framework are manifold, offering enhanced capabilities for monitoring, analysis, and decision support in water resources management (Kamyab et al., 2023). With AI-powered systems, water managers can detect patterns in data, predict future trends, and optimize operational strategies to meet evolving demands and environmental conditions. Moreover, the use of AI can enable proactive measures to mitigate risks, improve resource efficiency, and enhance overall resilience in water management practices (Gunasekaran & Boopathi, 2023).

Real-world applications of AI technologies in water resources management demonstrate their transformative potential in addressing key challenges such as water scarcity, pollution, and infrastructure optimization. From predictive modelling for drought forecasting to real-time monitoring of water quality, AI-driven systems offer innovative solutions to enhance decision-making processes and improve the sustainability of water management practices. Water managers can tackle complicated challenges with better precision, efficiency, and effectiveness by leveraging the combined power of AI technologies within a structured framework to maintain water resource viability in the future.

This chapter aims to offer a thorough examination of the envisioned structure for AI-powered decision support systems in the field of water resources management. It highlights the advantages and practical uses of combining diverse AI technologies to tackle the hurdles encountered in ensuring sustainable management of water resources. The chapter delves into the potential benefits and real-world applications of incorporating different AI technologies within the proposed framework. It showcases how this integration can facilitate the resolution of challenges in sustainable water resource management by providing advanced decision support capabilities. Through the utilization of AI-driven systems, water managers can leverage innovative solutions to optimize resource allocation, predict water availability and quality, identify patterns in data, and develop strategies for improved water management practices.

METHODOLOGY

The research design involved a comprehensive literature review to identify the current state of AI technologies in decision support systems for water resources management. The proposed framework was developed based on the integration of various AI methods and tools.

The analysis process included synthesizing information from various sources to identify the key AI methods and tools that could be integrated into the proposed framework. The analysis also involved evaluating the strengths and weaknesses of each AI method in the context of water management practices.

The proposed framework integrates various AI methods and tools, including expert systems, fuzzy logic, Gaussian processes, clustering algorithms, neural networks, random forest, and decision trees, to enhance decision-making processes in water management practices. This framework provides a roadmap for the development of intelligent decision support systems that can revolutionize the way decisions are made in water resources management (see in figure 1 below).

Figure 1. The framework for key AI tools to support decision for sustainable water management

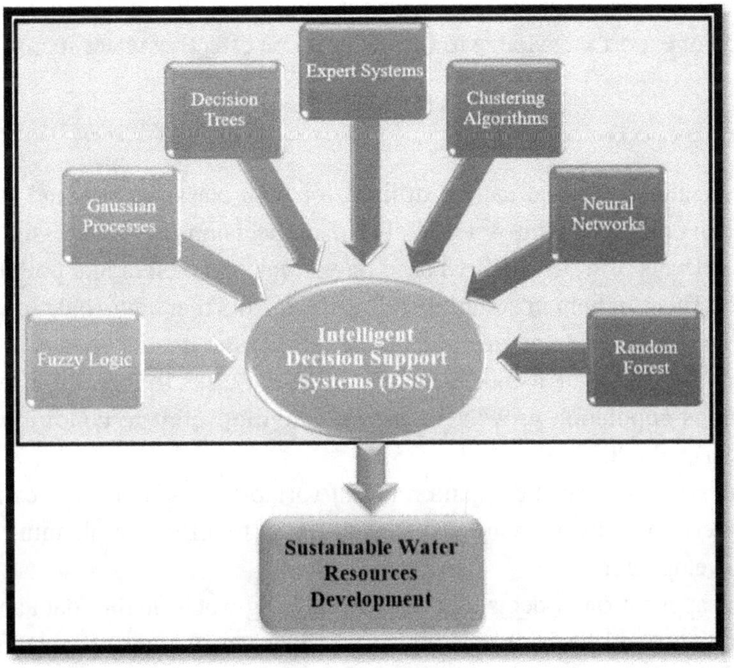

AI TOOLS TO SUPPORT DECISION IN WATER MANAGEMENT

Fuzzy Logic

Fuzzy logic serves as a valuable instrument for managing uncertainty in decision-making within water resources management. It offers a means to handle imprecise or uncertain information, making it a suitable tool for addressing the inherent variability and uncertainty within water resource systems (Mujumdar & Ghosh, 2008).

In the realm of water resources management, fuzzy logic plays a vital role in decision-making processes and action control when faced with uncertain or imprecise data. For instance, it can aid in adapting irrigation practices based on uncertain weather predictionsWater managers can make advanced irrigation scheduling decisions by using fuzzy logic to analyze and respond to weather forecasts, considering the forecast's uncertainties.

Moreover, fuzzy logic can be implemented in various areas of water resources management like flood control, reservoir operation, and water quality management. It facilitates the integration of expert knowledge and linguistic variables, which are often vague or subjective, into decision-making procedures (Mujumdar & Ghosh, 2008).

Furthermore, combining fuzzy logic with other modelling and decision support tools can offer a more comprehensive approach to addressing uncertainty in water resources systemsWater managers can improve decision-making quality in uncertain situations by combining fuzzy logic with mathematical models, optimization techniques, and data-driven approaches.

In essence, fuzzy logic presents an adaptive and nuanced strategy for managing uncertainty in decision-making within water resources management. It enables the consideration of imprecise data and the incorporation of expert knowledge to formulate more effective water management strategies.

Decision Trees

Decision trees are versatile tools that can be utilized for both classification and regression purposes in data analysis. When employed with water quality data, decision trees can assist in identifying and categorizing different levels of water quality based on various parameters like pollutant concentrations (Enayati et al., 2022). This can help in comprehending the factors that contribute to water pollution and in creating targeted strategies for managing and remediating water quality issues.

Furthermore, decision trees can also be used for regression tasks to forecast water demand by considering factors such as population growth, industrial activities, climate conditions, and agricultural requirements (Enayati et al., 2022).

Through the analysis of historical data and relevant variables, decision trees can offer valuable insights into future patterns of water demand, which is essential for effective planning of water resources and infrastructure development.

In conclusion, the application of decision trees in analysing water quality data and predicting water demand can support informed decision-making, proactive management of water resources, and the formulation of sustainable strategies for water management.

Expert Systems

Combining human expertise with AI algorithms, expert systems can offer invaluable water management decision suggestions, particularly beneficial for water managers and users lacking extensive technical expertise. These systems harness human experts' knowledge and experience alongside AI algorithms' computational power and analytical capabilities to provide practical and actionable guidance for water resource management (Guariso, 1988).

In the domain of water management, expert systems can be tailored to encapsulate and formalize the expertise of field specialists like hydrologists, water resource engineers, and environmental scientists. This expertise is then melded with AI algorithms to create a system capable of scrutinizing intricate data, interpreting ambiguous information, and furnishing customized recommendations for water management choices (Guariso, 1988)Individuals without specialized knowledge can use expert systems to receive distilled insights and ideas from domain experts in an easily accessible format, allowing them to make educated decisions about water use, conservation, and management.

Expert systems can span a broad array of water management functions, encompassing tasks such as irrigation scheduling, water distribution, drought mitigation, flood prevention, and water quality enhancement. These systems help water users optimize their water usage, reduce waste, and respond quickly to changing environmental conditions by providing specialized advice based on their individual local circumstances.

In conclusion, expert systems that amalgamate human wisdom with AI algorithms play a pivotal role in democratizing access to high-level water management guidance, empowering a wider spectrum of stakeholders to make informed choices about water resources. Simultaneously, these systems leverage AI capabilities to handle intricate data and uncertainties efficiently.

Random Forest

Random forest is a machine learning technique that combines multiple decision trees to enhance the accuracy and reliability of classification and regression tasks. In the field of water management and environmental engineering, random forest can be a useful tool for tasks such as flood prediction, water quality monitoring, and leak detection in water distribution systems (Alomani et al., 2022; Lučin et al., 2021).

For instance, in flood prediction, random forest can analyse historical data on factors like rainfall, river levels, land use, and soil characteristics to create models that can identify areas at risk of flooding. This information can be used to develop early warning systems and plan strategies for flood prevention. In water quality monitoring, random forest can analyse complex datasets to classify water quality levels and pinpoint potential sources of contamination (Esfandiari et al., 2020).

Additionally, in leak detection within water distribution systems, random forest can analyse various parameters like sensor data and flow rates to detect anomalies and locate potential leaks or pipe bursts (Lučin et al., 2021). The versatility of random forest in handling intricate data makes it a valuable tool for addressing challenges in water management and environmental engineering, ultimately aiding in more informed decision-making and sustainable resource management efforts.

Neural Networks

Neural networks, such as Convolutional Neural Networks (CNN) and Recurrent Neural Networks (RNN), are highly effective tools for various tasks in water management and environmental engineering. RNNs are particularly useful for analysing time-series water flow data, as they can capture temporal dependencies and identify patterns and anomalies (Wang et al., 2022).

In water quality prediction, neural networks can analyse relationships between different parameters to forecast future water quality levels, aiding in contamination risk assessment and treatment optimization (Wang et al., 2022).

Deep learning models like CNNs and RNNs are also valuable for flood forecasting by processing large amounts of spatial and temporal data to make accurate predictions, enhancing early warning systems.

Overall, these advanced technologies enable more precise predictions, better decision-making, and improved resource management in water management and environmental engineering.

Clustering Algorithms

Clustering algorithms such as K-means clustering or Density-Based Spatial Clustering of Applications with Noise (DBSCAN) can be utilized to group similar water quality samples or recognize spatial patterns within water resource data. K-means clustering, a well-known unsupervised learning technique, divides data into a predefined number of clusters based on their similarities (Vo-Van et al., 2020). In the context of water quality samples, K-means clustering can group samples sharing common traits, like comparable contaminant levels or overall water quality profiles. This aids in the identification of recurring patterns or irregularities in water quality data, which is crucial for understanding pollutant distribution or pinpointing potential contamination sources.

Density-Based Spatial Clustering of Applications with Noise (DBSCAN) presents another robust clustering algorithm useful for detecting spatial patterns in water resource data. DBSCAN excels in identifying clusters of various shapes and sizes while being resilient to data noise and outliers. Applied to water resource data, DBSCAN can pinpoint spatial clusters sharing similar water quality attributes, such as regions with consistently high or low levels of specific pollutants. This information proves valuable for environmental monitoring and management purposes (Mu et al., 2023).

Both K-means clustering and DBSCAN serve as valuable tools for conducting exploratory data analyses, offering insights into the spatial and temporal distribution of water quality attributes. They help to inform decision-making in water resource management and environmental engineering by offering a more in-depth understanding of these patterns.

Gaussian Processes

Gaussian processes (GPs) stand out as a potent instrument for quantifying uncertainty in water resource modelling and predictive assignments. They present a versatile and probabilistic method for modelling intricate systems, rendering them ideal for capturing and measuring uncertainty across various facets of water resource management and modelling.

Within the scope of water resource modelling, GPs can be harnessed to portray uncertain correlations between diverse variables, like the link between precipitation and streamflow or the spatial distribution of groundwater levels. Through the utilization of GPs in modelling these associations, one not only gains the ability to forecast future water resource conditions but also to assess the uncertainty linked with those forecasts (Sun et al., 2014).

Additionally, GPs prove to be invaluable in hydrological and environmental modelling by furnishing a probabilistic framework for assimilating observational data, model results, and expert insight. This framework enables the amalgamation of multiple information sources and the quantification of uncertainty in predictions and decision-making related to water resources. Moreover, GPs find utility in spatial interpolation and the mapping of water quality parameters, such as contaminant concentrations or pollutant levels, all while accommodating uncertainty relating to the spatial distribution of these parameters (Cui et al., 2021).

In essence, Gaussian processes present a pliable and robust method for uncertainty assessment in water resource modelling and prediction activities, rendering them a valuable resource for guiding decision-making processes in water resource management and environmental engineering.

WATER DATA ANALYSIS USING AI POWER

Big data analysis with AI technology is revolutionizing decision support systems for sustainable water resources management. By using the power of advanced algorithms and machine learning techniques, AI can process large volumes of data from varied sources, including sensors, satellites, and social media, to give important insights and forecasts that are critical for sustainable water management.

AI technologies for water solutions need to effectively combine different types of data from multiple sources, such as climate data, satellite images, IoT sensor data, and others. The capacity to handle and analyse large and varied datasets improves the precision and dependability of AI-generated insights aimed at addressing climate challenges.

One of the main advantages of leveraging AI in water resources management is its ability to handle large volumes of data with speed and precision (Sun & Scanlon, 2019). Traditional methods of data analysis often struggle to cope with the sheer volume and complexity of information generated by water infrastructure and environmental monitoring systems. AI, on the other hand, excels at processing and analysing this data, enabling water managers to extract meaningful patterns and trends that can inform decision-making processes. Moreover, AI technology can enhance the accuracy and reliability of predictive models used in water resources management. AI can accurately predict future water demand, quality, and availability by training algorithms on previous data and continuously updating them with real-time information.

These predictions are essential for planning and allocating water resources effectively, especially in regions facing water scarcity or rapid population growth.

Furthermore, AI can optimize the operation of water distribution networks by identifying inefficiencies and recommending adjustments to improve performanceAI can find anomalies or potential problems in water infrastructure in real time by analyzing data from sensors and other sources. This allows for proactive maintenance and risk management.

This proactive approach not only helps prevent costly failures but also ensures the reliability and efficiency of water supply systems, even in the face of changing environmental conditions.

In addition to optimizing water distribution networks, AI can also improve the efficiency of water treatment processes. Through analysing data from treatment plants and optimizing chemical dosages, energy consumption, and operational performance, AI can enhance the quality of treated water while reducing costs. This not only benefits public health and the environment but also contributes to the overall sustainability of water resources by ensuring that treatment processes are carried out efficiently and effectively.

Moreover, the integration of AI technology into decision support systems enables water managers to make more informed and timely decisions that are essential for sustainable water management practices. AI enables decision-makers to address challenges such as water scarcity, pollution, and infrastructure maintenance proactively by delivering real-time data and predictive analytics.

This proactive decision-making approach is crucial for mitigating risks, improving operational efficiency, and ensuring the long-term sustainability of water resources for future generations (Kamyab et al., 2023).

In conclusion, big data analysis with AI technology is a game-changer in decision support systems for sustainable water resources management. AI can provide useful insights and forecasts for effective water management by utilizing modern algorithms and machine learning techniques to process vast amounts of data from multiple sources.

As we continue to advance in this field, it is imperative that we embrace these technologies and harness their potential to create a more sustainable future for our planet and its inhabitants.

APPLICATIONS OF AI IN WATER MANAGEMENT MODELLING

AI integration into decision support systems for predictive modelling in water resources management is a ground-breaking advancement that holds immense potential for revolutionizing how we manage and safeguard our precious water resourcesWith leveraging the power of artificial intelligence, we can now obtain deep insights into water systems, predict future trends, and make informed decisions that maximize water distribution, usage, and conservation.

One of the key benefits of integrating AI into decision support systems for predictive modelling in water resources management is the ability to analyse vast amounts of data in real-time. AI algorithms can process data from various sources, such as sensors, satellites, weather forecasts, and historical records, to create accurate models that predict water availability, demand, quality, and other critical parameters (Kamyab et al., 2023). This predictive capability enables water managers to anticipate potential challenges, plan proactively, and implement effective strategies to ensure sustainable water management practices.

Moreover, AI-powered predictive modelling can help optimize decision-making processes by providing actionable insights and scenario planning capabilities. Decision-makers can use AI-generated predictions to evaluate different scenarios, assess the impact of various interventions, and choose the most effective course of action to achieve desired outcomes. Utilizing modeling various scenarios and predicting their outcomes, AI-powered decision support systems enable water managers to make informed decisions that maximize resource efficiency, reduce risks, and improve overall sustainability.

Furthermore, AI integration into decision support systems for predictive modelling in water resources management can enhance the resilience of water systems in the face of changing environmental conditions and uncertainties. Through continuously learning from new data inputs and updating models in real-time, AI algorithms can adapt to evolving circumstances and provide dynamic insights that help water managers respond effectively to unexpected events, such as extreme weather events, droughts, or pollution incidents. This adaptive capability enables decision-makers to mitigate risks, optimize resource allocation, and enhance the resilience of water systems against future challenges. Figure (2) illustrates a GeoAI application model for a smart IWRM support system.

Figure 2. A Geo AI model for smart decision support system for integrated water resources management (IWRM)
(Gonzales-Inca et al., 2022).

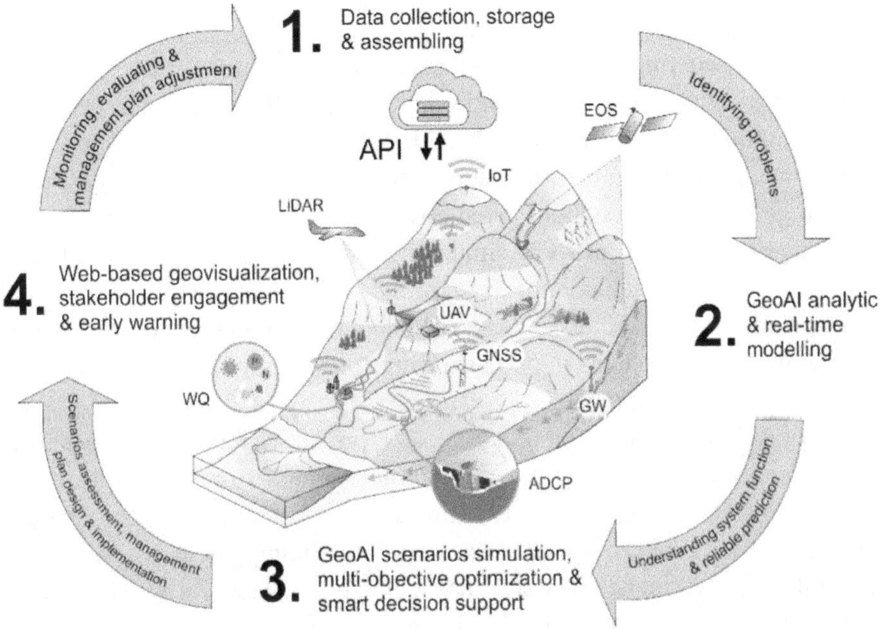

Below are some examples of predictive modelling applications in water resources management:

- **Water Supply Forecasting:** Predictive models can be used to forecast water availability in reservoirs, rivers, and groundwater sources based on historical data, weather patterns, and hydrological conditions (Kaur et al., 2021). These forecasts help water managers plan for future water supply needs, optimize reservoir operations, and make informed decisions about water allocation.
- **Drought Prediction:** Predictive models can help predict drought conditions by analysing historical precipitation data, soil moisture levels, and other relevant factors. Water managers can reduce the effects of water shortage on populations and ecosystems by forecasting drought episodes ahead of time (So et al., 2020).
- **Flood Risk Assessment:** Predictive modelling can be used to assess flood risk by simulating various scenarios and predicting potential flood events based on factors such as rainfall intensity, land use changes, and topography. These models help water managers identify high-risk areas, develop flood mitigation strategies, and improve emergency response planning (Han et al., 2007).
- **Water Quality Prediction:** Predictive models can forecast changes in water quality parameters such as nutrient levels, contaminants, and dissolved oxygen concentrations. When assessing historical data and environmental conditions, these models can help predict water quality trends, identify pollution sources, and guide decision-making to protect water resources and aquatic ecosystems (Alomani et al., 2022).

- **Groundwater Management:** Predictive modelling is essential for managing groundwater resources sustainably. Models can simulate groundwater flow, recharge rates, and contaminant transport to predict aquifer behaviour and assess the impact of pumping activities (Radhakrishnan & C A, 2023). Water managers can use these models to optimize groundwater extraction, prevent overexploitation, and protect groundwater quality.
- **Water Demand Forecasting:** Predictive models can forecast future water demand based on population growth, economic development, and other factors (Ristow et al., 2021). Predicting water demand patterns allows water utilities to plan infrastructure expenditures, improve water distribution networks, and provide reliable water supply to satisfy the needs of increasing communities.
- **Climate Change Impact Assessment:** Predictive modelling can help assess the potential impact of climate change on water resources by simulating future climate scenarios and predicting changes in precipitation patterns, temperature extremes, and hydrological processes (Schneider et al., 2023). These models enable water managers to adapt to changing conditions, develop resilience strategies, and mitigate the effects of climate change on water availability and quality.

The integration of AI into decision support systems for predictive modelling in water resources management represents a transformative innovation that has the potential to drive significant improvements in how we manage and protect our water resources. Through leveraging the power of artificial intelligence to analyse data, predict future trends, and optimize decision-making processes, we can enhance the efficiency, sustainability, and resilience of water management practices. As we continue to harness the capabilities of AI technology in this domain, we are poised to unlock new opportunities for achieving a more sustainable and secure water future for generations to come.

CHALLENGES AND OPPORTUNITIES

Decision support systems (DSS) play a crucial role in helping water resource managers make informed decisions based on available data and models. With the advancements in artificial intelligence (AI), there are both challenges and opportunities in integrating AI into DSS for predictive modeling in water resources management.

Limitations of this study include the lack of empirical validation of the proposed framework and the potential challenges in implementing AI technologies in real-world water management practices.

One of the main challenges of integrating AI into DSS for predictive modeling in water resources management is the complexity of the water system itself. Water systems are dynamic and influenced by a multitude of factors such as climate variability, land use changes, and human activities. AI algorithms need to be able to capture these complexities and uncertainties to provide accurate predictions. This requires a deep understanding of the underlying processes and interactions within the water system, as well as the ability to incorporate this knowledge into the AI models.

Another challenge is the availability and quality of data. AI algorithms rely on large amounts of data to learn patterns and make predictions. However, data in the water resources sector can be sparse, incomplete, or of varying quality. This can lead to biases and inaccuracies in the AI models, affecting the reliability of the predictions. Data collection and management are essential for successful integration of AI into DSS for water resources management.

Despite these challenges, there are also significant opportunities in integrating AI into DSS for predictive modeling in water resources management. AI has the potential to improve the accuracy and efficiency of predictions by analyzing large datasets and identifying patterns that may not be apparent to human analysts. This can help water resource managers make more informed decisions and optimize resource allocation.

AI can also enhance the adaptive capacity of DSS by continuously learning from new data and updating models in real-time. This can improve the resilience of water systems to changing conditions and uncertainties, such as climate change and extreme events. Implementing AI in DSS allows water resource managers to better anticipate and respond to future challenges, ensuring long-term water management.

Integrating AI into DSS for predictive modeling in water resources management presents both challenges and opportunities. Overcoming the challenges of complexity and data quality will require collaboration between water resource experts, data scientists, and AI specialists. Future research should focus on testing the effectiveness of the framework in practical applications.

CONCLUSION

In conclusion, the integration of artificial intelligence (AI) technologies in decision support systems for water resources management holds significant potential to enhance efficiency, accuracy, and sustainability in water management practices. Through a comprehensive literature review, key AI methods and tools have been identified for their applicability in addressing complex water resource challenges. The proposed framework developed in this study outlines a structured approach to leverage AI technologies such as machine learning, neural networks, and optimization algorithms for decision support in water resources management. Stakeholders may use AI capabilities to make better decisions, optimize resource allocation, and improve water quality and availability.

While the framework presents a promising pathway towards enhancing water management practices, there are challenges and limitations that need to be addressed. These include the need for empirical validation of the framework, overcoming implementation barriers, and ensuring ethical considerations related to data privacy and security. Future research should focus on testing the effectiveness of the proposed framework in real-world applications and exploring innovative ways to overcome the challenges associated with integrating AI technologies into water resources management. Water managers can better navigate the complicated field of water resource management and contribute to long-term water use by using AI-driven decision support tools.

REFERENCES

Alomani, S. M., Alhawiti, N. I., & Alhakamy, A. (2022). Prediction of Quality of Water According to a Random Forest Classifier. *International Journal of Advanced Computer Science and Applications*, *13*(6). doi:10.14569/IJACSA.2022.01306105

Cui, T., Pagendam, D., & Gilfedder, M. (2021). Gaussian process machine learning and Kriging for groundwater salinity interpolation. *Environmental Modelling & Software*, *144*, 105170. doi:10.1016/j.envsoft.2021.105170

Enayati, M., Bozorg-Haddad, O., Pourgholam-Amiji, M., Zolghadr-Asli, B., & Tahmasebi Nasab, M. (2022). Decision Tree (DT): A Valuable Tool for Water Resources Engineering (pp. 201–223). Springer. doi:10.1007/978-981-19-2519-1_10

EsfandiariM.JabariS.McGrathH.ColemanD. (2020). FLOOD MAPPING USING RANDOM FOREST AND IDENTIFYING THE ESSENTIAL CONDITIONING FACTORS; A CASE STUDY IN FREDERICTON, NEW BRUNSWICK, CANADA. *ISPRS Annals of the Photogrammetry, Remote Sensing and Spatial Information Sciences, V-3–2020*, 609–615. doi:10.5194/isprs-annals-V-3-2020-609-2020

Gonzales-Inca, C., Calle, M., Croghan, D., Torabi Haghighi, A., Marttila, H., Silander, J., & Alho, P. (2022). Geospatial Artificial Intelligence (GeoAI) in the Integrated Hydrological and Fluvial Systems Modeling: Review of Current Applications and Trends. *Water (Basel), 14*(14), 2211. doi:10.3390/w14142211

Guariso, G. (1988). Expert Systems in Water Resources. *IFAC Proceedings Volumes, 21*(17), 51–59. 10.1016/S1474-6670(17)54555-3

Gunasekaran, K., & Boopathi, S. (2023). *Artificial Intelligence in Water Treatments and Water Resource Assessments*. doi:10.4018/978-1-6684-6791-6.ch004

Han, D., Chan, L., & Zhu, N. (2007). Flood forecasting using support vector machines. *Journal of Hydroinformatics, 9*(4), 267–276. doi:10.2166/hydro.2007.027

Kamyab, H., Khademi, T., Chelliapan, S., SaberiKamarposhti, M., Rezania, S., Yusuf, M., Farajnezhad, M., Abbas, M., Hun Jeon, B., & Ahn, Y. (2023). The latest innovative avenues for the utilization of artificial Intelligence and big data analytics in water resource management. *Results in Engineering, 20*, 101566. doi:10.1016/j.rineng.2023.101566

Kaur, H., Alam, M. A., Mariyam, S., Alankar, B., Chauhan, R., Adnan, R. M., & Kisi, O. (2021). Predicting Water Availability in Water Bodies under the Influence of Precipitation and Water Management Actions Using VAR/VECM/LSTM. *Climate (Basel), 9*(9), 144. doi:10.3390/cli9090144

Lučin, I., Lučin, B., Čarija, Z., & Sikirica, A. (2021). Data-Driven Leak Localization in Urban Water Distribution Networks Using Big Data for Random Forest Classifier. *Mathematics, 9*(6), 672. doi:10.3390/math9060672

Mu, C., Hou, Y., Zhao, J., Wei, S., & Wu, Y. (2023). Stream-DBSCAN: A Streaming Distributed Clustering Model for Water Quality Monitoring. *Applied Sciences (Basel, Switzerland), 13*(9), 5408. doi:10.3390/app13095408

Mujumdar, P. P., & Ghosh, S. (2008). Fuzzy Logic-Based Approaches in Water Resource System Modelling. In *Practical Hydroinformatics* (pp. 165–176). Springer Berlin Heidelberg. doi:10.1007/978-3-540-79881-1_12

Radhakrishnan, R., & C A, L. D. (2023). Groundwater Level Prediction Using Support Vector Machine and M5 Model Tree-A Case Study. *SSRN Electronic Journal*. doi:10.2139/ssrn.4512253

Ristow, D. C. M., Henning, E., Kalbusch, A., & Petersen, C. E. (2021). Models for forecasting water demand using time series analysis: A case study in Southern Brazil. *Journal of Water, Sanitation, and Hygiene for Development : a Journal of the International Water Association*, *11*(2), 231–240. doi:10.2166/washdev.2021.208

Schneider, T., Behera, S., Boccaletti, G., Deser, C., Emanuel, K., Ferrari, R., Leung, L. R., Lin, N., Müller, T., Navarra, A., Ndiaye, O., Stuart, A., Tribbia, J., & Yamagata, T. (2023). Harnessing AI and computing to advance climate modelling and prediction. *Nature Climate Change*, *13*(9), 887–889. doi:10.1038/s41558-023-01769-3

So, J.-M., Lee, J.-H., & Bae, D.-H. (2020). Development of a Hydrological Drought Forecasting Model Using Weather Forecasting Data from GloSea5. *Water (Basel)*, *12*(10), 2785. doi:10.3390/w12102785

Sun, A. Y., & Scanlon, B. R. (2019). How can Big Data and machine learning benefit environment and water management: A survey of methods, applications, and future directions. *Environmental Research Letters*, *14*(7), 073001. doi:10.1088/1748-9326/ab1b7d

Sun, A. Y., Wang, D., & Xu, X. (2014). Monthly streamflow forecasting using Gaussian Process Regression. *Journal of Hydrology (Amsterdam)*, *511*, 72–81. doi:10.1016/j.jhydrol.2014.01.023

Vo-Van, T., Nguyen-Hai, A., Tat-Hong, M. V., & Nguyen-Trang, T. (2020). A New Clustering Algorithm and Its Application in Assessing the Quality of Underground Water. *Scientific Programming*, *2020*, 1–12. doi:10.1155/2020/6458576

Wang, G., Jia, Q.-S., Zhou, M., Bi, J., Qiao, J., & Abusorrah, A. (2022). Artificial neural networks for water quality soft-sensing in wastewater treatment: A review. *Artificial Intelligence Review*, *55*(1), 565–587. doi:10.1007/s10462-021-10038-8

Zeman, C., Rich, M., & Rose, J. (2006). World water resources: Trends, challenges, and solutions. *Reviews in Environmental Science and Biotechnology*, *5*(4), 333–346. doi:10.1007/s11157-005-4089-8

Chapter 16
Wise Agriculture for Sustainable Productivity and Healthy Ecosystems

Oumaima El Ghammad
Mohammed V University in Rabat, Morocco

Jamal Mabrouki
https://orcid.org/0000-0002-3841-7755
Mohammed V University in Rabat, Morocco

ABSTRACT

In the countryside of developing countries, the majority of people depend on agriculture. But economic crises, land maldistribution and climate concerns have hampered efforts to improve agricultural productivity and rural development. Technological advances have often pushed agriculture to a new stage, impacting plant diversity, harvests, and society. The idea of sustainable agriculture is gaining popularity because we want to reconcile productivity and environmental protection. Rich countries already adopted this concept in the 1990s. To generalize it, research is needed not only for large farms, but also for small farms in developing countries. Climate-smart agriculture preserves nature by using varied methods and planning well. The authors propose a new model based on data collection to better manage farms in a changing environment.

INTRODUCTION

The history of agriculture is marked by a series of revolutions which have considerably improved its productivity and yields. It all started with the agrarian revolution around 10,000 years ago, which allowed humanity to settle down and gave rise to the first societies and civilizations. Then, during the 1900s-1930s, mechanization came on the scene, making agricultural tasks much easier. Later, in the 1960s, the "green revolution" emerged, characterized by the development of resistant plant varieties and the

DOI: 10.4018/979-8-3693-3807-0.ch016

Copyright © 2024, IGI Global. Copying or distributing in print or electronic forms without written permission of IGI Global is prohibited.

use of agrochemicals, which led to a significant increase in food production. However, these advances have also had adverse consequences on human health and the environment.

More recently, we have seen the advent of the "digital agricultural revolution," an era where digital technologies are transforming agriculture. This revolution promises to sustain humanity and spur future prosperity (Smith and Jones, 2023).

Each agricultural revolution has brought its share of innovations and challenges. While mechanization made agricultural tasks more efficient, the green revolution increased yields but also posed health and environmental problems. The digital agricultural revolution appears to be the next step towards a more sustainable future for agriculture, harnessing digital technologies to solve modern challenges such as food security and environmental sustainability.

As part of the evolution towards agriculture 4.0, an advanced stage of agricultural technology, several crucial criteria are put forward. These include the need for climate-adaptive agriculture, long-term sustainable productivity through efficient crop and livestock practices, and optimal management of resources and nutrients. Moreover, this push towards digital agriculture offers varied prospects of ubiquitous connectivity and use of sovereign digital data (Gomez and Rodriguez, 2022).

Key research challenges include integrating diverse data sources, ranging from governance data to institutional data to end-user specific data, into a unified system. With the increasing integration of various sensors and digital platforms in agriculture, the diversity and volume of data continues to grow, opening new perspectives for analysis and innovation.

In many developing countries, agriculture is at the heart of progress, with small farmers playing a crucial role in global food production. For these farmers, monitoring and protecting their crops is a demanding task that takes up much of their time. This means they spend less time directly cultivating their land. Faced with this reality, it is increasingly evident that there is a need to adopt smarter agricultural practices to increase the quantity and quality of harvests while saving valuable resources such as labor and wastewater (Kumar and Singh, 2021).

A recent FAO study has highlighted the essential role of small farmers in global food production. Despite the modest size of their farms, they contribute significantly to feeding the world's population. However, it is alarming that so many people still suffer from malnutrition or deficiencies in essential nutrients. To reverse this trend, it is imperative to support these farmers by providing them with technical training on integrated agricultural methods, so that they can optimize their practices and guarantee food security for all (FAO, 2023).

Sustainability in agriculture represents a more circular and revitalizing model, capable of maintaining agricultural productivity indefinitely while taking an integrated approach to the preservation of natural resources and socio-economic equity. This sustainable approach emphasizes the use of favorable technologies, particularly favoring strategies that reduce dependence on non-renewable or environmentally harmful inputs. It encompasses various approaches such as eco-agriculture, permaculture, low-impact practices, and biodynamics, organic, ecological and environmentally friendly farming, as well as community-oriented and self-sufficiency methods.

Analytical skills, practical experience and a precise technological approach are essential to succeed in the field of sustainable agriculture. An integrated food system based on a digital chain depends on a deep understanding of ecosystems. Full adoption of advanced technologies such as Internet of Things, remote sensing, big data processing, and artificial intelligence and blockchain enables informed decisions at all levels of agriculture. This approach also accelerates industrial processes, reduces risks, provides real-time monitoring and optimizes costs, while preserving the quality of food products. To succeed in this complex system, it is essential to develop a new trans disciplinary understanding of the interactions between organisms and their environment (Gupta and Sharma, 2022).

The term "trans disciplinary" refers to ideas or methods that were originally developed in a specific field, but are now widely used in other disciplines. Figure 1 illustrates how a broad trans disciplinary approach can better organize data to improve understanding (Klien, 2004).

Advanced technologies offer new opportunities and significant benefits to developing countries, helping to reduce the use of agricultural inputs such as fertilizers, pesticides, energy and water, and associated environmental impacts. Renewable energy technologies and efficient energy conservation systems can drive technological progress and sustainable development.

The problem of post-harvest waste and crop losses constitutes a serious agricultural challenge, the scale of which will only increase in the years to come. Around a third of crops are abandoned without being consumed due to their short lifespan, thus causing significant losses. To mitigate this phenomenon, it is crucial to develop improved varieties of fruits, vegetables, herbs and ornamental plants adapted to supply chain constraints while ensuring optimal quality for consumers. This requires a holistic and collaborative approach, involving an integrated view both within disciplines, across disciplines, and beyond, in order to define strategies and encourage the adoption of best practices to promote food security and sustainable nutrition on a global scale (Godfray et al., 2010).

Agricultural practices are the largest consumers of water, but unfortunately, much of this valuable resource is wasted due to inefficient and unregulated use. Soil, this vital layer of the Earth, is essential to life as we know it. However, it is constantly threatened by pollution, which poses a danger to us humans and our environment. In arid and semi-arid regions, facing increasing water scarcity, recycling treated wastewater can offer a solution to improve soil fertility. However, it is crucial to note that using this wastewater can also pose risks. Indeed, effluents can contaminate soil and alter plant metabolism, thereby affecting groundwater quality and ultimately human health (Sato et al., 2013).

Climate change is disrupting weather and water patterns across the world, leading to water shortages in some regions and extreme weather events such as floods and cyclones in others. This situation, made worse by our current consumption habits, will only get worse. By 2025, two thirds of the world's population could suffer from water shortages, further endangering ecosystems (IPCC, 2014). Global warming will also affect growing seasons, altering photosynthetic processes and increasing the rate of respiration of crops, while disrupting essential bacterial populations in the soil. In addition, it alters the composition of nutrients, decreasing the effectiveness of fertilizers. At the same time, it intensifies soil evapotranspiration, thus depleting our natural resources. To counter these challenges, the circular economy is emerging as a promising solution, aiming to promote sustainable growth by maximizing the efficient use of resources and producing more with fewer inputs, in a more balanced system (Ellen MacArthur Foundation, 2015).

The fundamental objective of the research is to harness locally rooted knowledge to address the challenges facing agriculture, advocating sustainable intensification (SI). This approach is based on the optimal use of enabling technologies to manage aspects such as pests, nutrients, soil and water in an integrated and sustainable manner. Illuminatingly, we also explored how low-input regenerative agriculture, promoting the full participation of farmers at all stages of the development and expansion process, can help increase productivity. It is imperative to recognize that sustainable food systems, both in terms of production and consumption, cannot be ensured without considering the entire supply chain and the active engagement of all stakeholders at every stage. Developing digital chain technology, rooted in the value of a sustainable agri-food system, requires a deep understanding of ecosystems. Our ultimate goal is to build a paradigm based on spatio-temporal data to monitor, detect and guide agricultural ecosystems, improving market integration and productivity through technological advancements.

This text is structured into several distinct parts. It begins with an in-depth analysis of the relevant literature, highlighting the critical importance of the integrated agricultural approach to responding to global food security challenges. Next, the methodology used in this study is detailed. An in-depth exploration of the importance of sustainable agricultural intensification (SAI) is then presented. The results of the study and the resulting discussions are presented subsequently. Finally, the paper concludes by looking at future prospects.

Figure 1. From single-discipline to comprehensive transdisciplinary

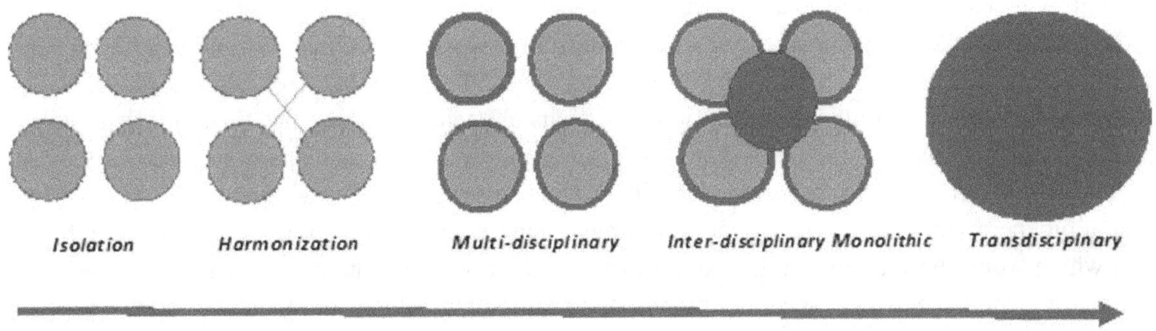

LITERATURE REVIEW

Sustainable agriculture can really help achieve many of the Sustainable Development Goals set by the United Nations for 2030. For example, it can help reduce poverty, ensure everyone has enough to eat, create decent jobs and grow the economy equitably. In addition, it can help reduce inequalities, encourage responsible consumption and production, and combat climate change. Many small farms, although often overlooked, play a crucial role in this sustainable agriculture. Despite predictions that these small farms will disappear, many still exist, and even their numbers are increasing in some areas (Lamb et al., 2021).

The current system that gets food from the farm to our plates is complicated. This is because the people who buy and sell the products, called brokers, are mostly concerned with their own profit, without really thinking about the farmers. Additionally, many farmers cannot read or write, which puts them at an even greater disadvantage.

To make the market more efficient and help farmers be more productive through technology, governments and international organizations should focus on two things: first, they should put in place rules to better control information about products. Next, they should help people, especially small farmers, to better understand and use computers and phones. This would allow farmers to be more included in the market and produce more (FAO, 2021).

It is crucial to develop environmentally friendly and economically viable agricultural practices, such as soil conservation, ecological agriculture, ecosystem management, small-scale irrigation, agroforestry and grazing control. In their study, Lal (2021) examined the benefits of leveraging big data and innovative technologies to make more informed decisions in agriculture. They also highlighted the importance of smart agricultural practices to improve the situation of the global food system.

Despite technological advances and the adoption of modern agricultural methods, persistent problems of environmental degradation remain. Developing nations already face stubborn challenges such as chronic poverty and food crises. Climate change is making this already precarious situation worse. According to the 2021 IPCC report, temperature variations in tropical areas generally have negative impacts on food production, with an estimate of food production in South Asia decreasing by around 30% by 2050 (Thornton and Herrero, 2021). Researchers Tukahirwa and Alavalapati, (2021) highlight that factors such as financing, training, expansion and access to markets have a positive effect on the adoption of climate-smart agricultural practices. A detailed study reveals that women farmers attach more importance to the nutritional benefits of food than their agricultural gains, thus pushing them to particularly focus on adopting smart agricultural practices to ensure a healthy diet.

Studies have looked at the impact of integrated farming systems on soil quality and residue management in combined rice crops. The results of the experiments indicated that to optimize nutrient utilization and promote soil fertility, an integration of various elements such as dairy farming, fish farming and poultry farming is necessary in agricultural systems in low-lying coastal areas altitude (Lopez-Carr et al., 2019). In addition, proposals for subsidies have been put forward, one aimed at farmers and the other at bioenergy producers. These programs aimed to increase the availability of biomass and bioenergy, which would have a positive impact on overall social welfare. These initiatives also considered the environmental benefits of bioenergy use for social welfare, as well as production uncertainties, risk aversion among farmers, and land capacity constraints in their analyzes (Altieri, 2018).

Soil is crucial to agriculture because it plays a central role in many processes essential to the agricultural ecosystem. Evaluating the health of the soil allows you to see if it is functioning well from a biological, chemical and physical point of view compared to its natural potential. Unfortunately, traditional soil testing methods are expensive, labor intensive, and can disturb the soil. This is where nanostructured sensors come into play. These small devices can help farmers quickly and accurately analyze soil water, nutrients, and moisture directly in the field, in addition to detecting plant pathogens (Antonacci et al., 2019).

To preserve biological diversity and improve the health of agricultural ecosystems, researchers have highlighted the importance of switching to organic farming and recommended a combined approach that brings local initiatives closer to government policies. Organic farming helps agricultural ecosystems be more resilient to climate change and can even reduce greenhouse gas emissions from agriculture (Rahman et al., 2019). To achieve better yields and soil quality, it is essential to 'have precise spatial data on the ground. This information is also crucial for combating global problems like soil erosion, malnutrition and desertification. Sensors used in smart agriculture are primarily used to directly measure soil chemical properties, such as pH, moisture, nutrients, and temperature. Sustainable integrated agriculture improves nutrient recycling and food production per unit of land and inputs, making the use of fertilizers and natural resources more efficient. The presence of animals and fish also alters the movement of nutrients through soil, plants and the atmosphere (Smith et al., 2020).

Some farmers choose to use organic fertilizers instead of chemical fertilizers. We look at different factors such as potential benefits, risks perceived by farmers, the environment and personal characteristics of farmers. Using ideas from psychology and behavior, researchers found that when farmers receive subsidies, they are more likely to opt for organic fertilizers. Thus, they suggest that the government should encourage farmers to join agricultural cooperation groups and receive organic fertilizer as an incentive, rather than money (Pretty, 2018).

Rural communities that have many strong social relationships have shown that they can better cope with food challenges during harsh weather conditions. Social capital, which represents these social connections, helps small farmers adapt to climate change by improving food security and nutrition. Additionally, being a member of local social groups, such as farmers' associations or religious groups, can help families access more financial resources (Huybrechts and Mondeleaers, 2021).

METHODOLOGY

The rapid expansion of urban areas is encroaching on land that could be used for agriculture or preserving biodiversity, reducing the space available to grow crops and raise livestock. Sustainable agricultural intensification aims to use environmentally friendly methods and technologies to increase food production without depleting natural resources. This approach meets the growing demand for food while helping to mitigate the effects of climate change and combat malnutrition.

After analyzing sustainable intensification (SI) in depth, several researchers have noticed that in the socio-scientific literature, there is a diversity of points of view and a transdisciplinary approach that predominates. This approach emphasizes the objectives to be achieved rather than the specific methods to be used. In other words, it does not dictate in advance which technologies or species combinations should be used. The idea behind the term "sustainable intensification" is that desired outcomes, such as increasing food production and improving the environmental sustainability of agroecosystems, can be achieved in different ways (Palosuo et al., 2021).

Sustainable agricultural systems tend to have a positive impact on the development of social, natural and human capital. In contrast, unsustainable systems that deplete these resources leave far fewer resources available for future generations. In summary, the emphasis is on finding agricultural methods that contribute to the long-term health of the environment and human communities, rather than approaches that deplete resources and compromise the future (Weltin et al., 2018).

AI-based applications are capable of analyzing a variety of data such as soil quality, weather conditions, crop status, and market trends. This up-to-date data is essential for making informed decisions regarding crop selection, optimal planting times, irrigation, fertilization and pest management.

Decision support systems, powered by machine learning algorithms, can provide instant recommendations based on the collected information. In summary, this approach uses data and technologies to optimize agricultural practices, improve productivity and minimize environmental impacts, while meeting consumer needs and preferences.

Figure 2. Utilizing intelligent sensing and monitoring to achieve strategic objectives

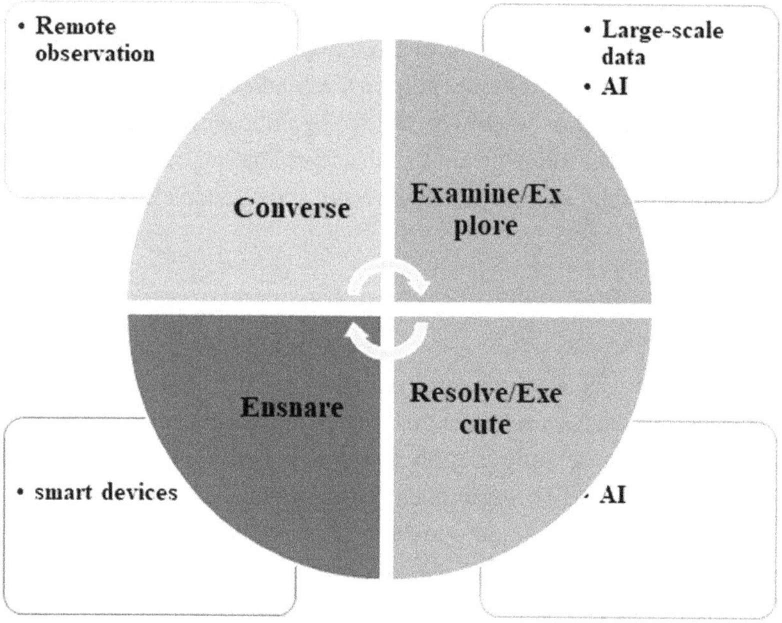

Technologies, ranging from efficient land management to climate information services, can only qualify as CSA if they manage to achieve the required goals in terms of food security and climate change adaptation and mitigation. To track food through the digital supply chain, it is necessary to have access to more than farm-to-table information, including data on shipping, packaging and local storage conditions up to 'to purchase by the consumer. This development promotes the transformation of fresh produce logistics into an intelligent cyber-physical system, driven by online monitoring and associated operational control, with the aim of improving freshness and food safety, reducing food waste, as well as improve the efficiency of transportation and distribution (Nantongo et al., 2018).

To meet this goal transparently, blockchain offers a decentralized, tamper-proof ledger that securely records all transactions and events along the supply chain. When it comes to sensing, monitoring and managing agricultural ecosystems through blockchain technology, it becomes more responsive as a means of confidential data storage. In recent times, the scope of blockchain's transformative potential as a tamper-proof ledger system has expanded, thereby providing a decentralized solution for ensuring privacy and security within large data systems (Azzi, 202).

Remote sensing involves using sensors installed on satellites, drones or planes to collect information about the Earth's surface. These sensors are very advanced and can record a wide variety of data, such as plant growth, temperature and precipitation, as well as other factors. This data is essential for understanding how agricultural fields and their environment are faring. Thanks to the increase in the use of wireless devices and technologies like artificial intelligence, researchers now have automated tools to develop agricultural systems that take all of this information into account (Neale and Maltese, 2019).

The modernization of agriculture, including the adoption of sustainable intensification technologies, will depend on ensuring public investments to support farmers, even in areas where yields are low. It has been noted that small farms have pushed farmers to favor more lucrative crops in order to increase their income. This is because the viability of primary agricultural production on these lands is uncertain, and off-farm opportunities are needed to meet financial needs. In contrast, large farms in sparsely populated regions have encouraged the adoption of mechanization to increase productivity while reducing reliance on labor (Wang et al., 2015).

Farmers engaged in collaborative learning gained a deep understanding of the connections between different facets of agriculture, leading to a more varied and intensive agricultural system. In addition to the distinctive testimonies of collaborative learning success stories, it has been noted that the current adjustment of capital resources and knowledge through an integrated approach to collaborative learning can effectively guide towards sustainable scale-up (Franzel, 2014).

SUSTAINABLE AGRICULTURAL INTENSIFICATION

Faced with many current challenges in agriculture, it is imperative to rapidly adopt climate-resilient agriculture on a global scale. This requires the implementation of appropriate policies, as well as widespread analysis and advocacy. It is clear that farmers' knowledge and adaptation to local resilience strategies are invaluable assets in steering modern agriculture towards a more sustainable and resilient development path. In this context, it is essential to explore different forms of agriculture, such as small-scale, organic or niche farming, and to develop methods aimed at preserving agricultural land, optimizing the use of available resources and to introduce new practices to promote sustainable agricultural production adapted to local physical, social and economic conditions (Newman, 2023).

Combined agriculture, also called agroecology, integrates different agricultural activities such as livestock breeding, poultry farming, floriculture, beekeeping and aquaculture. This model not only creates jobs, but also generates higher returns on investment throughout the year. For example, chickens can be raised in the upper area, where their droppings can be recycled. In the lower part, fish can be raised, and the remaining pond water can be reused for crop irrigation and fodder production. By taking into account both biosecurity and environmental concerns, different agricultural activities are closely connected in such a way that the by-products or wastes of one activity become valuable resources for another, thus ensuring efficient use optimal use of land and water resources on the farm. This approach results in diversified and maximum operational performance, while minimizing financial and labor costs (Altieri, 2018).

To solve the problem of losses and waste in horticulture, it is necessary to implement transdisciplinary solutions that integrate approaches from different fields such as engineering, biotechnology and socio-cultural sciences. Additionally, it is important to consider animal welfare in agricultural practices, particularly in grazing systems. Thus, the goal of climate-smart grazing practices is to maintain production levels while reducing herd sizes compared to well-organized and intensively farmed traditional pastures (Lipinski, 2013).

Studies have shown that rotational grazing can improve competitiveness by significantly reducing feed costs, leading to increased feed production and protein production, while reducing feed efficiency and crop intensity methane emissions. By adopting this approach, it is possible to optimize available resources and promote more sustainable and efficient food production (Souza Filho et al., 2019).

DISCUSSION

It is essential to recognize that despite the scale of manual labor required in agriculture, income from this activity remains very low, which constitutes a barrier for the people who depend on it. Thus, to ensure a prosperous future for small farmers, it is imperative to make them self-reliant through technical and professional training in integrated agriculture. It is equally crucial to provide similar training to other workers, especially women, in agriculture and related sectors, so that they can participate effectively in modern, sustainable and meaningful agricultural production. It is essential that the agricultural sector is prioritized in discussions on production and trade, ensuring that these activities take place in a fair and responsible manner (FAO, 2018).

Technological advances offer considerable potential to improve the management of our environment and stimulate a digitally inclusive and resource-efficient circular economy. The convergence of agricultural sensors, actuators, artificial intelligence, remote sensing and big data analytics via blockchain-integrated 5G networks is expected to increase the operational efficiency of the circular ecosystem. This could promote sustainable agriculture and resilient economic growth in the face of climate challenges. Potential applications of gene editing include creating drought-resistant crops and new antibiotics, using tools to insert, delete or modify an organism's genome. These technological advances, combined with a growing understanding of plant systems biology, will fundamentally transform how we breed new crop varieties. In the area of transparency and accountability in agricultural supply chains, blockchain offers the ability to ensure an immutable record from product origin to retail points of sale. This builds consumer confidence in the products they buy and gives producers the opportunity to be rewarded for adopting good agricultural practices. Ultimately, this would promote sustainable agricultural practices and responsible consumption (Nakagawa et al., 2019).

Agriculture is moving towards an increasingly knowledge-based and technologically advanced orientation. By combining data from various sources, we can obtain actionable insights. The fundamental elements of agriculture are soil and water. To promote integrated and sustainable agriculture, it is essential to adopt a cyclical and regenerative approach, using a diverse range of methods and technologies to manage pests, nutrients, soil and water synergistically. An appropriate combination of these elements guarantees satisfactory yields. However, due to poor agricultural practices, soil fertility declines, mainly due to inadequate nutrition compared to intensification of land exploitation (Tittonell and Giller, 2013).

Experienced smallholder farmers find themselves in a situation where they cannot fully exploit the benefits of agriculture due to their lack of technological skills and access to up-to-date information on government services. As a result, they are often forced to hand over their land to inexperienced people or rent it before moving from rural areas to cities. These new arrivals, unfamiliar with agricultural practices, resort to various chemical fertilizers and pesticides to increase productivity, leading to soil degradation, environmental pollution and risks to human health. This situation also contributes significantly to global warming. These challenges persist due to lack of agricultural knowledge adapted to climate change and limited use of information and communication technologies among farmers (Alston and Pardey, 2018).

The contemporary agricultural problem has two essential aspects. On the one hand, Sustainable Development Goal (SDG) 2.a.2 highlights the imperative to increase investments in various agricultural areas, such as rural infrastructure, technological development, agricultural research and agricultural services popularization. This initiative highlights the importance of international cooperation to strengthen these investments, particularly in developing and least developed countries, in order to strengthen their agricultural production capacity (Kumar et al., 2019).

On the other hand, there is the urban dimension of the agrarian crisis, which highlights the link between the degradation of agricultural and forest lands and migration to cities. This unplanned and unsustainable migration leads to adverse consequences such as increased pollution due to growing demand and expansion of services in urban areas. These developments are considered critical due to their potential to cause environmental deterioration and serious social problems (McGranahan et al., 2010).

CONCLUSION

Proposing a spatiotemporal data paradigm offers an approach to comprehensively and systematically assess the benefits and potential externalities of integrated agriculture. Small-scale farming plays a crucial role in local economic development and is closely linked to regional growth. Despite their importance as a pillar of agricultural activities, these farms often face challenges such as digital illiteracy among farmers, unsatisfactory levels of effectiveness and efficiency, lack of comprehensive integrated farming skills, sensitivity weather conditions, limited access to decentralized distributed chain technologies, inadequate infrastructure, as well as other external factors.

Therefore, it is imperative that international organizations, regional governments and policy makers place the highest priority on supporting small and marginal farmers, while encouraging the continuation of family farming to ensure that generations to come embrace this profession without hesitation. This will ensure that we only rely on agricultural production and not large factories. To achieve this, it is essential not only to increase agricultural production, but also to strengthen the resilience of rural populations by improving their living conditions. In terms of rural resilience, it is important to recognize that rurality and agriculture are closely linked, and any strategy to promote agriculture must take into account the multifunctional role it plays in rural reform.

If we are not all committed to preserving the environment and nature, we risk heading towards our own destruction. We will then be confronted with the devastating and disturbing consequences of nature with each change of season. Forests and the environment are closely linked to oxygen production. Trees and plants act as the green lungs of our planet, absorbing toxins from pollution and protecting us against many diseases. They are jewels of nature, producing oxygen and food necessary for our survival. Acting now to tackle the climate crisis is essential, because failure is simply not an option. The main barrier to overcome is that of will, not technology. Globally, strong and credible climate law is essential, encouraging climate action and supporting communities most affected by the impacts of climate change, particularly those historically most vulnerable.

Scientific transdisciplinarity must emanate from a broad understanding of diverse fields, including natural sciences, applied sciences, engineering, social sciences, landscape ecology, molecular biology and computational technology, reaching beyond the boundaries of traditional agronomy. Thus, throughout the development and expansion phases, low-input regenerative agriculture can improve productivity through farmers' knowledge and skills. However, reaching the level required by global and national food production systems only solves part of the food security challenge. It is essential to complement these advances with basic social, political, technical and economic conditions to ensure adequate access to nutrition and financial resources for all.

Ultimately, the integration of the aforementioned technologies within agricultural systems can equip relevant stakeholders with automated tools that improve decision-making, resource management and overall productivity. This fusion between technology and agriculture has the potential to drive the adoption of more sustainable and efficient agricultural practices, particularly through resource recycling. To achieve fundamental sustainability goals, it is imperative to foster the emergence of a transdisciplinary science deeply rooted in a broad and deep understanding of a broad spectrum of disciplines.

REFERENCES

Alston, J. M., & Pardey, P. G. (2018). Agriculture's role in meeting climate change challenges. *The Journal of Economic Perspectives*, *32*(4), 31–52.

Altieri, M. A. (2018). Agroecology and the design of climate change-resilient farming systems. *Agronomy for Sustainable Development*, *38*(4), 1–18.

Altieri, M. A. (2018). *Agroecology: The Science of Sustainable Agriculture*. CRC Press. doi:10.1201/9780429495465

Antonacci, P., Bertocchi, F., Coppola, G., & Fabbri, D. (2019). Nanostructured (bio)sensors for on-site detection of water, nutrients, and pathogens in soil: A review. *Journal of Agricultural and Food Chemistry*, *67*(8), 2073–2085.

Azzi, M. (2020). Blockchain technology in agriculture and food traceability. *Computers and Electronics in Agriculture*, *176*, 105681.

Ellen MacArthur Foundation. (2015). *Towards the Circular Economy: Accelerating the Scale-up Across Global Supply Chains*. Ellen MacArthur Foundation.

FAO. (2018). State of Food and Agriculture 2018: Migration, Agriculture and Rural Development. Food and Agriculture Organization of the United Nations.

FAO. (2021). *Digital Technologies in Agriculture and Rural Areas*. Food and Agriculture Organization of the United Nations.

FAO. (2023). *Smallholder Agriculture for Food Security: Investing in Sustainable Agricultural Practices*. Food and Agriculture Organization of the United Nations.

Franzel, S. (2014). Trees on farms: From opportunity costs to livelihood benefits. In J. C. Ingram, P. Ericksen, & D. Liverman (Eds.), *Food security and global environmental change* (pp. 265–272). Routledge.

Godfray, H. C. J., Beddington, J. R., Crute, I. R., Haddad, L., Lawrence, D., Muir, J. F., Pretty, J., Robinson, S., Thomas, S. M., & Toulmin, C. (2010). Food Security: The Challenge of Feeding 9 Billion People. *Science*, *327*(5967), 812–818. doi:10.1126/science.1185383 PMID:20110467

Gomez, M., & Rodriguez, A. (2022). Agriculture 4.0: A Comprehensive Review of Technological Advancements and Challenges. *Journal of Agricultural Engineering*, *10*(3), 45–62.

Gupta, S., & Sharma, A. (2022). Harnessing Advanced Technologies for Sustainable Agriculture: A Transdisciplinary Approach. *Journal of Sustainable Agriculture*, *18*(2), 215–230.

Huybrechts, B., & Mondelaers, K. (2021). Social capital and farmers' adaptive capacity to climate change: A systematic review. *Environmental Science & Policy*, *124*, 160–172.

IPCC (Intergovernmental Panel on Climate Change). (2014). *Climate Change 2014: Impacts, Adaptation, and Vulnerability. Contribution of Working Group II to the Fifth Assessment Report of the Intergovernmental Panel on Climate Change*. Cambridge University Press.

Klein, J. T. (2004). Prospects for Transdisciplinarity. *Futures, 36*(4), 515–526. doi:10.1016/j.futures.2003.10.007

Kumar, P., Prakash, A., & Bhartiya, A. (2019). Sustainable agriculture and its importance for food security. In Sustainable [). Springer, Cham.]. *Agricultural Reviews (Karnal), 37*, 53–68.

Kumar, S., & Singh, R. (2021). Smart Farming Solutions for Smallholder Farmers: Opportunities and Challenges. *International Journal of Agricultural and Biological Engineering, 14*(3), 78–94.

Lal, R. (2021). Soil health and climate change: Implications on agroecosystem services and food security. *Food and Energy Security, 10*(1), e252.

Lamb, A., Green, R., Bateman, I., Broadmeadow, M., Bruce, T., Burney, J., & Cisneros, E. (2021). The potential for land sparing to offset greenhouse gas emissions from agriculture. *Nature Sustainability, 4*(3), 192–199.

Lipinski, B., Hanson, C., Waite, R., Searchinger, T., & Lomax, J. (2013). *Reducing food loss and waste.*

López-Carr, D., Pricope, N. G., & Marlier, M. E. (2019). Integrated agriculture-aquaculture systems in coastal regions: Climate change mitigation and adaptation potential. *PLoS One, 14*(7), e0218986.

McGranahan, G., Tacoli, C., & Satterthwaite, D. (2010). Urbanization and its implications for food and farming. *Philosophical Transactions of the Royal Society of London. Series B, Biological Sciences, 365*(1554), 2809–2820. doi:10.1098/rstb.2010.0136 PMID:20713386

Nakagawa, M., Nakamura, Y., Honda, H., & Yamagishi, H. (2019). Blockchain technology: Principles and applications. *Journal of Cryptographic Engineering, 9*(2), 95–108.

Nantongo, H. (2023). Combining remote sensing and crop growth models to assess the impacts of climate change on maize yield in Uganda. *International Journal of Applied Earth Observation and Geoinformation, 98*, 102456.

Neale, C. M., & Maltese, A. (2019, December). Remote Sensing for Agriculture, Ecosystems, and Hydrology XXI. In *Proc. of SPIE Vol* (Vol. 11149, pp. 1114901-1).

Newman, J. (2023). *Sustainable Agriculture: Concepts, Practices, and Future Directions*. Springer.

Palosuo, T. (2021). Sustainable intensification in a northern European context: A review. *Agronomy for Sustainable Development, 41*(1), 1–17.

Pretty, J. (2018). Agricultural sustainability: Concepts, principles and evidence. *Philosophical Transactions of the Royal Society of London. Series B, Biological Sciences, 363*(1491), 447–465. doi:10.1098/rstb.2007.2163 PMID:17652074

Rahman, S., Rahman, K., & Rahman, M. (2019). Impact of organic farming on environmental sustainability. *Agriculture, Ecosystems & Environment, 284*, 106589.

Sato, T., Qadir, M., Yamamoto, S., Endo, T., & Zahoor, A. (2013). Global, Regional, and Country Level Need for Data on Wastewater Generation, Treatment, and Use. *Agricultural Water Management, 130*, 1–13. doi:10.1016/j.agwat.2013.08.007

Smith, P., Soussana, J.-F., Angers, D., Schipper, L., Chenu, C., Rasse, D. P., Batjes, N. H., van Egmond, F., McNeill, S., Kuhnert, M., Arias-Navarro, C., Olesen, J. E., Chirinda, N., Fornara, D., Wollenberg, E., Álvaro-Fuentes, J., Sanz-Cobena, A., & Klumpp, K. (2020). How to measure, report and verify soil carbon change to realize the potential of soil carbon sequestration for atmospheric greenhouse gas removal. *Global Change Biology*, *26*(1), 219–241. doi:10.1111/gcb.14815 PMID:31469216

Smith, T., & Jones, R. (2023). Digital Agriculture: Transforming the Future of Farming. *Agricultural Technology (Thailand)*, *15*(2), 87–104.

Souza Filho, W. (2019). Economic and Energetic Assessment of Integrated Crop-Livestock-Forest Systems Under Different Management Scenarios in the Brazilian Cerrado. *Agricultural Systems*, *170*, 31–42.

Thornton, P. K., & Herrero, M. (2021). Adaptation of food systems to climate change: Challenges and options. *Environmental Research Letters*, *16*(5), 053001.

Tittonell, P., & Giller, K. E. (2013). When yield gaps are poverty traps: The paradigm of ecological intensification in African smallholder agriculture. *Field Crops Research*, *143*, 76–90. doi:10.1016/j.fcr.2012.10.007

Tukahirwa, J. T., & Alavalapati, J. R. R. (2021). Women's involvement in climate-smart agriculture (CSA): A case study of the banana systems in Uganda. *Women's Studies International Forum*, *86*, 102487.

Wang, S. L., Heisey, P., Schimmelpfennig, D., & Ball, V. E. (2015). *Agricultural productivity growth in the United States: Measurement, trends, and drivers*. Economic Research Service, Paper No. Err-189.

Weltin, M. (2018). Sustainable intensification of agriculture: The role of grazing and forage in supporting food security. *Sustainability*, *10*(11), 4044.

Compilation of References

Abbot, J., & Marohasy, J. (2012). Application of Artificial Neural Networks to rainfall forecasting in Queensland, Australia. *Advances in Atmospheric Sciences*, *29*(4), 717–730. doi:10.1007/s00376-012-1259-9

Abdollahi, M., Zuki, A. B. Z., Goh, Y. M., Rezaeizadeh, A., & Noordin, M. M. (2010). The effects of Momordica charantia on the liver in streptozotocin-induced diabetes in neonatal rats. *African Journal of Biotechnology*, *9*(31), 5004–5012.

Abounahia, N. M., El-Sayed, A. M. A., Saleem, H., & Zaidi, S. J. (2023). An overview on the progress in produced water desalination by membrane-based technology. *Journal of Water Process Engineering*, *51*, 103479. doi:10.1016/j.jwpe.2022.103479

Abou-Shady, A., Siddique, M. S., & Yu, W. (2023). A Critical Review of Recent Progress in Global Water Reuse during 2019–2021 and Perspectives to Overcome Future Water Crisis. *Environments (Basel, Switzerland)*, *10*(9), 159. doi:10.3390/environments10090159

Adamchuk, V. I., & Hummel, J. W. (2016). Remote sensing and imaging sensors for site-specific management in precision agriculture. *Journal of Applied Remote Sensing*, *10*(1), 015009.

Adnan, R. (2016). *A 3-hours river water level rainfall prediction model using NNARX with improves modelling strategy*. 2021 6th International Conference on System Engineering and Technology (ICSET), Bandung. 10.1109/ICSEngT.2016.7849617

Aissaoui, A., Zizi, S., Israili, Z. H., & Lyoussi, B. (2011). Hypoglycemic and hypolipidemic effects of Coriandrum sativum L. in Meriones shawi rats. *Journal of Ethnopharmacology*, *137*(1), 652–661. doi:10.1016/j.jep.2011.06.019 PMID:21718774

Akhtar, N., Rehman, A., Hussnain, M., Rohail, S., Missen, M. S., Nasir, M., Hayder, A., Salamat, N., & Pasha, M. (2019). Hierarchical Coloured Petri-Net Based Multi-Agent System for Rainfall Monitoring, Prediction, and Rescue (FMPR). *IEEE Access : Practical Innovations, Open Solutions*, *7*, 180544–180557. doi:10.1109/ACCESS.2019.2958258

Alboaneen, D., Pranggono, B., Alshammari, D., Alqahtani, N., & Alyaffer, R. (2020). Predicting the epidemiological outbreak of the coronavirus disease 2019 (COVID-19) in Saudi Arabia. *International Journal of Environmental Research and Public Health*, *17*(4568), 4568. doi:10.3390/ijerph17124568 PMID:32630363

Algapani, D. E., Qiao, W., Su, M., di Pumpo, F., Wandera, S. M., Adani, F., & Dong, R. (2016). Bio-hydrolysis and biohydrogen production from food waste by thermophilic and hyperthermophilic anaerobic process. *Bioresource Technology*, *216*, 768–777. doi:10.1016/j.biortech.2016.06.016 PMID:27295255

Alimadadi, A., Aryal, S., Manandhar, I., Munroe, P. B., Joe, B., & Cheng, X. (2020). Artificial intelligence and machine learning to fight COVID-19. *Physiological Genomics*, *52*(4), 200–202. doi:10.1152/physiolgenomics.00029.2020 PMID:32216577

Aljohani, N. S., Kavil, Y. N., Al-Farawati, R. K., Aljohani, N. H., Orif, M. I., Ghandourah, M. A., Bahshwan, S. M., Aloufi, F. A., Halawani, R. F., & Salam, M. A. (2023). The assessment of environmental parameter along the desalination plants in the Kingdom of Saudi Arabia. *Open Chemistry*, *21*(1), 20220274. doi:10.1515/chem-2022-0274

Alleau, T., T Veziroglu, N., & Lequeux, G. (2000). *The hydrogen : A clean and durable energy; L'hydrogene : Une energie propre et durable.*

Allen, R. G., Pereira, L. S., Raes, D., & Smith, M. (1998). Crop evapotranspiration: Guidelines for computing crop water requirements. *FAO Irrigation and Drainage Paper*, 56.

Almgren, R., & Skobelev, D. (2020). Evolution of technology and technology governance. *Journal of Open Innovation*, *6*(2), 22. doi:10.3390/joitmc6020022

Alomani, S. M., Alhawiti, N. I., & Alhakamy, A. (2022). Prediction of Quality of Water According to a Random Forest Classifier. *International Journal of Advanced Computer Science and Applications*, *13*(6). doi:10.14569/IJACSA.2022.01306105

Alston, J. M., & Pardey, P. G. (2018). Agriculture's role in meeting climate change challenges. *The Journal of Economic Perspectives*, *32*(4), 31–52.

Altieri, M. A. (2018). Agroecology and the design of climate change-resilient farming systems. *Agronomy for Sustainable Development*, *38*(4), 1–18.

Altieri, M. A. (2018). *Agroecology: The Science of Sustainable Agriculture*. CRC Press. doi:10.1201/9780429495465

Alwashmi, M. F. (2020). The Use of Digital Health in the Detection and Management of COVID-19. *International Journal of Environmental Research and Public Health*, *17*(8), 2906. doi:10.3390/ijerph17082906 PMID:32340107

Antonacci, P., Bertocchi, F., Coppola, G., & Fabbri, D. (2019). Nanostructured (bio)sensors for on-site detection of water, nutrients, and pathogens in soil: A review. *Journal of Agricultural and Food Chemistry*, *67*(8), 2073–2085.

Appenroth, K.-J., Sree, K. S., Böhm, V., Hammann, S., Vetter, W., Leiterer, M., & Jahreis, G. (2017, February). Nutritional value of duckweeds (Lemnaceae) as human food. *Food Chemistry*, *217*, 266–273. doi:10.1016/j.foodchem.2016.08.116 PMID:27664634

Arenas-Sánchez, A., Rico, A., & Vighi, M. (2016). Effects of water scarcity and chemical pollution in aquatic ecosystems: State of the art. *The Science of the Total Environment*, *572*, 390–403. doi:10.1016/j.scitotenv.2016.07.211 PMID:27513735

Arık, S. O., Li, X., & Menon, V. (2021). A prospective evaluation of AI-augmented epidemiology to forecast COVID-19 in the USA and Japan. *NPJ Digital Medicine*, *4*(96), 146. doi:10.1038/s41746-021-00511-7 PMID:34625656

Arimbrathodi, S., Javed, M. A., Hamouda, M. A., Aly Hassan, A., & Ahmed, M. E. (2023). BioH2 Production Using Microalgae : Highlights on Recent Ad-vancements from a Bibliometric Analysis. *Water (Basel)*, *15*(1), 185. doi:10.3390/w15010185

Arslan, O., & Colkesen, I. (2021). Precision Farming Applications and Future Perspectives: A Review. *Computers and Electronics in Agriculture*, *190*, 106335.

Aruna, M. S., & Pradesh, M. (2012). *A Review: The application of Remote Sensing*. GIS and GPS in Precision Agriculture.

Asafo-Agyei, T., Appau, Y., Barimah, K. B., & Asase, A. (2023). Medicinal plants used for management of diabetes and hypertension in Ghana. *Heliyon*, *9*(12), e22977. doi:10.1016/j.heliyon.2023.e22977 PMID:38076168

Asseng, S., Martre, P., Maiorano, A., Rötter, R. P., O'Leary, G. J., Fitzgerald, G. J., Girousse, C., Motzo, R., Giunta, F., Babar, M. A., Reynolds, M. P., Kheir, A. M. S., Thorburn, P. J., Waha, K., Ruane, A. C., Aggarwal, P. K., Ahmed, M., Balkovič, J., Basso, B., & Ewert, F. (2019). Climate change impact and adaptation for wheat protein. *Global Change Biology*, *25*(1), 155–173. doi:10.1111/gcb.14481 PMID:30549200

Ayer, N. W. (2019). Life cycle assessment of a novel low-input agricultural system for nutrient management. *Journal of Cleaner Production*, *212*, 401–410.

Ayuso, M., Carpena, M., Taofiq, O., Albuquerque, T. G., Simal-Gandara, J., Oliveira, M. B. P. P., & Barros, L. (2022). Fig "Ficus carica L." and its by-products: A decade evidence of their health-promoting benefits towards the development of novel food formulations. *Trends in Food Science & Technology*, *127*, 1–13. doi:10.1016/j.tifs.2022.06.010

Ayyoubzadeh, S. M., Zahedi, H., Ahmadi, M., & Niakan Kalhori, R. S. (2020). Predicting COVID-19 incidence through analysis of Google Trends data in Iran: Data mining and deep learning pilot study. *JMIR Public Health and Surveillance*, *6*(2, e18828), e18828. doi:10.2196/18828 PMID:32234709

Azzi, M. (2020). Blockchain technology in agriculture and food traceability. *Computers and Electronics in Agriculture*, *176*, 105681.

Backer, J. A., Klinkenberg, D., & Wallinga, J. (2020). Incubation period of 2019 novel coronavirus (2019-nCoV) infections among travellers from Wuhan, China, 20–28 January 2020. *Eurosurveillance*, *25*(2000062). doi:10.2807/1560-7917.ES.2020.25.5.2000062 PMID:32046819

Badri, W., Dari, M. B., Agunaou, M., & Bounakhla, M. (2016). Atmospheric Pollution Research in Morocco: An Overview. *Journal of Materials & Environmental Sciences*, *7*(6), 2178–2185.

Barile, S., Orecchini, F., Saviano, M., & Farioli, F. (2018). People, technology, and governance for sustainability: The contribution of systems and cyber-systemic thinking. *Sustainability Science*, *13*(5), 1197–1208. doi:10.1007/s11625-018-0621-y

Barkaoui, M., Katiri, A., Boubaker, H., & Msanda, F. (2017). Ethnobotanical survey of medicinal plants used in the traditional treatment of diabetes in Chtouka Ait Baha and Tiznit (Western Anti-Atlas). *Journal of Ethnopharmacology*, *198*, 338–350. doi:10.1016/j.jep.2017.01.023 PMID:28109915

Barzman, M., Bàrberi, P., Birch, A. N. E., Boonekamp, P., Dachbrodt-Saaydeh, S., Graf, B., Hommel, B., Jensen, J. E., Kiss, J., Kudsk, P., Lamichhane, J. R., Messéan, A., Moonen, A.-C., Ratnadass, A., Ricci, P., Sarah, J.-L., & Sattin, M. (2015). Eight principles of integrated pest management. *Agronomy for Sustainable Development*, *35*(4), 1199–1215. doi:10.1007/s13593-015-0327-9

Bassem, S. M. (2020). *Water pollution and aquatic biodiversity*. https://doi.org/ doi:10.15406/bij.2020.04.00159

BatistaM. (2020). Estimation of the final size of the COVID-19 epidemic. medRxiv. doi:10.1101/2020.02.16.20023606

Bekkar, A., Hssina, B., Douzi, S., & Douzi, K. (2022). Air Quality Forecasting using decision trees algorithms. In *2022 2nd International Conference on Innovative Research in Applied Science, Engineering and Technology (IRASET)* (pp. 1-4). IEEE. 10.1109/IRASET52964.2022.9737814

Bekkar, A., Hssina, B., Douzi, S., & Douzi, K. (2023). *Transformer-Based Model for Multi-Horizon Forecasting Ozone in Marrakech city, Morocco.* 2023 14th International Conference on Intelligent Systems: Theories and Applications (SITA), Casablanca, Morocco. 10.1109/SITA60746.2023.10373756

Bekkar, A., Hssina, B., Douzi, S., & Douzi, K. (2022). Spatiotemporal Prediction of Concentrations Based on IoT Sensors. In Z. Boulouard, M. Ouaissa, M. Ouaissa, & S. El Himer (Eds.), *AI and IoT for Sustainable Development in Emerging Countries. Lecture Notes on Data Engineering and Communications Technologies* (Vol. 105). Springer. doi:10.1007/978-3-030-90618-4_10

Bennett, G., & Dixon, J. (Eds.). (2017). *Farming Systems and Food Security in Africa: Priorities for Science and Policy Under Global Change*. Routledge.

Berg, C. C. (2003). Flora Malesiana precursor for the treatment of Moraceae 1: the main subdivision of Ficus: the subgenera Blumea-Biodiversity. *Blumea*, *48*(1), 166–177. doi:10.3767/000651903X686132

Bernholt, S. (2021). Neglected and Underutilized Crop Species: A Multidimensional Analysis of Data Availability. *Sustainability, 13*(14), 7787.

Bhardwaj, A., Dagar, V., Khan, M. O., Aggarwal, A., Alvarado, R., Kumar, M., Irfan, M., & Proshad, R. (2022). Smart IoT and Machine Learning-based Framework for Water Quality Assessment and Device Component Monitoring. *Environmental Science and Pollution Research International*, *29*(30), 46018–46036. doi:10.1007/s11356-022-19014-3 PMID:35165843

Bhat, S., Kaushal, P., & Sharma, H. K. (2014). Coriander (*Coriandrum sativum* L.): Processing, nutritional and functional aspects. *African Journal of Plant Science*, *8*(1), 25–33. doi:10.5897/AJPS2013.1118

Bhattacharya, S., & Sen, S. (2022) *In West Bengal, ambitious efforts to plant mangroves yield limited results.* Scroll.in. https://scroll.in/article/1032297/in-west-bengal-ambitious-efforts-to-plant-mangroves-yield-limited-results

Bhatt, S., Gething, P. W., Brady, O. J., Messina, J. P., Farlow, A. W., Moyes, C. L., Drake, J. M., Brownstein, J. S., Hoen, A. G., Sankoh, O., Myers, M. F., George, D. B., Jaenisch, T., Wint, G. R. W., Simmons, C. P., Scott, T. W., Farrar, J. J., & Hay, S. I. (2013). The global distribution and burden of dengue. *Nature*, *496*(7446), 504–507. doi:10.1038/nature12060 PMID:23563266

Boni, G. (2016, June). A Prototype System for Rainfall Monitoring Based on Rainfall Forecast Combined With COSMO-SkyMed and Sentinel-1 Data. *IEEE Journal of Selected Topics in Applied Earth Observations and Remote Sensing*, *9*(6), 2794–2805. doi:10.1109/JSTARS.2016.2514402

Boote, K. J., Jones, J. W., White, J. W., Asseng, S., & Lizaso, J. I. (2019). Putting mechanisms into crop production models: Challenges and opportunities. *Agronomy Journal*, *111*(2), 535–555.

Botanical Online. (n.d.). *Cultivation of the Prickly Pear*. Botanical Onlin..

Bouma, J. (2016). Soil science contributions to precision agriculture. In *Soil Mapping and Process Modeling for Sustainable Land Use Management* (pp. 1–19). CRC Press.

Box, G. E. P., Jenkins, G. M., & Reinsel, G. C. (1994). *Time Series Analysis: Forecasting and Control* (3rd ed.). Prentice-Hall.

Breiman, L. (2001). Random Forests. *Machine Learning*, *45*(1), 5–32. doi:10.1023/A:1010933404324

Butcher, J., & Beridze, I. (2019). What is the state of artificial intelligence governance globally? *RUSI Journal*, *164*(5-6), 88–96. doi:10.1080/03071847.2019.1694260

Büyüksalih, G. (2021). Challenges and Opportunities of Geospatial Technologies in Agriculture. In Handbook of Research on Smart Technology Integration in the Agricultural Sector (pp. 1-16). IGI Global.

Cai, X., Wang, Y. F., Mao, F. F., Lan, Y. U., Liu, C. X., Zhang, H., & Wei, X. L. (2011). Hypoglycemic and hyperglycemia-prevention effects of crude tea flower polysaccharide. Modern [in Chinese]. *Food Science and Technology (Campinas), 27*(3), 262–266.

Calicioglu, O., & Brennan, R. A. (2018, June). Sequential ethanol fermentation and anaerobic digestion increases bioenergy yields from duckweed. *Bioresource Technology, 257*, 344–348. doi:10.1016/j.biortech.2018.02.053 PMID:29605287

Calzadilla, A., Rehdanz, K., & Tol, R. S. J. (2010). The economic impact of more sustainable water use in agriculture: A computable general equilibrium analysis. *Journal of Hydrology (Amsterdam), 384*(3–4), 292–305. doi:10.1016/j.jhydrol.2009.12.012

Carter, T. R., Jones, R. N., & Lu, X. (2007). Assessing Climate Change Impacts on Agriculture: The Value of Climate Analogues. In Ecosystems and Human Well-Being: Current State and Trends (pp. 253-263). Island Press.

Ceylan, Z. (2020). Estimation of COVID-19 prevalence in Italy, Spain, and France. *The Science of the Total Environment, 729*(138817), 138817. doi:10.1016/j.scitotenv.2020.138817 PMID:32360907

Chacko, S. M., Thambi, P. T., Kuttan, R., & Nishigaki, I. (2010). Beneficial effects of green tea: A literature review. *Chinese Medicine, 5*(1), 13. doi:10.1186/1749-8546-5-13 PMID:20370896

Chakraborti, S. (2020, October 16). New Town gets one-stop waste-to-wealth store: Kolkata news - times of India. *The Times of India*. https://timesofindia.indiatimes.com/city/kolkata/new-town-gets-one-stop-waste-to-wealth-store/articleshow/78689888.cms (Accessed: 10 June 2023).

Chakraborty, D., Ghosh, S., & Joshi, P. K. (2012). Application of geospatial technology in precision agriculture: A review. *Journal of Agricultural Physics, 12*(1), 1–15.

Chandolias, K., Pardaev, S., & Taherzadeh, M. J. (2016). Biohydrogen and carboxylic acids production from wheat straw hydrolysate. *Bioresource Technology, 216*, 1093–1097. doi:10.1016/j.biortech.2016.05.119 PMID:27268482

Chattopadhyay, S., & Bandyopadhyay, G. (2007). Artificial neural network with backpropagation learning to predict mean monthly total ozone in Arosa, Switzerland. *International Journal of Remote Sensing, 28*(20), 4471–4482. doi:10.1080/01431160701250440

Chen, C., Yang, Y., Graham, N. J., Li, Z., Yang, X., Wang, Z., Farhat, N., Vrouwenvelder, J. S., & Hou, L. (2024). A comprehensive evaluation of the temporal and spatial fouling characteristics of RO membranes in a full-scale seawater desalination plant. *Water Research, 249*, 120914. doi:10.1016/j.watres.2023.120914 PMID:38007899

Cheng, H.-H., Yu, W.-S., Tseng, S.-C., Wu, Y.-J., Hsieh, C.-L., Lin, S.-S., Chu, C.-P., Huang, Y.-D., Chen, W.-R., Lin, T.-F., & Whang, L.-M. (2023). Reclaimed water in Taiwan: Current status and future prospects. *Sustainable Environment Research, 33*(1), 16. doi:10.1186/s42834-023-00177-8

Cheng, J. J., & Stomp, A. (2009, January). Growing Duckweed to Recover Nutrients from Wastewaters and for Production of Fuel Ethanol and Animal Feed. *Clean (Weinheim), 37*(1), 17–26. doi:10.1002/clen.200800210

Chen, T., & Guestrin, C. (2016). XGBoost: A Scalable Tree Boosting System. *Proceedings of the 22nd ACM SIGKDD International Conference on Knowledge Discovery and Data Mining*. ACM. 10.1145/2939672.2939785

Chen, Z. (2020). Geospatial big data-driven smart agriculture: A review. *Computers and Electronics in Agriculture, 176*, 105612. doi:10.1016/j.compag.2020.105612

Chew, A. W. Z., Pan, Y., Wang, Y., & Zhang, L. (2021). Hybrid deep learning of social media big data for predicting the evolution of COVID-19 transmission. *Knowledge-Based Systems, 233*, 107417. doi:10.1016/j.knosys.2021.107417 PMID:34690447

Chisti, Y. (2008). Biodiesel from microalgae beats bioethanol. *Trends in Biotechnology*, *26*(3), 126–131. doi:10.1016/j.tibtech.2007.12.002 PMID:18221809

Chithra, V., & Leelamma, S. (1999). *Coriandrum sativum* — Mechanism of hypoglycemic action. *Food Chemistry*, *67*(3), 229–231. doi:10.1016/S0308-8146(99)00113-2

Chouit, E. M., Rachdi, M., Bellafkih, M., & Raouyane, B. (2023). *Morocco's COVID-19 instance as a potential epidemiological scenario*. 2023 3rd International Conference on Innovative Research in Applied Science, Engineering and Technology (IRASET), Mohammedia, Morocco. 10.1109/IRASET57153.2023.10152923

Chouit, E.M. (2022). Forecasting of the epidemiological situation: Case of COVID-19 in Morocco. *Journal of the Nigerian Society of Physical Sciences*.

Chouit, E. M., Rachdi, M., Bellafkih, M., & Raouyane, B. (2022). Forecasting of the epidemiological situation: Case of COVID-19 in Morocco. *Journal of the Nigerian Society of Physical Sciences*, *4*(4), 843. doi:10.46481/jnsps.2022.843

Christodoulou, M., Tchoumtchoua, J., Skaltsounis, A., Scorilas, A., & Halabalaki, M. (2019). Natural alkaloids intervening the insulin pathway: New hopes for anti-diabetic agents? *Current Medicinal Chemistry*, *26*(32), 5982–6015. doi:10.2174/0929867325666180430152618 PMID:29714135

Chukwuma, C. I., Mopuri, R., Nagiah, S., Chuturgoon, A. A., & Islam, M. S. (2017). Erythritol reduces small intestinal glucose absorption, increases muscle glucose uptake, improves glucose metabolic enzymes activities and increases expression of Glut-4 and IRS-1 in type 2 diabetic rats. *European Journal of Nutrition*. PMID:28770335

Chum, H. L., & Overend, R. P. (2001). Biomass and renewable fuels. *Fuel Processing Technology*, *71*(1-3), 187–195. doi:10.1016/S0378-3820(01)00146-1

Colt, J., Schuur, A. M., Weaver, D., & Semmens, K. (2022). Engineering design of aquaponics systems. *Reviews in Fisheries Science & Aquaculture*, *30*(1), 33–80. doi:10.1080/23308249.2021.1886240

Comrie, A. C. (1997). Comparing neural networks and regression models for ozone forecasting. *Journal of the Air & Waste Management Association*, *47*(6), 653–663. doi:10.1080/10473289.1997.10463925

Conn, S. J., Hocking, B., Dayod, M., Xu, B., Athman, A., Henderson, S., Aukett, L., Conn, V., Shearer, M. K., Fuentes, S., Tyerman, S. D., & Gilliham, M. (2013). Protocol: Optimising hydroponic growth systems for nutritional and physiological analysis of Arabidopsis thaliana and other plants. *Plant Methods*, *9*(1), 1–11. doi:10.1186/1746-4811-9-4 PMID:23379342

Correspondent, H. T. (2023, June 6). *West Bengal govt launches buses with air purifiers in Kolkata to beat pollution*. Hindustan Times. https://www.hindustantimes.com/cities/kolkata-news/west-bengal-govt-launches-buses-with-air-purifiers-in-kolkata-to-beat-pollution-101686042102914.html

Cortes, C., & Vapnik, V. (1995). Support-Vector Networks. *Machine Learning*, *20*(3), 273–297. doi:10.1007/BF00994018

Cui, T., Pagendam, D., & Gilfedder, M. (2021). Gaussian process machine learning and Kriging for groundwater salinity interpolation. *Environmental Modelling & Software*, *144*, 105170. doi:10.1016/j.envsoft.2021.105170

Dabas, M., Pandey, S. K., & Singh, S. (2020). Precision Farming Technologies for Sustainable Agriculture: A Review. *IOP Conference Series. Earth and Environmental Science*, *431*(1), 012057.

Darkrim-Lamari, F., & Malbrunot, P. (2006). *Combustible hydrogène-Production*.

Das, D., & Veziroglu, T. N. (2008). Advances in biological hydrogen production processes. *International Journal of Hydrogen Energy*, *33*(21), 6046–6057. doi:10.1016/j.ijhydene.2008.07.098

Davis, K. E., Kingsbury, B., & Merry, S. E. (2012). Indicators as a technology of global governance. *Law & Society Review*, *46*(1), 71–104. doi:10.1111/j.1540-5893.2012.00473.x

de Nicolás, A. P., Molina-García, Á., García-Bermejo, J. T., & Vera-García, F. (2023). Desalination, minimal and zero liquid discharge powered by renewable energy sources: Current status and future perspectives. *Renewable & Sustainable Energy Reviews*, *187*, 113733. doi:10.1016/j.rser.2023.113733

de Ruiter, H. (2018). The Potential of Multi-Criteria Decision Analysis for Evaluating Sustainable Agriculture at the Farm Level. *Sustainability 10*(11).

Debry, E., & Mallet, V. (2014). Ensemble forecasting with machine learning algorithms for ozone, nitrogen dioxide and PM10 on the Prev'Air platform. *Atmospheric Environment*, *91*, 71–84. doi:10.1016/j.atmosenv.2014.03.049

DeFelice, N. B., Little, E., Campbell, S. R., & Shaman, J. (2018). Ensemble forecast of human West Nile virus cases and mosquito infection rates. *Nature Communications*, *9*(1), 1–8. PMID:29317637

Demirci, U. B., & Miele, P. (2010). Les méthodes de production d'hydrogène sont-elles vertes. *L'Actualité Chimique*, *341*, 5–12.

Deshpande, R. R. (2012). On the rainfall time series prediction using Multilayer Perceptron Artificial Neural Network. *International Journal of Emerging Technology and Advanced Engineering*, *2*(1), 148–153.

Dhakal, N., Salinas-Rodriguez, S. G., Hamdani, J., Abushaban, A., Sawalha, H., Schippers, J. C., & Kennedy, M. D. (2022). Is Desalination a Solution to Freshwater Scarcity in Developing Countries? *Membranes*, *12*(4), 381. doi:10.3390/membranes12040381 PMID:35448351

Dias Canedo, E., Morais do Vale, A. P., Patrão, R. L., Camargo de Souza, L., Machado Gravina, R., & Eloy dos Reis, V., & T. de Sousa Jr, R. (2020). Information and Communication Technology (ICT) Governance Processes: A Case Study. *Information (Basel)*, *11*(10), 462. doi:10.3390/info11100462

Dineshkumar, B., Mitra, A., & Manjunatha, M. (2010). A comparative study of alpha amylase inhibitory activities of common anti-diabetic plants at Kharagpur 1 block Int. *J. Green Pharm*, *4*(2).

Doorn, N. (2021). Artificial intelligence in the water domain: Opportunities for responsible use. *Science of The Total Environment*, *755*, 142561. doi:10.1016/j.scitotenv.2020.142561

Drobenko, B. (2021). *Introduction au droit de l'eau*. Johanet.

Duxbury, C. L., Dixon, D. G., & Greenberg, B. M. (1997, August). Effects of simulated solar radiation on the bioaccumulation of polycyclic aromatic hydrocarbons by the duckweed Lemna gibba. *Environmental Toxicology and Chemistry*, *16*(8), 1739–1748. doi:10.1002/etc.5620160824

Eakin, H. (2021). The Global South's Contribution to Global Crop Diversity and the Role of International Agricultural Research. *Environmental Science & Policy*, *118*, 236–243.

Ebert, A. W. (2022). Neglected and underutilized crop species: An overview on prospects for improving nutrition and food security. *Sustainability*, *14*.

Eidi, M., Eidi, A., Saeidi, A., Molanaei, S., Sadeghipour, A., Bahar, M., & Bahar, K. (2009). Effect of coriander seed (Coriandrum sativum L.) ethanol extract on insulin release from pancreatic beta cells in streptozotocin-induced diabetic rats. *Phytotherapy Research*, *23*(3), 404–406. doi:10.1002/ptr.2642 PMID:19003941

Ekins, P., & Hughes, N. (2009). The prospects for a hydrogen economy (1) : Hydrogen futures. *Technology Analysis and Strategic Management*, *21*(7), 783–803. doi:10.1080/09537320903182264

Ellen MacArthur Foundation. (2015). *Towards the Circular Economy: Accelerating the Scale-up Across Global Supply Chains*. Ellen MacArthur Foundation.

El-Moustaqim, K., El Bakraoui, H., Mabrouki, J., Fouad, Y., Slaoui, M., Hmouni, D., Benyeogor, M. S., & Igbigbi, T. L. (2023). Combination of Microalgae Method, Decantation, and Filtration for Domestic Wastewater Treatment. *Sustainability (Basel), 15*(22), 22. doi:10.3390/su152216110

El-Moustaqim, K., Mabrouki, J., & Azrour, M. (2023). *Monitoring of Water Toxicity Through the Internet of Things to Protect the Health of the Population.*, doi:10.1201/9781003430735-7

El-Moustaqim, K., Mabrouki, J., & Hmouni, D. (2024). *Applied Sciences Green Microalgae for Future Biomass Development.*, doi:10.1007/978-3-031-50860-8_12

Elmulthum, N. A., Zeineldin, F. I., Al-Khateeb, S. A., Al-Barrak, K. M., Mohammed, T. A., Sattar, M. N., & Mohmand, A. S. (2023). Water Use Efficiency and Economic Evaluation of the Hydroponic versus Conventional Cultivation Systems for Green Fodder Production in Saudi Arabia. *Sustainability (Basel), 15*(1), 822. doi:10.3390/su15010822

Elshafai, S., Elgohary, F., Nasr, F., Petervandersteen, N., & Gijzen, H. (2007, March). Nutrient recovery from domestic wastewater using a UASB-duckweed ponds system. *Bioresource Technology, 98*(4), 798–807. doi:10.1016/j.biortech.2006.03.011 PMID:16713255

Eltoukhy, A. E. E., Shaban, I. A., Chan, F. T. S., & Abdel-Aal, M. A. M. (2020). *Data analytics for predicting COVID-19 cases in top affected countries: Observations and recommendations*. Research in Public Health.

Enayati, M., Bozorg-Haddad, O., Pourgholam-Amiji, M., Zolghadr-Asli, B., & Tahmasebi Nasab, M. (2022). Decision Tree (DT): A Valuable Tool for Water Resources Engineering (pp. 201–223). doi:10.1007/978-981-19-2519-1_10

Encinar, J. M., Beltrán, F. J., Bernalte, A., Ramiro, A., & González, J. F. (1996). Pyrolysis of two agricultural residues : Olive and grape bagasse. Influence of particle size and temperature. *Biomass and Bioenergy, 11*(5), 397–409. doi:10.1016/S0961-9534(96)00029-3

EsfandiariM.JabariS.McGrathH.ColemanD. (2020). FLOOD MAPPING USING RANDOM FOREST AND IDENTIFYING THE ESSENTIAL CONDITIONING FACTORS; A CASE STUDY IN FREDERICTON, NEW BRUNSWICK, CANADA. *ISPRS Annals of the Photogrammetry, Remote Sensing and Spatial Information Sciences, V-3–2020*, 609–615. doi:10.5194/isprs-annals-V-3-2020-609-2020

Express News Service. (2023, May 16). *Experts brainstorm on strategies to improve air quality in Delhi.* The Indian Express. https://www.newindianexpress.com/cities/delhi/2023/May/15/experts-brainstorm-on-strategies-to-improve-air-quality-in-delhi-2575552.html

Facebook, Inc. (2017). *Prophet: Automatic forecasting procedure [Computer software].* GitHub repository. https://github.com/facebook/prophet

FAO. (2017). *Land-use planning: Making sustainable land-use decisions at the local level.* Food and Agriculture Organization of the United Nations.

FAO. (2018). State of Food and Agriculture 2018: Migration, Agriculture and Rural Development. Food and Agriculture Organization of the United Nations.

FAO. (2021). *Digital Technologies in Agriculture and Rural Areas.* Food and Agriculture Organization of the United Nations.

FAO. (2021). *Sustainable Crop Production.* FAO Agricultural Development Economics Briefs.

FAO. (2023). *Smallholder Agriculture for Food Security: Investing in Sustainable Agricultural Practices*. Food and Agriculture Organization of the United Nations.

Faraji, M., & Saidi, M. (2021). Hydrogen-rich syngas production via integrated configuration of pyrolysis and air gasification processes of various algal biomass : Process simulation and evaluation using Aspen Plus software. *International Journal of Hydrogen Energy*, *46*(36), 18844–18856. doi:10.1016/j.ijhydene.2021.03.047

Farzaei, F., Morovati, M. R., Farjadmand, F., & Farzaei, M. H. (2017). A mechanistic review on medicinal plants used for diabetes mellitus in traditional Persian medicine. *Journal of Evidence-Based Complementary & Alternative Medicine*, *22*(4), 944–955. doi:10.1177/2156587216686461 PMID:29228789

Fathima, S., & Hossain, M. E. (2021). Application of XgBoost algorithm for epidemic outbreak prediction. *Journal of Healthcare Engineering*, 2021.

Fathima, T., & Sudarson, J. (2020). Analysis of dengue infection based on Raman spectroscopy and machine learning. *IEEE Access : Practical Innovations, Open Solutions*, *8*, 76577–76585.

Ferrari, E., Verda, D., Pinna, N., & Muselli, M. (2023). Optimizing Water Distribution through Explainable AI and Rule-Based Control. *Computers*, *12*(6), 123. doi:10.3390/computers12060123

Finger, M., & Pécoud, G. (2003). From e-Government to e-Governance? Towards a model of e-Governance. *Electronic. Journal of E-Government*, *1*(1), 52–62.

Foley, J. A. (2021). Our share of the planetary pie. *Proceedings of the National Academy of Sciences of the United States of America*, *118*, 26. PMID:17646656

Food and Agriculture Organization of the United Nations (FAO). (2021). *FAO*. Ecology, Cultivation and Uses of the Prickly Pear.

Food and Agriculture Organization of the United Nations. (2022). *The 2022 Statistical Yearbook of the FAO*. FAO. www.fao.org

Ford, J. D., Berrang-Ford, L., Lesnikowski, A., Barrera, M., & Heymann, J. (2013). How to track adaptation to climate change: A typology of approaches for national-level application. *Ecology and Society*, *18*(3), 40. doi:10.5751/ES-05732-180340

Fotheringham, A. S., Brunsdon, C., & Charlton, M. (2017). *Geographically Weighted Regression: The Analysis of Spatially Varying Relationships*. Wiley.

Franzel, S. (2014). Trees on farms: From opportunity costs to livelihood benefits. In J. C. Ingram, P. Ericksen, & D. Liverman (Eds.), *Food security and global environmental change* (pp. 265–272). Routledge.

Frelat, R., Lopez-Ridaura, S., Giller, K. E., Herrero, M., Douxchamps, S., Djurfeldt, A. A., Erenstein, O., Henderson, B., Kassie, M., Paul, B. K., Rigolot, C., Ritzema, R. S., Rodriguez, D., van Asten, P. J. A., & van Wijk, M. T. (2016). Drivers of household food availability in sub-Saharan Africa based on big data from small farms. *Proceedings of the National Academy of Sciences of the United States of America*, *113*(2), 458–463. doi:10.1073/pnas.1518384112 PMID:26712016

Fu, L., Ding, Z., Kumpeangkeaw, A., Tan, D., Han, B., Sun, X., & Zhang, J. (2020, January). De novo assembly, transcriptome characterization, and simple sequence repeat marker development in duckweed Lemna gibba. *Physiology and Molecular Biology of Plants*, *26*(1), 133–142. doi:10.1007/s12298-019-00726-9 PMID:32158126

Furtak, K., & Wolińska, A. (2023). The impact of extreme weather events as a consequence of climate change on the soil moisture and on the quality of the soil environment and agriculture – A review. *Catena*, *231*, 107378. doi:10.1016/j.catena.2023.107378

Gade, D. S. (2021). Reinventing Smart Water Management System through ICT and IoT Driven Solution for Smart Cities. *International Journal of Applied Engineering and Management Letters*, 132–151. doi:10.47992/IJAEML.2581.7000.0109

Gao, F., Luo, Y., & Zhao, C. (2023). Effects of Climate and Land-Use Change on the Supply and Demand Relationship of Water Provision Services in the Yellow River Basin. *Land (Basel)*, *12*(12), 2089. doi:10.3390/land12122089

García, S., Luengo, J., & Herrera, F. (2015). *Data Preprocessing in Data Mining*. Springer. doi:10.1007/978-3-319-10247-4

Gardner, M. W., & Dorling, S. R. (1999). Neural network modelling and prediction of hourly NOx and NO2 concentrations in urban air in London. *Atmospheric Environment*, *33*(5), 709–719. doi:10.1016/S1352-2310(98)00230-1

Garnier-Crussard, A., Forestier, E., Gilbert, T., & Krolak-Salmon, P. (2020). Novel coronavirus (COVID-19) epidemic: What are the risks for older patients? *Journal of the American Geriatrics Society*, *68*(939–940), 939–940. doi:10.1111/jgs.16407 PMID:32162679

Gastón, M. S., Cid, M. P., Vázquez, A. M., Decarlini, M. F., Demmel, I. G., Rossi, L. I., Aimar, M. L., & Salvatierra, N. A. (2016). Sedative Effect of Central Administration of Coriandrum Sativum Essential Oil and Its Major Component Linalool in Neonatal Chicks. *Pharmaceutical Biology*, *54*(10), 1954–1961. doi:10.3109/13880209.2015.1137602 PMID:26911626

Gebbers, R., & Adamchuk, V. I. (2010). Precision agriculture and food security. *Science*, *327*(5967), 828–831. doi:10.1126/science.1183899 PMID:20150492

Geetha & Selvaraj. (2011). Prediction of monthly rainfall in Chennai using Back Propagation Neural Network model. *Int. J. of Eng. Sci. and Technology, 3*(1), 211-213.

Geng, X., & Rasmussen, J. (2021). Overcoming barriers to precision agriculture adoption: A systematic review. *Agricultural Systems*, *188*, 103015.

Ge, Y., & Thomasson, J. A. (2017). *Geospatial techniques for managing agriculture and natural resources*. CRC Press.

Goddek, S., Delaide, B., Mankasingh, U., Ragnarsdottir, K. V., Jijakli, H., & Thorarinsdottir, R. (2015). Challenges of sustainable and commercial aquaponics. *Sustainability (Basel)*, *7*(4), 4199–4224. doi:10.3390/su7044199

Godfray, H. C. J., Beddington, J. R., Crute, I. R., Haddad, L., Lawrence, D., Muir, J. F., Pretty, J., Robinson, S., Thomas, S. M., & Toulmin, C. (2010). Food Security: The Challenge of Feeding 9 Billion People. *Science*, *327*(5967), 812–818. doi:10.1126/science.1185383 PMID:20110467

Goh, S. Y., & Cooper, M. E. (2008). Clinical review: The role of advanced glycation end products in progression and complications of diabetes. *The Journal of Clinical Endocrinology and Metabolism*, *93*(4), 1143–1152. doi:10.1210/jc.2007-1817 PMID:18182449

Gomez, M., & Rodriguez, A. (2022). Agriculture 4.0: A Comprehensive Review of Technological Advancements and Challenges. *Journal of Agricultural Engineering*, *10*(3), 45–62.

Gonzales-Inca, C., Calle, M., Croghan, D., Torabi Haghighi, A., Marttila, H., Silander, J., & Alho, P. (2022). Geospatial Artificial Intelligence (GeoAI) in the Integrated Hydrological and Fluvial Systems Modeling: Review of Current Applications and Trends. *Water (Basel)*, *14*(14), 2211. doi:10.3390/w14142211

Goria, K., Singh, H. M., Singh, A., Kothari, R., & Tyagi, V. V. (2024). Insights into biohydrogen production from algal biomass : Challenges, recent advancements and future directions. *International Journal of Hydrogen Energy*, *52*, 127–151. doi:10.1016/j.ijhydene.2023.03.174

Gourchane F. (2023). Bioadsorption of CI Reactive Blue 203 dye by duckweed Lemna gibba's powder. *Moroccan Journal of Chemistry*. doi:10.48317/IMIST.PRSM/morjchem-v11i3.40166

Gourchane, F., Al-Jadabi, N., Benrahmane, L., & El Alami, M. (2022). A.bouyahya, M.laaouan, A. EL hourch « Influence of climatic conditions and aquatic environment on the growth of Lemna gibba in the pilot lagoon treatment plant of Bouregreg RABAT-Morocco ». *International Journal of Chemical and Biochemical Science*, 176–182. https://www.iscientific.org/volume-22-2022/

Grafton, R. Q., Pittock, J., Davis, R., Williams, J., Fu, G., Warburton, M., Udall, B., McKenzie, R., Yu, X., Che, N., Connell, D., Jiang, Q., Kompas, T., Lynch, A., Norris, R., Possingham, H., & Quiggin, J. (2013). Global insights into water resources, climate change and governance. *Nature Climate Change*, *3*(4), 315–321. doi:10.1038/nclimate1746

Granovskii, M., Dincer, I., & Rosen, M. A. (2007). Air pollution reduction via use of green energy sources for electricity and hydrogen production. *Atmospheric Environment*, *41*(8), 1777–1783. doi:10.1016/j.atmosenv.2006.10.023

Grieve, M. (1971). *A modern herbal Dover Publications.*

Groot, R. S. (2021). Neglected and Underutilized Crop Species: The Key to Sustainable Food and Nutrition Security in a Changing World? *Sustainability, 19*, 10777.

Guariso, G. (1988). Expert Systems in Water Resources. *IFAC Proceedings Volumes, 21*(17), 51–59. 10.1016/S1474-6670(17)54555-3

Guhathakurta, P. (2008). Long lead monsoon rainfall prediction using deterministic Artificial Neural Network model. *Meteorology and Atmospheric Physics*, *101*, 93–108. doi:10.1007/s00703-008-0335-2

Gunasekaran, K., & Boopathi, S. (2023). *Artificial Intelligence in Water Treatments and Water Resource Assessments.*, doi:10.4018/978-1-6684-6791-6.ch004

Gunda, N. S., Gautam, S., & Mitra, S. (2018). Artificial Intelligence for Water Quality Monitoring. *ECS Meeting Abstracts, MA2018-02*(56), 1997–1997. 10.1149/MA2018-02/56/1997

Gupta, U. (2023, June 15). India, France discuss ISA priorities for accelerating global energy transition. *PV Magazine*. https://www.pv-magazine-india.com/2023/06/15/india-france-discuss-isa-priorities-for-accelerating-global-energy-transition/

Gupta, S. V. K., & Suhas. (2009). Application of low-cost adsorbents for dye removal – A review. *Journal of Environmental Management*, *90*(8), 2313–2342. doi:10.1016/j.jenvman.2008.11.017

Gupta, S., & Sharma, A. (2022). Harnessing Advanced Technologies for Sustainable Agriculture: A Transdisciplinary Approach. *Journal of Sustainable Agriculture*, *18*(2), 215–230.

Hajam, T. A., & Saleem, H. (2022). Phytochemistry, biological activities, industrial and traditional uses of fig (*Ficus carica*): A review. *Chemico-Biological Interactions*, *368*, 110237. doi:10.1016/j.cbi.2022.110237 PMID:36288779

Hajiya Hasan, A. H., Al-Bader, D. A., Woodward, S., Peters, A. F., & Küpper, F. C. (2023). Ecophysiology of Kuwaiti macroalgae with special emphasis on temperature and salinity tolerance related to the conditions at desalination plant outfalls. *Botanica Marina*, *66*(5), 373–390. doi:10.1515/bot-2022-0049

Han, Q., Yu, Q.Y., Shi, J., Xiong, C.Y., Ling, Z.J., & He, P.M. (2011). Molecular characterization and hypoglycemic activity of a novel water-soluble polysaccharide from tea (Camellia sinensis) flower. *Carbohydrate Polymers, 86* (2), 797-805.

Han, D., Chan, L., & Zhu, N. (2007). Flood forecasting using support vector machines. *Journal of Hydroinformatics*, *9*(4), 267–276. doi:10.2166/hydro.2007.027

Haq, S., Boz, I., & Shahbaz, P. (2021). Adoption of climate-smart agriculture practices and differentiated nutritional outcome among rural households: A case of Punjab province, Pakistan. *Food Security, 13*(4), 913–931. doi:10.1007/s12571-021-01161-z

Hartemink, A. E. (2008). Geostatistics and soil mapping. *Geoderma, 146*(1-2), 90–99.

Hasnaoui, S. (2021). *Optimisation de la production du bio-hydrogène par Spirulina sp via un photo-bioréacteur électrochimique (EPBR) : Application au traitement des eaux de lac de Réghaia.*

Hattink, J., Wolterbeek, H. Th., & De Goeij, J. J. M. (2001, May). Influence of salinity and eutrophication on bioaccumulation of 99 technetium in duckweed. *Environmental Toxicology and Chemistry, 20*(5), 996–1002. doi:10.1002/etc.5620200509 PMID:11337889

Hemschemeier, A., Melis, A., & Happe, T. (2009). Analytical approaches to photobiological hydrogen production in unicellular green algae. *Photosynthesis Research, 102*(2-3), 523–540. doi:10.1007/s11120-009-9415-5 PMID:19291418

Hengl, T., & Heuvelink, G. B. M. (2013). Soil mapping and soil monitoring: A review. *International Journal of Applied Earth Observation and Geoinformation, 23*, 1–14.

Hii, Y. L., Zhu, H., Ng, N., Ng, L. C., & Rocklöv, J. (2012). Forecast of dengue incidence using temperature and rainfall. *PLoS Neglected Tropical Diseases, 6*(11), e1908. doi:10.1371/journal.pntd.0001908 PMID:23209852

Hiroi & Kawaguchi. (2016). RainfallEye: Real-time flash rainfall prediction system for urban complex water flow. 2021 IEEE SENSORS, 1-3. . doi:10.1109/ICSENS.2016.7808626

Hoffmann, H., Zhao, G., Asseng, S., Bindi, M., Biernath, C., Constantin, J., Coucheney, E., Dechow, R., Doro, L., Eckersten, H., Gaiser, T., Grosz, B., Heinlein, F., Kassie, B. T., Kersebaum, K.-C., Klein, C., Kuhnert, M., Lewan, E., Moriondo, M., & Ewert, F. (2016). Impact of spatial soil and climate input data aggregation on regional yield simulations. *PLoS One, 11*(8), e0151782. doi:10.1371/journal.pone.0151782 PMID:27055028

Hoi, K. I., Yuen, K. V., & Mok, K. M. (2014). Prediction of daily maximum ozone concentrations by seasonally integrated time series models. *Atmospheric Environment, 89*, 168–175.

Holzworth, D. P., Huth, N. I., deVoil, P. G., Zurcher, E. J., Herrmann, N. I., McLean, G., & Snow, V. O. (2014). APSIM - Evolution towards a new generation of agricultural systems simulation. *Environmental Modelling & Software, 62*, 327–350. doi:10.1016/j.envsoft.2014.07.009

Ho, S.-H., Huang, S.-W., Chen, C.-Y., Hasunuma, T., Kondo, A., & Chang, J.-S. (2013). Bioethanol production using carbohydrate-rich microalgae biomass as feedstock. *Bioresource Technology, 135*, 191–198. doi:10.1016/j.biortech.2012.10.015 PMID:23116819

Htike, K. K., & Khalifa, O. O. (2010). Rainfall forecasting models using Focused Time-Delay Neural Networks. Comput. and Commun. Eng. (ICCCE), Int. Conf. on IEEE. 10.1109/ICCCE.2010.5556806

Huang, J., Yang, X., Rozelle, S., Li, N., & Zhang, J. (2010). Irrigation, agricultural performance and poverty reduction in China. *Food Policy, 35*(6), 455–464.

Hütten, M. (2019). The soft spot of hard code: Blockchain technology, network governance and pitfalls of technological utopianism. *Global Networks, 19*(3), 329–348. doi:10.1111/glob.12217

Hu, Y., Sun, X., Nie, X., Li, Y., & Liu, L. (2019). An enhanced LSTM for trend following of time series. *IEEE Access : Practical Innovations, Open Solutions, 7*, 34020–34030. doi:10.1109/ACCESS.2019.2896621

Huybrechts, B., & Mondelaers, K. (2021). Social capital and farmers' adaptive capacity to climate change: A systematic review. *Environmental Science & Policy*, *124*, 160–172.

Hu, Z., Gao, F., Qin, L., Yang, Y., & Xu, H. (2019). A case-control study on risk factors and their Interactions with prediabetes among the elderly in rural communities of yiyang city, hunan province. *Journal of Diabetes Research*, *2019*, 1386048. doi:10.1155/2019/1386048 PMID:30911549

Ienca, M., & Vayena, E. (2020). On the responsible use of digital data to tackle the COVID-19 pandemic. *Nature Medicine*, *26*(4), 463–464. doi:10.1038/s41591-020-0832-5 PMID:32284619

Ingrao, C., Strippoli, R., Lagioia, G., & Huisingh, D. (2023). Water scarcity in agriculture: An overview of causes, impacts and approaches for reducing the risks. *Heliyon*, *9*(8), e18507. doi:10.1016/j.heliyon.2023.e18507 PMID:37534016

IPCC. (2014). *Climate Change 2014: Impacts, Adaptation, and Vulnerability. Contribution of Working Group II to the Fifth Assessment Report of the Intergovernmental Panel on Climate Change*. Cambridge University Press.

Islam, S. M. R., Kwak, D., Kabir, M. H., Hossain, M., & Kwak, K.-S. (2015). The Internet of Things for Health Care: A Comprehensive Survey. *IEEE Access : Practical Innovations, Open Solutions*, *3*, 678–708. doi:10.1109/ACCESS.2015.2437951

Ivorra, B., Ferrández, M. R., Vela-Pérez, M., & Ramos, A. M. (2020). Mathematical modeling of the spread of the coronavirus disease 2019 (COVID-19) taking into account the undetected infections. The case of China. *Communications in Nonlinear Science and Numerical Simulation*, *88*(105303), 105303. doi:10.1016/j.cnsns.2020.105303 PMID:32355435

Janssen, S., Porter, C. H., Moore, A. D., Athanasiadis, I. N., Foster, I., Jones, J. W., & Antle, J. M. (2017). Towards a new generation of agricultural system data, models and knowledge products: Design and improvement. *Agricultural Systems*, *155*, 255–268. doi:10.1016/j.agsy.2016.09.017 PMID:28701817

Javed, M. A., Alyammahi, S. M. A., Alshehhi, F. A. A., Aljneibi, S. A. A., Alketbi, S. A. O., & Aly Hassan, A. (2024). A novel two-stage immobilized bioreactor for biohydrogen production using a partial microalgal-bacterial (*Chlorella vulgaris* and wastewater activated sludge) co-culture. *Sustainable Energy Technologies and Assessments*, *62*, 103624. doi:10.1016/j.seta.2024.103624

Jayawardena, A. W. (n.d.). *CHALLENGES FOR SUSTAINABLE WATER MANAGEMENT*. BBC. http://news.bbc.co.uk/2/hi/science/nature/7329799.stm

Jiao, H., Tsigkou, K., Elsamahy, T., Pispas, K., Sun, J., Manthos, G., Schagerl, M., Sventzouri, E., Al-Tohamy, R., Kornaros, M., & Ali, S. S. (2024). Recent advances in sustainable hydrogen production from microalgae : Mechanisms, challenges, and future perspectives. *Ecotoxicology and Environmental Safety*, *270*, 115908. doi:10.1016/j.ecoenv.2023.115908 PMID:38171102

John Deere. (2021). *Precision Agriculture: The Future of Farming*. John Deere. https://www.deere.com/en/technology-products/precision-agriculture/

Johns Hopkins, C. S. S. E. (2020). *2019 Novel Coronavirus COVID-19 (2019-nCoV)*. Data Repository by Johns Hopkins CSSE. https://github.com/CSSEGISandData/COVID-19

Johnson, L. R. (2021). Forecasting West Nile virus outbreaks with decision tree-based models. *Diseases (Basel, Switzerland)*, *9*(2), 41. PMID:34200868

Jones, J. W., Antle, J. M., Basso, B., Boote, K. J., Conant, R. T., Foster, I., & Porter, C. H. (2017). Toward a new generation of agricultural system data, models, and knowledge products: State of agricultural systems science. *Agricultural Systems*, *155*, 269–288. doi:10.1016/j.agsy.2016.09.021 PMID:28701818

Juiz, C., Guerrero, C., & Lera, I. (2014). *Implementing good governance principles for the public sector in information technology governance frameworks.* Open Journal of Accounting. doi:10.4236/ojacct.2014.31003

Jumin, E., Zaini, N., Ahmed, A. N., Abdullah, S., Ismail, M., Sherif, M., Sefelnasr, A., & El-Shafie, A. (2020). Machine learning versus linear regression modelling approach for accurate ozone concentrations prediction. *Engineering Applications of Computational Fluid Mechanics*, *14*(1), 713–725. doi:10.1080/19942060.2020.1758792

Kalogiannidis, S., Kalfas, D., Giannarakis, G., & Paschalidou, M. (2023). Integration of Water Resources Management Strategies in Land Use Planning towards Environmental Conservation. *Sustainability (Basel)*, *15*(21), 15242. doi:10.3390/su152115242

Kamyab, H., Khademi, T., Chelliapan, S., SaberiKamarposhti, M., Rezania, S., Yusuf, M., Farajnezhad, M., Abbas, M., Hun Jeon, B., & Ahn, Y. (2023). The latest innovative avenues for the utilization of artificial Intelligence and big data analytics in water resource management. *Results in Engineering*, *20*, 101566. doi:10.1016/j.rineng.2023.101566

Kannan & Ghosh. (2010). Prediction of daily rainfall state in a river basin using statistical downscaling from GCM output. Springer-Verlag.

Kannan, M., Prabhakaran, S., & Ramachandran, P. (2010). Rainfall Forecasting Using Data Mining Technique. *IACSIT International Journal of Engineering and Technology*, *2*(6), 397–401.

Kaur, H. (2020). Challenges and opportunities in the use of remote sensing and geospatial technology for sustainable agriculture. *Journal of Spatial Science*, *65*(2), 241–263.

Kaur, H., Alam, M. A., Mariyam, S., Alankar, B., Chauhan, R., Adnan, R. M., & Kisi, O. (2021). Predicting Water Availability in Water Bodies under the Influence of Precipitation and Water Management Actions Using VAR/VECM/LSTM. *Climate (Basel)*, *9*(9), 144. doi:10.3390/cli9090144

Khajah, M., Ahmed, M.E., & Al-Matouq, A. (2023). *Characterization of reverse osmosis reject wastewater generated from Sulaibiya Wastewater Treatment Plant.*

Khalil, S., & Belitski, M. (2020). Dynamic capabilities for firm performance under the information technology governance framework. *European Business Review*, *32*(2), 129–157. doi:10.1108/EBR-05-2018-0102

Khan, M. S. (2021). Role of remote sensing in sustainable water resources management in agriculture: A review. *Agricultural Water Management*, *248*, 106770.

Khan, Z. H., Bhat, S. A., Khare, D., & Wani, S. H. (2020). Precision Farming in Agriculture: A Review. *Journal of Pure & Applied Microbiology*, *14*(3), 1925–1937.

Khosla, R., & Reich, R. M. (2016). The role of geospatial techniques in precision agriculture. In M. Gebbers & A. Adamchuk (Eds.), *Precision agriculture basics* (pp. 17–29). Springer.

Kim, D., Park, J., Han, H., Lee, H., Kim, H. S., & Kim, S. (2023). Application of AI-Based Models for Flood Water Level Forecasting and Flood Risk Classification. *KSCE Journal of Civil Engineering*, *27*(7), 3163–3174. doi:10.1007/s12205-023-2175-5

Klein, J. T. (2004). Prospects for Transdisciplinarity. *Futures*, *36*(4), 515–526. doi:10.1016/j.futures.2003.10.007

Kocaman, S., Demircioglu, M. B., & Yilmaz, E. (2019). The use of geospatial techniques in climate change research. *Journal of Environmental Science and Engineering B*, *8*(6), 173–185. doi:10.17265/2162-5263/2019.06.002

Kohavi, R. (1995). A study of cross-validation and bootstrap for accuracy estimation and model selection. *IJCAI (United States)*, *14*(2), 1137–1145.

Kolesnik, A. A., Kakhniashvili, T. A., Zherebin, Y. L., Golubev, V. N., & Pilipenko, L. N. (1986). Lipids of the fruit of Ficus carica. *Chemistry of Natural Compounds*, *22*(4), 394–397. doi:10.1007/BF00579808

Korak, J. A., Mungan, A. L., & Watts, L. T. (2023). Critical review of waste brine management strategies for drinking water treatment using strong base ion exchange. *Journal of Hazardous Materials*, *441*, 129473. doi:10.1016/j.jhazmat.2022.129473

Kumar, M. (2022). State Pollution Control Boards in India neither have enough staff nor expertise. *Scroll.in*. https://scroll.in/article/1036752/state-pollution-control-boards-in-india-neither-have-enough-staff-nor-expertise

Kumar, A., Kumar, A., Ranjan, R., & Kumar, S. (2012). *A rainfall prediction model using artificial neural network. In Control and Syst. Graduate Research Colloq*. ICSGRC.

Kumar, B. M. (2018). Agroforestry: A pathway to sustainable land use and climate resilient farming. *Current Opinion in Environmental Science & Health*, *5*, 106–113.

Kumar, G., Mudhoo, A., Sivagurunathan, P., Nagarajan, D., Ghimire, A., Lay, C.-H., Lin, C.-Y., Lee, D.-J., & Chang, J.-S. (2016). Recent insights into the cell immobilization technology applied for dark fermentative hydrogen production. *Bioresource Technology*, *219*, 725–737. doi:10.1016/j.biortech.2016.08.065 PMID:27561626

Kumari, U., Swamy, K., Gupta, A., Karri, R. R., & Meikap, B. C. (2021). Global water challenge and future perspective. In *Green Technologies for the Defluoridation of Water* (pp. 197–212). Elsevier. doi:10.1016/B978-0-323-85768-0.00002-6

Kumar, P., Morawska, L., Martani, C., Biskos, G., Neophytou, M., Di Sabatino, S., & Norford, L. (2019). The rise of low-cost sensing for managing air pollution in cities. *Environment International*, *75*, 199–205. doi:10.1016/j.envint.2014.11.019 PMID:25483836

Kumar, P., Prakash, A., & Bhartiya, A. (2019). Sustainable agriculture and its importance for food security. In Sustainable []. Springer, Cham.]. *Agricultural Reviews (Karnal)*, *37*, 53–68.

Kumar, S., & Singh, R. (2021). Smart Farming Solutions for Smallholder Farmers: Opportunities and Challenges. *International Journal of Agricultural and Biological Engineering*, *14*(3), 78–94.

Kutcher, H. R. (2020). The Role of Alternative Crops in the Long-term Resilience of Prairie Agricultural Systems: Implications for Food Security. *Frontiers in Sustainable Food Systems*, *4*, 146.

Lal, N. (2023, January 17). *Mumbai slum residents stand up against Adani's redevelopment plan*. Nikkei Asia. https://asia.nikkei.com/Spotlight/Asia-Insight/Mumbai-slum-residents-stand-up-against-Adani-s-redevelopment-plan

Lalmuanawma, S., Hussain, J., & Chhakchhuak, L. (2020). Applications of machine learning and artificial intelligence for Covid-19 (SARS-CoV-2) pandemic: A review. *Chaos, Solitons, and Fractals*, *139*, 110059. doi:10.1016/j.chaos.2020.110059 PMID:32834612

Lal, R. (2021). Soil health and climate change: Implications on agroecosystem services and food security. *Food and Energy Security*, *10*(1), e252.

Lamb, A., Green, R., Bateman, I., Broadmeadow, M., Bruce, T., Burney, J., & Cisneros, E. (2021). The potential for land sparing to offset greenhouse gas emissions from agriculture. *Nature Sustainability*, *4*(3), 192–199.

Langa, M. (2022, October 20). PM Modi launches Mission LiFE in the presence of U.N. Secretary General. *The Hindu*. https://www.thehindu.com/news/national/pm-modi-launch-mission-life-presence-u-n-secretary-general-antonio-guterres/article66035847.ece

Laribi, B., Kouki, K., M'Hamdi, M., & Bettaieb, T. (2015). Coriander (Coriandrum sativum L.) and its bioactive constituents. *Fitoterapia*, *103*, 9–26. doi:10.1016/j.fitote.2015.03.012 PMID:25776008

Lauer, S. A., Grantz, K. H., Bi, Q., Jones, F. K., Zheng, Q., Meredith, H. R., Azman, A. S., Reich, N. G., & Lessler, J. (2020). The incubation period of coronavirus disease 2019 (COVID-19) from publicly reported confirmed cases: Estimation and application. *Annals of Internal Medicine*, *172*(577–582), 577–582. doi:10.7326/M20-0504 PMID:32150748

LeCun, Y., Bengio, Y., & Hinton, G. (2015). Deep learning. *Nature*, *521*(7553), 436–444. doi:10.1038/nature14539 PMID:26017442

Lee, C., Ho, H., Chen, W., & Iizuka, A. (2024). Total Resource Circulation of Desalination Brine: A Review. *Advanced Sustainable Systems*, 2300460. doi:10.1002/adsu.202300460

Lee, H., & Tan, T. P. (2016). Singapore's experience with reclaimed water: NEWater. *International Journal of Water Resources Development*, *32*(4), 611–621. doi:10.1080/07900627.2015.1120188

Leong, Y. K., Su, R.-H., Varjani, S., Lee, D.-J., & Chang, J.-S. (2024). Integrated biohydrogen production and dairy manure wastewater treatment via a microalgae platform. *International Journal of Hydrogen Energy*, *52*, 404–417. doi:10.1016/j.ijhydene.2023.06.308

León, L. F. A., & Rosen, J. (2020). Technology as ideology in urban governance. *Annals of the American Association of Geographers*, *110*(2), 497–506. doi:10.1080/24694452.2019.1660139

Liang, S. (2020). Applications of remote sensing in precision agriculture: A review. *The Science of the Total Environment*, 142063.

Li, C., & Fang, H. H. P. (2007). Fermentative Hydrogen Production From Wastewater and Solid Wastes by Mixed Cultures. *Critical Reviews in Environmental Science and Technology*, *37*(1), 1–39. doi:10.1080/10643380600729071

Li, C., Peng, J., Wang, H., & Yang, S. X. (2017). A rainfall prediction method based on streaming big data processing. *2022 IEEE International Conference on Information and Automation (ICIA)*, 898-902. 10.1109/ICInfA.2017.8079030

Li, J., Xu, Q., Cuomo, R., Purushothaman, V., & Mackey, T. (2020). Data mining and content analysis of the Chinese social media platform Weibo during the early COVID-19 outbreak: Retrospective observational infoveillance study. *JMIR Public Health and Surveillance*, *6*(2, e18700), e18700. doi:10.2196/18700 PMID:32293582

Limpisathian, P. (2011). *Geographic Information System in Agriculture and Precision Farming*. Academic Press.

Linden, M. & van der, (2024). *Environmental effects of brine disposal and seawater usage for offshore green hydrogen production and storage in the Dutch North Sea.*

Li, P., Ma, F., & Cheng, L. (2013). Primary and secondary metabolism in the sun-exposed peel and the shaded peel of apple fruit. *Physiologia Plantarum*, *148*(1), 9–24. doi:10.1111/j.1399-3054.2012.01692.x PMID:22989296

Lipinski, B., Hanson, C., Waite, R., Searchinger, T., & Lomax, J. (2013). *Reducing food loss and waste.*

Lipper, L. (2020). Climate-smart agriculture for food security. *Nature Climate Change*, *10*(12), 1074–1076. PMID:33262808

Lipsitch, M., Finelli, L., Heffernan, R. T., Leung, G. M., & Redd, S. C. (2016). Improving the evidence base for decision making during a pandemic: The example of 2009 influenza A/H1N1. *Biosecurity and Bioterrorism*, *9*(2), 89–115. PMID:21612363

Li, S., Li, F., & Ho, S.-H. (2021). Biohydrogen production from microalgae for environmental sustainability. *Chemosphere*, *291*, 132717. doi:10.1016/j.chemosphere.2021.132717 PMID:34757051

Liu, B. (2016). Statistical models for yield prediction in agriculture: Opportunities and challenges. *Journal of Agricultural Biological & Environmental Statistics*, *21*(3), 361–380.

Lobell, D. B. (2014). Remote sensing and geospatial technologies for agriculture. In G. Backhaus & R. J. Hussey (Eds.), *The Routledge handbook of environment and communication* (pp. 250–258). Routledge.

Lobell, D. B. (2019). The use of satellite data for crop yield gap analysis. *Field Crops Research*, *237*, 104–110.

López-Carr, D., Pricope, N. G., & Marlier, M. E. (2019). Integrated agriculture-aquaculture systems in coastal regions: Climate change mitigation and adaptation potential. *PLoS One*, *14*(7), e0218986.

Los Angeles Department of Water and Power in partnership with TreePeople. (2015). *Stormwater Capture Master Plan August 2015*. Los Angeles Department of Water and Power in Partnership with TreePeople.

Lu, B., & Xie, W. (2020). Geospatial technology applications in autonomous farming: A review. *Computers and Electronics in Agriculture*, *175*, 105565.

Lučin, I., Lučin, B., Čarija, Z., & Sikirica, A. (2021). Data-Driven Leak Localization in Urban Water Distribution Networks Using Big Data for Random Forest Classifier. *Mathematics*, *9*(6), 672. doi:10.3390/math9060672

Lu, F. S., Hattab, M. W., Clemente, C. L., Biggerstaff, M., & Santillana, M. (2019). Improved state-level influenza nowcasting in the United States leveraging Internet-based data and network approaches. *Nature Communications*, *10*(1), 1–10. doi:10.1038/s41467-018-08082-0 PMID:30635558

Lu, X., Zhang, X., Li, J., & Zhang, H. (2021). Application of geospatial techniques in climate change research: A review. *Remote Sensing*, *13*(3), 541. doi:10.3390/rs13030541

Luzi, F., Del Buono, D., Orfei, B., Moretti, C., Buonaurio, R., Torre, L., & Puglia, D. (2022, June). Lemna minor aqueous extract as a natural ingredient incorporated in poly (vinyl alcohol)-based films for active food packaging systems. *Food Packaging and Shelf Life*, *32*, 100822. doi:10.1016/j.fpsl.2022.100822

Malinga, R., & Totolo, O. (2017). The role of land use planning in achieving sustainable land management in rural Botswana. *Journal of Environmental Planning and Management*, *60*(10), 1861–1881.

Maneeruttanarungroj, C., Lindblad, P., & Incharoensakdi, A. (2010). A newly isolated green alga, *Tetraspora* sp. CU2551, from Thailand with efficient hydrogen production. *International Journal of Hydrogen Energy*, *35*(24), 13193–13199. doi:10.1016/j.ijhydene.2010.08.096

Manson, N. C., & O'Neill, O. (2007). *Rethinking Informed Consent in Bioethics*. Cambridge University Press. doi:10.1017/CBO9780511814600

Maoka, T. (2020). Carotenoids as natural functional pigments. *Journal of Natural Medicines*, *74*(1), 1–16. doi:10.1007/s11418-019-01364-x PMID:31588965

Marbán, G., & Valdés-Solís, T. (2007). Towards the hydrogen economy? *International Journal of Hydrogen Energy*, *32*(12), 1625–1637. doi:10.1016/j.ijhydene.2006.12.017

Martin-Carrasco, F., Garrote, L., Iglesias, A., & Mediero, L. (2013). Diagnosing Causes of Water Scarcity in Complex Water Resources Systems and Identifying Risk Management Actions. *Water Resources Management*, *27*(6), 1693–1705. doi:10.1007/s11269-012-0081-6

Marwat, S. K., Khan, M. A., Khan, M. A., Ahmad, M., Zafar, M., Rehman, F., & Sultana, S. (2009). Fruit plant species mentioned in the Holy Qura'n and Ahadith and their ethno medicinal importance. *American-Eurasian Journal of Agricultural & Environmental Sciences*, *5*, 284–295.

Matondo, J. I. (2002). A comparison between conventional and integrated water resources planning and management. *Physics and Chemistry of the Earth, Parts A/B/C, 27*(11–22), 831–838. doi:10.1016/S1474-7065(02)00072-4

Matson, P. A. (2021). Agriculture's contribution to biodiversity conservation. *Science, 371*(6531), 239–240.

Matsuda, H., Hamao, M., Nakamura, S., Kon'i, H., Murata, M., & Yoshikawa, M. (2012). Medicinal Flowers. XXXIII. Anti-hyperlipidemic and anti-hyperglycemic effects of chakasaponins I-III and structure of chakasaponin IV from flower buds of Chinese tea plant (Camellia sinensis). *Chemical & Pharmaceutical Bulletin, 60*(5), 674–680. doi:10.1248/cpb.60.674 PMID:22689407

Mbow, C. (2019). Agroforestry solutions to address food security and climate change challenges in Africa. *Current Opinion in Environmental Sustainability, 41*, 61–67.

McBratney, A. B., Mendonça Santos, M. L., & Minasny, B. (2003). On digital soil mapping. *Geoderma, 117*(1-2), 3–52. doi:10.1016/S0016-7061(03)00223-4

Mcgranahan, G. (2002). *Demand-Side Water Strategies and the Urban Poor.*

McGranahan, G., Tacoli, C., & Satterthwaite, D. (2010). Urbanization and its implications for food and farming. *Philosophical Transactions of the Royal Society of London. Series B, Biological Sciences, 365*(1554), 2809–2820. doi:10.1098/rstb.2010.0136 PMID:20713386

McKenzie, N. J., Grundy, M. J., Webster, R., & Minasny, B. (2019). *Digital Soil Mapping Encyclopedia of Soil Science* (3rd ed.).

Meher Kotay, S., & Das, D. (2008). Biohydrogen as a renewable energy resource—Prospects and potentials. *International Journal of Hydrogen Energy, 33*(1), 258–263. doi:10.1016/j.ijhydene.2007.07.031

Mehta, M., Majumdar, A., & Shukla, H. (2020). Harnessing the power of machine learning and artificial intelligence for epidemic preparedness: Trends and future directions. *Journal of Global Health, 10*(1), 010318.

Meléndez-martínez, A. J., Mandić, A. I., Bantis, F., Borge, G. I. A., Brnčić, M., Bysted, A., Cano, M. P., Dias, G., Elgersma, A., Fikselová, M., García-alonso, J., Gonçalves, V. S. S., Hornero-méndez, D., Lavelli, V., Manganaris, G. A., Mapelli-brahm, P., Olmedilla-alonso, B., Periago-castón, M. J., Sheehan, J. J., & Šaponjac, V. T. (2022). A comprehensive review on carotenoids in foods and feeds: Status quo, applications, patents, and research needs. *Critical Reviews in Food Science and Nutrition, 62*(8), 1999–2049. doi:10.1080/10408398.2020.1867959 PMID:33399015

Meran, G., Siehlow, M., & von Hirschhausen, C. (2021). Integrated Water Resource Management: Principles and Applications (pp. 23–121). doi:10.1007/978-3-030-48485-9_3

Meziant, L., Bachir-bey, M., Bensouici, C., Saci, F., Boutiche, M., & Louaileche, H. (2021). Assessment of inhibitory properties of flavonoid-rich fig (Ficus carica L.) peel extracts against tyrosinase, α-glucosidase, urease and cholinesterases enzymes, and relationship with antioxidant activity. *European Journal of Integrative Medicine, 43*, 101272. doi:10.1016/j.eujim.2020.101272

Minasny, B. (2017). Soil sensing for precision agriculture. In *Precision Agriculture* (pp. 223–241). Springer.

Ministry of Agriculture, Fisheries, Rural Development, Water and Forests, Kingdom of Morocco. (2019). *Sugar Sector.* Ministère de l'agriculture.

Mishra, A. K., Raghuwanshi, N. S., & Singh, R. (2018). Geospatial techniques in agriculture: An overview. *Current Science, 114*(3), 575–586.

Mittal, P., & Kaur, A. (2013). E-governance: A challenge for India. *International Journal of Advanced Research in Computer Engineering and Technology*, *2*(3).

Mittelstadt, B. (2017). Ethics of the Health-related Internet of Things: A Narrative Review. *Ethics and Information Technology*, *19*(3), 157–175. doi:10.1007/s10676-017-9426-4

Mkhabela, M. (2020). An overview of precision farming technologies used in the agricultural industry. *Agrotechnology*, *9*(2), 1–8.

Mohamed, W. A., Mansour, M. M. A., & Salem, M. Z. M. (2019, May). Lemna gibba and Eichhornia crassipes extracts: Clean alternatives for deacidification, antioxidation and fungicidal treatment of historical paper. *Journal of Cleaner Production*, *219*, 846–855. doi:10.1016/j.jclepro.2019.02.097

Mohammadi, M., Fazeli, F., & Baghsheikhi, M. (2024). Enhancing water desalination and power generation in microbial desalination cells: A comprehensive review of effective parameters and structural types for optimal performance. *Journal of Power Sources*, *602*, 234267. doi:10.1016/j.jpowsour.2024.234267

Mohan, A. K., & Tiwari, A. (2023) Making sense of bengaluru's Messy Urban Development Data. *Citizen Matters, Bengaluru*. https://bengaluru.citizenmatters.in/making-sense-of-bengalurus-messy-urban-development-data-117710

Mohan, V. (2023, February 2). Govt budgets for 'green growth' but experts call it inadequate to tackle air pollution. *The Times of India*. https://timesofindia.indiatimes.com/business/budget/govt-budgets-for-green-growth-but-experts-call-it-inadequate-to-tackle-air-pollution/articleshow/97559795.cms

Molden, D. J., Oweis, T., Steduto, P., Bindraban, P. S., Hanjra, M. A., & Kijne, J. W. (Eds.). (2010). Improving water productivity in agriculture: Editor's overview. *Agricultural Water Management*, *97*(4), 517–523. doi:10.1016/j.agwat.2009.03.023

Moreno-Garcia, L., Adjallé, K., Barnabé, S., & Raghavan, G. S. V. (2017). Microalgae biomass production for a bio-refinery system : Recent advances and the way towards sustainability. *Renewable & Sustainable Energy Reviews*, *76*, 493–506. doi:10.1016/j.rser.2017.03.024

Mort, M., Finch, T., & May, C. (2009). Making and unmaking telepatients: Identity and governance in new health technologies. *Science, Technology & Human Values*, *34*(1), 9–33. doi:10.1177/0162243907311274

Mu, C., Hou, Y., Zhao, J., Wei, S., & Wu, Y. (2023). Stream-DBSCAN: A Streaming Distributed Clustering Model for Water Quality Monitoring. *Applied Sciences (Basel, Switzerland)*, *13*(9), 5408. doi:10.3390/app13095408

Mujumdar, P. P., & Ghosh, S. (2008). Fuzzy Logic-Based Approaches in Water Resource System Modelling. In *Practical Hydroinformatics* (pp. 165–176). Springer Berlin Heidelberg. doi:10.1007/978-3-540-79881-1_12

Mulligan, D. K., & Bamberger, K. A. (2018). Saving governance-by-design. *California Law Review*, *106*(3), 697–784.

Musso, J., Weare, C., & Hale, M. (2000). Designing web technologies for local governance reform: Good management or good democracy? *Political Communication*, *17*(1), 1–19. doi:10.1080/105846000198486

Nakagawa, M., Nakamura, Y., Honda, H., & Yamagishi, H. (2019). Blockchain technology: Principles and applications. *Journal of Cryptographic Engineering*, *9*(2), 95–108.

Nantongo, H. (2023). Combining remote sensing and crop growth models to assess the impacts of climate change on maize yield in Uganda. *International Journal of Applied Earth Observation and Geoinformation*, *98*, 102456.

Neale, C. M., & Maltese, A. (2019, December). Remote Sensing for Agriculture, Ecosystems, and Hydrology XXI. In *Proc. of SPIE Vol* (Vol. 11149, pp. 1114901-1).

Newman, J. (2023). *Sustainable Agriculture: Concepts, Practices, and Future Directions*. Springer.

News Desk. (2023, May 29). *Global Recognition to Namami Gange Programme.* The Avenue Mail. https://avenuemail.in/global-recognition-to-namami-gange-programme/

Nikhil Binoy, C., Arjun, N., Keerthi, C., Sreerag, S., & Nair, A. H. (2019). Rainfall Prediction Using Flow and Depth Measurement with Artificial Neural Network in Canals. 2019 3rd International Conference on Computing Methodologies and Communication (ICCMC), 798-801. 10.1109/ICCMC.2019.8819807

Nova, K. (2023). AI-Enabled Water Management Systems: An Analysis of System Components and Interdependencies for Water. *Conservation*, *8*(1).

Nunes, A. R., Alves, M. G., Moreira, P. I., Oliveira, P. F., & Silva, B. M. (2014). Can tea consumption be a safe and effective therapy against diabetes mellitus-induced neurodegeneration? *Current Neuropharmacology*, *12*(6), 475–489. doi:10.2174/1570159X13666141204220539 PMID:25977676

Nunes, B., Veiga, V., Frankenbach, S., Serôdio, J., & Pinto, G. (2019, November). Evaluation of physiological changes induced by the fluoroquinolone antibiotic ciprofloxacin in the freshwater macrophyte species Lemna minor and Lemna gibba. *Environmental Toxicology and Pharmacology*, *72*, 103242. doi:10.1016/j.etap.2019.103242 PMID:31473558

Obirikorang, K. A., Sekey, W., Gyampoh, B. A., Ashiagbor, G., & Asante, W. (2021). Aquaponics for improved food security in Africa: A review. *Frontiers in Sustainable Food Systems*, *5*, 705549. doi:10.3389/fsufs.2021.705549

OECD/FAO. (2020). *OECD-FAO Agricultural Outlook 2020-2029, Sugar*. OECD iLibrary.

Okello, C., Tomasello, B., Greggio, N., Wambiji, N., & Antonellini, M. (2015). Impact of Population Growth and Climate Change on the Freshwater Resources of Lamu Island, Kenya. *Water (Basel)*, *7*(12), 1264–1290. doi:10.3390/w7031264

Omer, S., Saravanan, P., Kumar, P., Rajeshkannan, R., Manivasagan, R., & Shanmugam, V. (2024). Insights into renewable biohydrogen production from algal biomass : Technical hurdles and economic analysis. *Biomass Conversion and Biorefinery*, 1–20. doi:10.1007/s13399-023-05263-w

Oron, G., Porath, D., & Jansen, H. (1987, February). Performance of the duckweed species Lemna gibba on municipal wastewater for effluent renovation and protein production. *Biotechnology and Bioengineering*, *29*(2), 258–268. doi:10.1002/bit.260290217 PMID:18576384

Oumsalem, M. N. I., & Zanoun, L. (2019). *Culture de microalgue verte halophile pour production de biodiesel* [PhD Thesis].

Pagliuso, D., Grandis, A., Fortirer, J. S., Camargo, P., Iochevet, E., Floh, S., & Buckeridge, M. S. (2022). *Duckweeds as Promising Food Feedstocks Globally*., doi:10.3390/agronomy12040796

Palosuo, T. (2021). Sustainable intensification in a northern European context: A review. *Agronomy for Sustainable Development*, *41*(1), 1–17.

Pandit, P. R., & Fulekar, M. H. (2019). Biodiesel production from microalgal biomass using CaO catalyst synthesized from natural waste material. *Renewable Energy*, *136*, 837–845. doi:10.1016/j.renene.2019.01.047

Pantanella, E. (2012). Nutrition and quality of aquaponic systems.

Pantanella, E. (2012). Nutrition and quality of aquaponic systems. *Sustainable Agriculture.*

Pantanella, E. (2018). Aquaponics production, practices and opportunities. *Sustainable Aquaculture*, 191-248.

Paolacci, S., Stejskal, V., Toner, D., & Jansen, M. A. K. (2022). Wastewater valorisation in an integrated multitrophic aquaculture system; assessing nutrient removal and biomass production by duckweed species. *Environmental Pollution, 302*, 119059. doi:10.1016/j.envpol.2022.119059 PMID:35227845

Papastefanopoulos, V., Linardatos, P., & Kotsiantis, S. (2020). COVID-19: A comparison of time series methods to forecast percentage of active cases per population. *Applied Sciences (Basel, Switzerland), 10*(3880), 3880. doi:10.3390/app10113880

Parmesan, C., & Yohe, G. (2003). A globally coherent fingerprint of climate change impacts across natural systems. *Nature, 421*(6918), 37–42. doi:10.1038/nature01286 PMID:12511946

Pathan, S., Sood, A., & Ahmad, T. (2021). A Review on Precision Agriculture: Techniques, Challenges, and Opportunities. *Computers and Electronics in Agriculture, 185*, 106131.

Pei, S., Kandula, S., Yang, W., & Shaman, J. (2018). Forecasting the spatial transmission of influenza in the United States. *Proceedings of the National Academy of Sciences of the United States of America, 115*(11), 2752–2757. doi:10.1073/pnas.1708856115 PMID:29483256

Peng, L., Yang, W., Zhang, D., Zhuge, C., & Hong, L. (2020). Epidemic analysis of COVID-19 in China by dynamical modeling. arXiv preprint arXiv:2002.06563. doi:10.1101/2020.02.16.20023465

Pereira, L. S. (2019). Improving irrigation water use efficiency in agriculture: A review. *Agronomy for Sustainable Development, 39*(1), 1–22. PMID:30881486

Pereira, L. S., Cordery, I., & Iacovides, I. (2012). Coping with water scarcity: An action framework for agriculture and food security. *Agriculture, Ecosystems & Environment, 144*, 54–65.

Pham, Q.-V., Nguyen, D. C., Huynh-The, T., Hwang, W.-J., & Pathirana, P. N. (2021). Artificial intelligence (AI) and big data for coronavirus (COVID-19) pandemic: A survey on the state-of-the-arts. *IEEE Access : Practical Innovations, Open Solutions, 8*, 130820–130839. doi:10.1109/ACCESS.2020.3009328 PMID:34812339

Piwek, L., Ellis, D. A., Andrews, S., & Joinson, A. (2016). The rise of consumer health wearables: Promises and barriers. *PLoS Medicine, 13*(2), e1001953. doi:10.1371/journal.pmed.1001953 PMID:26836780

Ploug, T., & Holm, S. (2020). The four dimensions of contestable AI diagnostics – A patient-centric approach to explainable AI. *Artificial Intelligence in Medicine, 104*, 101820. doi:10.1016/j.artmed.2020.101901 PMID:32828448

Podmore, C. (2009). *Irrigation salinity – causes and impacts*. DPI. https://www.dpi.nsw.gov.au/__data/assets/pdf_file/0018/310365/Irrigation-salinity-causes-and-impacts.pdf

Poirier, K., Lotfi, M., Garg, K., Patchigolla, K., Anthony, E. J., Faisal, N. H., Mulgundmath, V., Sahith, J. K., Jadhawar, P., Koh, L., Morosuk, T., & Al Mhanna, N. (2023). A comprehensive review of pre-and post-treatment approaches to achieve sustainable desalination for different water streams. *Desalination, 566*, 116944. doi:10.1016/j.desal.2023.116944

Politano, A., Al-Juboori, R.A., Alnajdi, S., Alsaati, A., Athanassiou, A., Bar-Sadan, M., Naderi Beni, A., Campi, D., Cupolillo, A., & D'Olimpio, G. (2024). 2024 Roadmap on membrane desalination technology at the water-energy nexus. *Journal of Physics: Energy.*

Pontius, J., & McIntosh, A. (2024). Water Scarcity. In *Environmental Problem Solving in an Age of Climate Change, Springer Textbooks in Earth Sciences, Geography and Environment* (pp. 87–103). Springer International Publishing. doi:10.1007/978-3-031-48762-0_8

Porter, J. R. (2014). Food security and food production systems. In *Climate Change 2014: Impacts, Adaptation, and Vulnerability. Part A: Global and Sectoral Aspects. Contribution of Working Group II to the Fifth Assessment Report of the Intergovernmental Panel on Climate Change*. Cambridge University Press.

Pouriyeh, S., Vahid, S., Sannino, G., De Pietro, G., Arabnia, H., & Gutierrez, J. (2017). A comprehensive investigation and comparison of machine learning techniques in the domain of heart disease. In *2017 IEEE Symposium on Computers and Communications (ISCC)* (pp. 204-207). IEEE. 10.1109/ISCC.2017.8024530

Prabhu, T. N. (2020). *Population by country - 2020*. Kaggle. https://www.kaggle.com/tanuprabhu/population-by-country-2020

Prachayasittikul, V., Prachayasittikul, S., Ruchirawat, S., & Prachayasittikul, V. (2018). Coriander (*Coriandrum sativum*): A promising functional food toward the well-being. *Food Research International*, *105*, 305–323. doi:10.1016/j.foodres.2017.11.019 PMID:29433220

Pradhan, A., Patel, A. B., & Singh, S. K. (2019, June). Evaluation of live duckweed, Wolffia globosa as an allochthonous feed for Labeo rohita fry during nursery rearing. *Aquaculture Research*, *50*(6), 1557–1563. doi:10.1111/are.14025

Pradhan, A., & Srinivasan, V. (2022). Do dams improve water security in India? A review of post facto assessments. *Water Security*, *15*, 100112. doi:10.1016/j.wasec.2022.100112

Prasher, G. (2023). Bengaluru, we have a problem: It's our Lakes. *Bangalore Mirror*. https://bangaloremirror.indiatimes.com/bangalore/civic/bengaluru-we-have-a-problem-its-our-lakes/articleshow/97289067.cms

Pratley, R. E. (2013). The early treatment of type 2 diabetes. *The American Journal of Medicine*, *126*(9), S2–S9. doi:10.1016/j.amjmed.2013.06.007 PMID:23953075

Pretty, J. (2018). Agricultural sustainability: Concepts, principles and evidence. *Philosophical Transactions of the Royal Society of London. Series B, Biological Sciences*, *363*(1491), 447–465. doi:10.1098/rstb.2007.2163 PMID:17652074

Priyan, K. (2021). Issues and Challenges of Groundwater and Surface Water Management in Semi-Arid Regions. In *Groundwater Resources Development and Planning in the Semi-Arid Region* (pp. 1–17). Springer International Publishing. doi:10.1007/978-3-030-68124-1_1

Punetha, D., Tewari, G., & Pande, C. (2018). Compositional variability in inflorescence essential oil of Coriandrum sativum from North India. *Journal of Essential Oil Research, 30*(2), 113-119.

Qian, X., Liu, X., Mao, Z., Abdulai, T., Dong, X., Tu, R., Wang, Y., Liu, X., Luo, Z., Qiao, D., Wang, C., & Li, Y. (2020). Fruit and vegetable intake modifies the associations between suppressor of cytokine signaling 3 genetic variants and type 2 diabetes. *European Journal of Nutrition*, *59*(8), 3441–3449. doi:10.1007/s00394-020-02178-4 PMID:31927672

Qin, L., Sun, Q., Wang, Y., Wu, K. F., Chen, M., Shia, B. C., & Wu, S. Y. (2020). Prediction of number of cases of 2019 novel coronavirus (COVID-19) using social media search index. *International Journal of Environmental Research and Public Health*, *17*(2365). PMID:32244425

Qureshi, M. F., Hussain, M., Shabbir, R. N., & Baig, I. A. (2020). Precision Agriculture: A Review. *Advances in Horticultural Science*, *34*(3), 393–410.

Radhakrishnan, R., & C A, L. D. (2023). Groundwater Level Prediction Using Support Vector Machine and M5 Model Tree-A Case Study. *SSRN Electronic Journal*. doi:10.2139/ssrn.4512253

Rahman, S., Rahman, K., & Rahman, M. (2019). Impact of organic farming on environmental sustainability. *Agriculture, Ecosystems & Environment*, *284*, 106589.

Rai, R., Sharma, A., & Sinha, A. (2018). Challenges in the adoption of geospatial technology for agricultural decision-making: A review. *International Journal of Remote Sensing & Geoscience*, *7*(4), 2482–2489.

Rajamanickam, R., & Selvasembian, R. (2024). A sustainable resource recovery approach through micro-algae integrated brine management in minimal liquid discharge system. *Desalination*, *578*, 117424. doi:10.1016/j.desal.2024.117424

Rakocy, J. E. (2012). Aquaponics—integrating fish and plant culture. *Aquaculture production systems*, 344-386.

Rakocy, J. E. (2012). Aquaponics—integrating fish and plant culture. *Aquaculture production systems*.

Ramadan, M.F. (2023). *Fig* (Ficus carica): *Production, processing, and properties*. Springer.

Rashidi, B., Malekzadeh, M., Goodarzi, M., Masoudifar, A., & Mirzaei, H. (2017). Green tea and its anti-angiogenesis effects. *Biomedicine and Pharmacotherapy*, *89*, 949–956. doi:10.1016/j.biopha.2017.01.161 PMID:28292023

Razouki, S., Kousksou, T., Allaoui, T., Rhafiki, T. E. M., & Jamil, A. (2018). Air quality indicators in Morocco - Application to urban, rural, and industrial areas. *Energy Procedia*, *139*, 696–701.

Remuzzi, A., & Remuzzi, G. (2020). COVID-19 and Italy: What next? *Lancet*, *395*(1225–1228). PMID:32178769

Requena-Leal, I., Carvela, M., Fernández-Marchante, C. M., Lobato, J., & Rodrigo, M. A. (2024). On the use of chlor-alkali technology to power environmental electrochemical treatment technologies. *Current Opinion in Electrochemistry*, *45*, 101461. doi:10.1016/j.coelec.2024.101461

Rezaei, E. E., Siebert, S., Ewert, F., Webber, H., Gaiser, T., Naumann, G., & Zhao, G. (2019). Assessing yield gaps at the country level using an adapted FAO EcoCrop model. *European Journal of Agronomy*, *103*, 79–88.

Ribeiro, M. H. D. M., da Silva, R. G., Larcher, J. H. K., Mariani, V. C., & Coelho, L. D. S. (2022). Ensemble learning models coupled with urban mobility information applied to predict COVID-19 incidence cases. *Modeling, Control and Drug Development for COVID-19 Outbreak Prevention, 24*, 821–858.

Ristow, D. C. M., Henning, E., Kalbusch, A., & Petersen, C. E. (2021). Models for forecasting water demand using time series analysis: A case study in Southern Brazil. *Journal of Water, Sanitation, and Hygiene for Development : a Journal of the International Water Association*, *11*(2), 231–240. doi:10.2166/washdev.2021.208

Rivero-Falcón, Á., Peñate Suárez, B., & Melián-Martel, N. (2023). SWRO Brine Characterisation and Critical Analysis of Its Industrial Valorisation: A Case Study in the Canary Islands (Spain). *Water (Basel)*, *15*(8), 1600. doi:10.3390/w15081600

Robeson, S. M., & Steyn, D. G. (1990). Evaluation and comparison of statistical forecast models for daily maximum ozone concentrations. *Atmospheric Environment. Part B, Urban Atmosphere*, *24*(2), 303–312. doi:10.1016/0957-1272(90)90036-T

Rocher, L., Hendrickx, J. M., & de Montjoye, Y. A. (2019). Estimating the success of re-identifications in incomplete datasets using generative models. *Nature Communications*, *10*(1), 3069. doi:10.1038/s41467-019-10933-3 PMID:31337762

Roco, M. C. (2008). Possibilities for global governance of converging technologies. *Journal of Nanoparticle Research*, *10*(1), 11–29. doi:10.1007/s11051-007-9269-8

Roman, B., & Brennan, R. A. (2019). A beneficial by-product of ecological wastewater treatment: An evaluation of wastewater-grown duckweed as a protein supplement for sustainable agriculture. *Ecological Engineering*, *142*, 100004. doi:10.1016/j.ecoena.2019.100004

Ross, R., & Hamer, W. H. (2013). *The SIR model and the foundations of public health*. Materials Matematics.

Ruiz-Ramos, M. (2019). Adaptation to climate change through the choice of cropping system and sowing date in northern Spain. *European Journal of Agronomy*, *103*, 97–109.

Ruslan, F. A., Samad, A. M., & Adnan, R. (2020). Modelling of rainfall prediction system using hybrid NNARX and Extended Kalman Filter. *2017 IEEE 13th International Colloquium on Signal Processing & its Applications (CSPA)*, 149-152. 10.1109/CSPA.2017.8064941

Ruslan, F. A., Samad, A. M., Tajjudin, M., & Adnan, R. (2016). 7 hours rainfall prediction modelling using NNARX structure: Case study Terengganu. 2019 IEEE 12th International Colloquium on Signal Processing & Its Applications (CSPA), 263-268. 10.1109/CSPA.2016.7515843

Russo, T., Alfredo, K., & Fisher, J. (2014). Sustainable Water Management in Urban, Agricultural, and Natural Systems. *Water (Basel)*, *6*(12), 3934–3956. doi:10.3390/w6123934

Sachdeva, S. (2002). e-Governance strategy in India. *White Paper on e-Governance strategy in India.*

Sadiq, A. (2016). Air quality monitoring stations data - Marrakech - Morocco. figshare. Dataset. doi:10.6084/m9.figshare.3113515.v1

Sahota, S., Kumar, S., & Lombardi, L. (2024). Biohythane, Biogas, and Biohydrogen Production from Food Waste : Recent Advancements, Technical Bottlenecks, and Prospects. *Energies*, *17*(3), 3. Advance online publication. doi:10.3390/en17030666

Saidi, A., Chaouchi-Abada, B., & Chergui-Bouafia, S. (2007). *PRODUIRE DE L'HYDROGENE À PARTIR D'EAU ET DE LUMIERE GRACE AUX MICROORGANISMES.*

Sankaranath, S., Ms, K., Sehar, S., & Suganya, E. (2023). *Challenges and Opportunities for Green Energy Harvesting in Sustainable IoT Systems.*, doi:10.1007/978-3-031-35965-1_7

Santos, E., Carvalho, M., & Martins, S. (2023). Sustainable Water Management: Understanding the Socioeconomic and Cultural Dimensions. *Sustainability (Basel)*, *15*(17), 13074. doi:10.3390/su151713074

Sato, T., Qadir, M., Yamamoto, S., Endo, T., & Zahoor, A. (2013). Global, Regional, and Country Level Need for Data on Wastewater Generation, Treatment, and Use. *Agricultural Water Management*, *130*, 1–13. doi:10.1016/j.agwat.2013.08.007

Scharf, P. C. (2016). Precision nutrient management. In *Precision Agriculture* (pp. 399–432). Springer.

Scharlemann, J. P., & Laurance, W. F. (2008). How green are biofuels? *Science*, *319*(5859), 43–44. doi:10.1126/science.1153103 PMID:18174426

Schneider, T., Behera, S., Boccaletti, G., Deser, C., Emanuel, K., Ferrari, R., Leung, L. R., Lin, N., Müller, T., Navarra, A., Ndiaye, O., Stuart, A., Tribbia, J., & Yamagata, T. (2023). Harnessing AI and computing to advance climate modelling and prediction. *Nature Climate Change*, *13*(9), 887–889. doi:10.1038/s41558-023-01769-3

Scopus. (2020). *Analyze Search Results, TITLE-ABS-KEY ("COVID-19" or "CoViD-19" or "Hubei Pneumonia" or "Novel Corona Virus").* Scopus. https://www.scopus.com

Sghaier, N., & Labbene, M. (2016). *PROJET MENA-DELP*. Middle East & North Africa - Desert Ecosystems and Livelihoods Program. March 2016.

Sharma, B. R., & Minhas, P. S. (2010). Water management in agriculture: A global perspective. *Journal of Applied and Natural Science*, *2*(1), 88–94.

Sharma, D. (2021). Rainwater harvesting: A boon for sustainable agriculture. *Journal of Water Supply: Research & Technology - Aqua*, 70.

Sharma, M. A., & Singh, J. B. (2011). *Comparative Study of rainfall forecasting models*. New York Sci. J.

Sharma, N., Acharya, S., Kumar, K., Singh, N., & Chaurasia, O. P. (2018). Hydroponics as an advanced technique for vegetable production: An overview. *Journal of Soil and Water Conservation*, *17*(4), 364–371. doi:10.5958/2455-7145.2018.00056.5

Shemer, H., Wald, S., & Semiat, R. (2023). Challenges and Solutions for Global Water Scarcity. *Membranes*, *13*(6), 612. doi:10.3390/membranes13060612 PMID:37367816

Shen, C., Chen, A., Luo, C., Zhang, J., Feng, B., & Liao, W. (2020). Using reports of symptoms and diagnoses on social media to predict COVID-19 case counts in mainland China: Observational infoveillance study. *Journal of Medical Internet Research*, *22*(5, e19421), e19421. doi:10.2196/19421 PMID:32452804

Shen, K., Yang, Y., Wang, T., Zhao, D., Jiang, Y., Jin, R., Zheng, Y., Xu, B., Xie, Z., Lin, L., Shang, Y., Lu, X., Shu, S., Bai, Y., Deng, J., Lu, M., Ye, L., Wang, X., Wang, Y., & Gao, L. (2020). Diagnosis, treatment, and prevention of 2019 novel coronavirus infection in children: Experts' consensus statement. *World Journal of Pediatrics*, *16*(3), 223–231. doi:10.1007/s12519-020-00343-7 PMID:32034659

Sillero, L., Sganzerla, W. G., Forster-Carneiro, T., Solera, R., & Perez, M. (2022). A bibliometric analysis of the hydrogen production from dark fermentation. *International Journal of Hydrogen Energy*, *47*(64), 27397–27420. doi:10.1016/j.ijhydene.2022.06.083

Silovská, T., Matoušek, J., Fajstavr, D., Švorčík, V., & Kolská, Z. (2020). Antimicrobial effect of polymers grafted with cinnamaldehyde. *Materials Letters*, *277*, 128274. doi:10.1016/j.matlet.2020.128274

Singh, D., Singh, B., & Goel, R. K. (2011). Traditional uses, phytochemistry and pharmacology of Ficus religiosa: A review. *Journal of Ethnopharmacology*, *134*(3), 565–583. doi:10.1016/j.jep.2011.01.046 PMID:21296646

Singh, K. P., Gupta, S., & Rai, P. (2013). Identifying pollution sources and predicting urban air quality using ensemble learning methods. *Atmospheric Environment*, *80*, 426–437. doi:10.1016/j.atmosenv.2013.08.023

Singh, V. P., & Frevert, D. K. (2018). *Handbook of Applied Hydrology* (2nd ed.). McGraw-Hill Education.

Sirota, R., Winters, G., Levy, O., Marques, J., Paytan, A., Silverman, J., Sisma-Ventura, G., Rahav, E., Antler, G., & Bar-Zeev, E. (2024). Impacts of Desalination Brine Discharge on Benthic Ecosystems. *Environmental Science & Technology*, *58*(13), 5631–5645. doi:10.1021/acs.est.3c07748 PMID:38516811

Smith, G., & Nguyen, T. (2022). Utilizing FbProphet for forecasting infectious diseases: A COVID-19 case study. *Health Informatics Journal*, *28*(1).

Smith, P., Soussana, J.-F., Angers, D., Schipper, L., Chenu, C., Rasse, D. P., Batjes, N. H., van Egmond, F., McNeill, S., Kuhnert, M., Arias-Navarro, C., Olesen, J. E., Chirinda, N., Fornara, D., Wollenberg, E., Álvaro-Fuentes, J., Sanz-Cobena, A., & Klumpp, K. (2020). How to measure, report and verify soil carbon change to realize the potential of soil carbon sequestration for atmospheric greenhouse gas removal. *Global Change Biology*, *26*(1), 219–241. doi:10.1111/gcb.14815 PMID:31469216

Smith, T., & Jones, R. (2023). Digital Agriculture: Transforming the Future of Farming. *Agricultural Technology (Thailand)*, *15*(2), 87–104.

So, J.-M., Lee, J.-H., & Bae, D.-H. (2020). Development of a Hydrological Drought Forecasting Model Using Weather Forecasting Data from GloSea5. *Water (Basel)*, *12*(10), 2785. doi:10.3390/w12102785

Souto, L. R. F., Silva, I. F. D., Ninow, J. L., Collins, S. R. A., Elliston, A., & Waldron, K. W. (2019, August). Effect of hydrothermal pre-treatment on duckweed (Landoltia punctata) biomass for simultaneous saccharification and fermentation process. *Biomass and Bioenergy*, *127*, 105259. doi:10.1016/j.biombioe.2019.105259

Souza Filho, W. (2019). Economic and Energetic Assessment of Integrated Crop-Livestock-Forest Systems Under Different Management Scenarios in the Brazilian Cerrado. *Agricultural Systems*, *170*, 31–42.

Spitzer, V., Sanson, F., Feliot, P., Gurevsky, E., & Rossi, A. (s. d.). *Optimisation de la production d'hydrogène vert renouvelable sous incertitudes.*

Sridhar, A., Ponnuchamy, M., Senthil Kumar, P., Kapoor, A., & Xiao, L. (2022). Progress in the production of hydrogen energy from food waste : A bibliometric analysis. *International Journal of Hydrogen Energy*, *47*(62), 26326–26354. doi:10.1016/j.ijhydene.2021.09.258

Srivastava, A. (2020). A GIS-based approach for crop yield estimation using multiple regression analysis. *Geocarto International*, *35*(3), 225–239.

Stagnari, F., Ramazzotti, S., & Pisante, M. (2009). Conservation Agriculture: A Different Approach for Crop Production Through Sustainable Soil and Water Management. *RE:view*, *1*, 55–83. doi:10.1007/978-1-4020-9654-9_5

Suastes-Rivas, J. K., Romero-Pineda, M. J., Monje-Ramírez, I., Velasquez-Orta, S. B., Velasco, A., & Orta-Ledesma, M. T. (2024). Direct biohydrogen production from chitosan harvested microalgae biomass and an isolated yeast. *Energy Conversion and Management*, *302*, 118081. doi:10.1016/j.enconman.2024.118081

Sun, A. Y., & Scanlon, B. R. (2019). How can Big Data and machine learning benefit environment and water management: A survey of methods, applications, and future directions. *Environmental Research Letters*, *14*(7), 073001. doi:10.1088/1748-9326/ab1b7d

Sun, A. Y., Wang, D., & Xu, X. (2014). Monthly streamflow forecasting using Gaussian Process Regression. *Journal of Hydrology (Amsterdam)*, *511*, 72–81. doi:10.1016/j.jhydrol.2014.01.023

Swain, D., Roy, P., & Panda, S. S. (2020). Challenges in the adoption of geospatial technology for precision agriculture in India: A review. *International Journal of Scientific Research in Science, Engineering and Technology*, *6*(4), 373–381.

Sweeney, L. (2002). k-anonymity: A model for protecting privacy. *International Journal of Uncertainty, Fuzziness and Knowledge-based Systems*, *10*(05), 557–570. doi:10.1142/S0218488502001648

Tadesse, T. W. (2021). Adaptation to climate change in agriculture: Analysis of potential determinants for investment in Ethiopia. *World Development*, *140*, 105365.

Taylor, S. J., & Letham, B. (2017). *Prophet: Forecasting at scale [Computer software manual].* Github. https://facebook.github.io/prophet/

Taylor, S. J., & Letham, B. (2018). Forecasting at scale. PeerJ Preprints, 5:e3190v2. https://peerj.com/preprints/3190/

Taylor, S. J., & Letham, B. (2018). Forecasting at scale. *The American Statistician*, *72*(1), 37–45. doi:10.1080/00031305.2017.1380080

Tetteh Kwasi Nuer, A., Agbeko, D., Worlali, S., Mwangi Thiga, M., Ndogo Ndungu, S., Wangari Mutiga, M., & Babcock, L. H. (2018). *Why Invest in ICTs for agriculture?* CTA.

The Nature Conservancy. (2019). Latin America 2019 Annual Report. *The Nature Conservancy*, 1–25.

Thenkabail, P. S. (2019). Advancements in remote sensing of agriculture: Context description, existing operational monitoring systems, and major information needs. *Remote Sensing*, *11*(18), 2107.

Thompson, J. A. (2017). Geospatial applications in precision agriculture: An overview. *Applied Geomatics*, *9*(1), 67–74.

Thornton, P. K., & Herrero, M. (2021). Adaptation of food systems to climate change: Challenges and options. *Environmental Research Letters*, *16*(5), 053001.

Tittonell, P., & Giller, K. E. (2013). When yield gaps are poverty traps: The paradigm of ecological intensification in African smallholder agriculture. *Field Crops Research*, *143*, 76–90. doi:10.1016/j.fcr.2012.10.007

TNN. (2023, June 2). How planning and development of Mumbai can involve citizens. *The Times of India*. https://timesofindia.indiatimes.com/city/mumbai/how-planning-and-development-of-mumbai-can-involve-citizens/articleshow/100691710.cms

Torous, J., Rodriguez, J., & Powell, A. (2017). The new digital divide for digital biomarkers. *Digital Biomarkers*, *1*(1), 87–91. doi:10.1159/000477382 PMID:29104959

Torres-León, C., & Rebolledo, R. F. (2023). Article. *Saudi Pharmaceut. J.*, *31*(1), 21–28.

Trejo-Téllez, L. I. (2012). Nutrient solutions for hydroponic systems. *Hydroponics-a standard methodology for plant biological researches*.

Tropogo. (n.d.). *Application of drones in acriculture in India*. Toropogo. https://tropogo.com/blogs/application-of-drones-in-agriculture-in-india

Tuite, A. R., Bogoch, I. I., Sherbo, R., Watts, A., Fisman, D., & Khan, K. (2020). Estimation of coronavirus disease 2019 (COVID-19) burden and potential for international dissemination of infection from Iran. *Annals of Internal Medicine*, *172*(699–701), 699–701. doi:10.7326/M20-0696 PMID:32176272

Tuite, A. R., Ng, V., Rees, E., & Fisman, D. (2020). Estimation of COVID-19 outbreak size in Italy. *The Lancet. Infectious Diseases*, *20*(537), 537. doi:10.1016/S1473-3099(20)30227-9 PMID:32199494

Tukahirwa, J. T., & Alavalapati, J. R. R. (2021). Women's involvement in climate-smart agriculture (CSA): A case study of the banana systems in Uganda. *Women's Studies International Forum*, *86*, 102487.

Tun, P. P. P., & Sein, M. M. (2018). Rainfall Prediction System for Middle Region of Myanmar. *2018 IEEE 7th Global Conference on Consumer Electronics (GCCE)*, 769-770. 10.1109/GCCE.2018.8574757

Tu, W. H., Zhao, Y., Chan, W. P., & Lisak, G. (2024). Reclaimed seawater discharge–Desalination brine treatment and resource recovery system. *Water Research*, *251*, 121096. doi:10.1016/j.watres.2023.121096 PMID:38184912

Ugulu, I., Baslar, S., Yorek, N., & Dogan, Y. (2009). The investigation and quantitative ethnobotanical evaluation of medicinal plants used around Izmir province. *Turkey Journal of Medicinal Plant Research*, *3*, 345–367.

Umoh, A., Nwasike, C., Tula, O., Adekoya, O., & Gidiagba, J. (2024). A REVIEW OF SMART GREEN BUILDING TECHNOLOGIES : INVESTIGATING THE INTEGRATION AND IMPACT OF AI AND IOT IN SUSTAINABLE BUILDING DESIGNS. *Computer Science & IT Research Journal*, *5*(1), 141–165. doi:10.51594/csitrj.v5i1.715

UNFCCC. (2012). *Technical Guidelines for the National Adaptation Plan Process*. United Nations Framework Convention on Climate Change.

Valdeyron, G. & Lloyd, D.G. (1979). Sex differences and flowering phenology in the common fig. *Ficus carica L Evolution, 33*, 673-685.

van der Voet, H. (2019). Methodology for the systematic analysis of trade-offs in sustainability assessment: make fewer assumptions. *Sustainability, 11*, 499.

Veinot, T. C., Mitchell, H., & Ancker, J. S. (2018). Good intentions are not enough: How informatics interventions can worsen inequality. *Journal of the American Medical Informatics Association : JAMIA*, *25*(8), 1080–1088. doi:10.1093/jamia/ocy052 PMID:29788380

Velazquez-Gonzalez, R. S., Garcia-Garcia, A. L., Ventura-Zapata, E., Barceinas-Sanchez, J. D. O., & Sosa-Savedra, J. C. (2022). A review on hydroponics and the technologies associated for medium-and small-scale operations. *Agriculture*, *12*(5), 646. doi:10.3390/agriculture12050646

Vellore Vettalakar, A.B. (2023). *THE PRODUCTION BRINE WATER PROBLEMS AND RESOURCES.*

Verma, S., Suresh, S., Mili, M., Diwedi, P., Bajpai, H., Khan, M., Hashmi, S., Rathore, S., & Srivastava, A. (2023). Future process options for brine sludge management from the chlor-alkali industry. *International Journal of Environmental Science and Technology*, *20*(8), 9291–9310. doi:10.1007/s13762-022-04433-z

Vidisha, S. (2023). Mumbai slum residents stand up against Adani's redevelopment plan. *Nikkei Asia*. https://asia.nikkei.com/Spotlight/Asia-Insight/Mumbai-slum-residents-stand-up-against-Adani-s-redevelopment-plan

Vinayagam, M., Grace, L. K. J., Ranjit, P. S., Parikh, S., Venkatesh, R., Soudagar, M. E. M., Obaid, S. A., Alharbi, S. A., & Ramachandaramurthy, V. K. (2024). Synergistic development of nonvascular species microalgae for bio-hydrogen production featured with nano titanium dioxide additive. *Process Safety and Environmental Protection*. doi:10.1016/j.psep.2024.02.080

Virtanen, P., Gommers, R., Oliphant, T. E., Haberland, M., Reddy, T., Cournapeau, D., Burovski, E., Peterson, P., Weckesser, W., Bright, J., van der Walt, S. J., Brett, M., Wilson, J., Millman, K. J., Mayorov, N., Nelson, A. R. J., Jones, E., Kern, R., Larson, E., & van Mulbregt, P. (2020). SciPy 1.0: Fundamental algorithms for scientific computing in Python. *Nature Methods*, *17*(3), 261–272. doi:10.1038/s41592-019-0686-2 PMID:32015543

Viscarra Rossel, R. A., & Behrens, T. (2010). Using data mining to model and interpret soil diffuse reflectance spectra. *Geoderma*, *158*(1-2), 46–54. doi:10.1016/j.geoderma.2009.12.025

Vo-Van, T., Nguyen-Hai, A., Tat-Hong, M. V., & Nguyen-Trang, T. (2020). A New Clustering Algorithm and Its Application in Assessing the Quality of Underground Water. *Scientific Programming*, *2020*, 1–12. doi:10.1155/2020/6458576

Waltner-Law, M. E., Wang, X. L., Law, B. K., Hall, R. K., Nawano, M., & Granner, D. K. (2002). Epigallocatechin gallate, a constituent of green tea, represses hepatic glucose production. *The Journal of Biological Chemistry*, *277*(38), 34933–34940. doi:10.1074/jbc.M204672200 PMID:12118006

Wang, S. L., Heisey, P., Schimmelpfennig, D., & Ball, V. E. (2015). *Agricultural productivity growth in the United States: Measurement, trends, and drivers*. Economic Research Service, Paper No. Err-189.

Wang, D., Hu, B., Hu, C., Zhu, F., Liu, X., Zhang, J., Wang, B., Xiang, H., Cheng, Z., Xiong, Y., Zhao, Y., Li, Y., Wang, X., & Peng, Z. (2020). Clinical characteristics of 138 hospitalized patients with 2019 novel coronavirus-infected pneumonia in Wuhan, China. *Journal of the American Medical Association*, *323*(11), 1061–1069. doi:10.1001/jama.2020.1585 PMID:32031570

Wang, G., Jia, Q.-S., Zhou, M., Bi, J., Qiao, J., & Abusorrah, A. (2022). Artificial neural networks for water quality soft-sensing in wastewater treatment: A review. *Artificial Intelligence Review*, *55*(1), 565–587. doi:10.1007/s10462-021-10038-8

Wang, L. (2021). Neural network-based models for influenza predictions in public health. *Frontiers in Public Health*, 9.

Wang, L., Ning, S., Zheng, W., Guo, J., Li, Y., Li, Y., Chen, X., Ben-Gal, A., & Wei, X. (2023). Performance analysis of two typical greenhouse lettuce production systems: commercial hydroponic production and traditional soil cultivation. Frontiers in Plant Science, 14.

Wang, L., Ning, S., Zheng, W., Guo, J., Li, Y., Li, Y., Chen, X., Ben-Gal, A., & Wei, X. (2023). Performance analysis of two typical greenhouse lettuce production systems: commercial hydroponic production and traditional soil cultivation. *Frontiers in Plant Science*.

Wang, W., Tang, J., & Wei, F. (2020). Updated understanding of the outbreak of 2019 novel coronavirus (2019-nCoV) in Wuhan, China. *Journal of Medical Virology*, *92*(4), 441–447. doi:10.1002/jmv.25689 PMID:31994742

Wang, Y. F., Yu, L., & Wei, X. L. (2012). Monosaccharide composition and bioactivity of tea flower polysaccharides obtained by ethanol fractional precipitation and stepwise precipitation. *CYTA: Journal of Food*, *10*(1), 1–4. doi:10.1080/19476337.2010.523901

Wang, Z., Ma, D., Guo, W., & Zhang, Y. (2019). Crop yield estimation based on machine learning: A review. *Sensors (Basel)*, *19*(19), 4302. PMID:31590234

Wei, J.-N., Liu, Z.-H., Zhao, Y.-P., Zhao, L.-L., Xue, T.-K., & Lan, Q.-K. (2019). Phytochemical and bioactive profile of Coriandrum sativum L. *Food Chemistry*, *286*(juillet), 260–267. doi:10.1016/j.foodchem.2019.01.171 PMID:30827604

Wei, M., Yuan, J., Liu, Y., Fu, T., Yu, X., & Zhang, Z.-J. (2020). Novel coronavirus infection in hospitalized infants under 1 year of age in China. *Journal of the American Medical Association*, *323*(1313 1314), 1313. doi:10.1001/jama.2020.2131 PMID:32058570

Wei, X. L., Chen, M. A., Xiao, J. B., Liu, Y., Yu, L., Zhang, H., & Wang, Y. F. (2010). Composition and bioactivity of tea flower polysaccharides obtained by different methods. *Carbohydrate Polymers*, *79*(2), 418–422. doi:10.1016/j.carbpol.2009.08.030

Weldon, L., McBride, M. F., & Earley, R. (2018). Challenges to implementing precision agriculture techniques: A case study of farmer attitudes in Ireland. *Land Use Policy*, *70*, 439–446.

Weltin, M. (2018). Sustainable intensification of agriculture: The role of grazing and forage in supporting food security. *Sustainability*, *10*(11), 4044.

West Bengal Govt launches buses with Air Purifiers in Kolkata to beat pollution . (2023) *Hindustan Times*. https://www.hindustantimes.com/cities/kolkata-news/west-bengal-govt-launches-buses-with-air-purifiers-in-kolkata-to-beat-pollution-101686042102914.html (Accessed: 11 June 2023).

White, J. W. (2011). Field-based approaches for developing maize germplasm adapted to drought and heat stress. *Crop Science*, *51*(3), 1–15.

Whiting, D. R., Guariguata, L., Weil, C., & Shaw, J. (2011). J. Shaw IDF diabetes atlas: Global estimates of the prevalence of diabetes for 2011 and 2030. *Diabetes Research and Clinical Practice*, *94*(3), 311–321. doi:10.1016/j.diabres.2011.10.029 PMID:22079683

Widiasari, Nugroho, & Widyawan. (2017). Deep learning multilayer perceptron (MLP) for rainfall prediction model using wireless sensor network based hydrology time series data mining. 2019 International Conference on Innovative and Creative Information Technology (ICITech), 1-5. . doi:10.1109/INNOCIT.2017.8319150

Wiens, J., & Shenoy, E. S. (2018). Machine learning for healthcare: On the verge of a major shift in healthcare epidemiology. *Clinical Infectious Diseases*, *66*(1), 149–153. doi:10.1093/cid/cix731 PMID:29020316

Wollenberg, E. (2016). Climate change mitigation and adaptation in the land sector: From complementarity to synergy. *Environmental Research Letters*, *11*(3), 033001.

Woodland, D. W. (1997). *Contemporary Plant Systematics* (2nd ed.). Andrews University Press.

World Health Organization. (2013). *WHO Traditional Medicine Strategy*. World Health Organization.

World Health Organization. (2021). *Infectious Diseases*. WHO.

Wu, C. L., Chau, K. W., & Fan, C. (2010). Prediction of rainfall time series using Modular Artificial Neural Networks coupled with data preprocessing techniques. *Journal of Hydrology (Amsterdam)*, *389*(1), 146–167. doi:10.1016/j.jhydrol.2010.05.040

Xing, L., Zhang, H., Qi, R., Tsao, R., & Mine, Y. (2019). Recent advances in the understanding of the health benefits and molecular mechanisms associated with green tea polyphenols. *Journal of Agricultural and Food Chemistry*, *67*(4), 1029–1043. doi:10.1021/acs.jafc.8b06146 PMID:30653316

Xu, J., Zhao, H., Stomp, A., & Cheng, J. J. (2012). The production of duckweed as a source of biofuels. Future Science. doi:10.4155/bfs.12.31

Xu, J., Shen, Y., Zheng, Y., Smith, G., Sun, X. S., Wang, D., Zhao, Y., Zhang, W., & Li, Y. (2023, August). Duckweed (Lemnaceae) for potentially nutritious human food: A review. *Food Reviews International*, *39*(7), 3620–3634. doi:10.1080/87559129.2021.2012800

Yadav, N., & Singh, V. B. (2013). E-governance: past, present and future in India. *arXiv preprint arXiv:1308.3323. Experts brainstorm on strategies to improve air quality in Delhi.* https://www.newindianexpress.com/cities/delhi/2023/may/16/experts-brainstorm-on-strategies-to-improve-air-quality-in-delhi-2575552.html

Yang, W., Kandula, S., Huynh, M., Greene, S. K., Van Wye, G., Li, W., & Shaman, J. (2020). Estimating the direct and indirect effects of vaccinations using epidemiological and economic data, and the implications for public health policy. *Journal of the Royal Society, Interface*, *17*(168), 20200144. PMID:32693748

Yarkent, Ç., Oral, İ., Öncel, D. Ş., & Oncel, S. S. (2024). Focusing on a new biohydrogen production strategy (HS-HS) using *Chlamydomonas reinhardtii* mutants : The effects of mutation points on biohydrogen production capacity. *International Journal of Hydrogen Energy*, *52*, 88–102. doi:10.1016/j.ijhydene.2023.02.034

Yep, B., Gale, N. V., & Zheng, Y. (2020). Comparing hydroponic and aquaponic rootzones on the growth of two drug-type Cannabis sativa L. cultivars during the flowering stage. *Industrial Crops and Products*, *157*, 112881. doi:10.1016/j.indcrop.2020.112881

Yılmaz, A. (2021). Ozone level prediction with machine learning algorithms. *Journal of Aeronautics and Space Technologies*, *14*(2), 177–183.

Yoshikawa, M., Wang, T., Sugimoto, S., Nakamura, S., & Nagatomo, A. (2008). Functional saponins in tea flower (Flower buds of Camellia sinensis): Gastroprotective and hypoglycemic effects of floratheasaponins and qualitative and quantitative analysis using HPLC Yakugaku. *Zasshi-Journal of the Pharmaceutical Society of Japan*, *128*(1) 141-151.

Yu, K. L., Chen, W.-H., Sheen, H.-K., Chang, J.-S., Lin, C.-S., Ong, H. C., Show, P. L., & Ling, T. C. (2020). Bioethanol production from acid pretreated microalgal hydrolysate using microwave-assisted heating wet torrefaction. *Fuel*, *279*, 118435. doi:10.1016/j.fuel.2020.118435

Yu, X., Xiao, J., Chen, S., Yu, Y., Ma, J., Lin, Y., Li, R., Lin, J., Fu, Z., Zhou, Q., Chao, Q., Chen, L., Yang, L., & Liu, R. (2020). Metabolite signatures of diverse Camellia sinensis tea populations. *Nature Communications*, *11*(1), 5586–5600. doi:10.1038/s41467-020-19441-1 PMID:33149146

Zeman, C., Rich, M., & Rose, J. (2006). World water resources: Trends, challenges, and solutions. *Reviews in Environmental Science and Biotechnology*, *5*(4), 333–346. doi:10.1007/s11157-005-4089-8

Zhang, Y., Bocquet, M., Mallet, V., Seigneur, C., & Baklanov, A. (2012). Real-time air quality forecasting, part I: History, techniques, and current status. *Atmospheric Environment*, *60*, 632–655. doi:10.1016/j.atmosenv.2012.06.031

Zhang, Y., Chen, J., Zeng, Y., Huang, D., & Xu, Q. (2019). Involvement of AMPK activation in the inhibition of hepatic gluconeogenesis by Ficus carica leaf extract in diabetic mice and HepG2 cells. *Biomedicine and Pharmacotherapy*, *109*, 188–194. doi:10.1016/j.biopha.2018.10.077 PMID:30396076

Zhang, Y., Liu, X., Liang, S., & Yang, X. (2019). Applications of remote sensing in precision agriculture: A review. *Journal of Agricultural Science*, *11*(2), 42–54.

Zhang, Z. B., Xiong, T., Chen, J. H., Ye, F., Cao, J. J., Chen, Y. R., Zhao, Z. W., & Luo, T. (2023). Understanding the origin and evolution of tea (*Camellia sinensis* [L.]): Genomic advances in tea. *Journal of Molecular Evolution*, *91*(2), 156–168. doi:10.1007/s00239-023-10099-z PMID:36859501

Zhao, C., Li, S., Li, C., Wang, T., Tian, Y., & Li, X. (2021). Flavonoids from Fig (*Ficus carica* Linn.) Leaves: The development of a new extraction method and identification by UPLC-QTOF-MS/MS. *Applied Sciences (Basel, Switzerland)*, *11*(16), 7718. doi:10.3390/app11167718

Zhao, L., Sun, Z. F., Zhang, C. C., Nan, J., Ren, N. Q., Lee, D. J., & Chen, C. (2022). Advances in pretreatment of lignocellulosic biomass for bioenergy production : Challenges and perspectives. *Bioresource Technology*, *343*, 126123. doi:10.1016/j.biortech.2021.126123 PMID:34653621

Zhuang, Z., Zhao, S., Lin, Q., Cao, P., Lou, Y., Yang, L., & He, D. (2020). Preliminary estimation of the basic reproduction number of novel coronavirus (2019-nCoV) in China, from 2019 to 2020: A data-driven analysis in the early phase of the outbreak. *International Journal of Infectious Diseases*, *94*(29–31). PMID:32088335

Zhuang, Z., Zhao, S., Lin, Q., Cao, P., Lou, Y., Yang, L., & He, D. (2020). Preliminary estimation of the novel coronavirus disease (COVID-19) cases in Iran: A modelling analysis based on overseas cases and air travel data. *International Journal of Infectious Diseases*, *94*(29–31), 29–31. doi:10.1016/j.ijid.2020.03.019 PMID:32171951

Zhu, Y., & Shyy, G. (2017). *Sub-Daily Rainfall Forecasting by Using Facebook Prophet.* In *Proceedings of the 2017 International Conference on Computational Science and Computational Intelligence (CSCI)*, Las Vegas, USA.

Related References

To continue our tradition of advancing academic research, we have compiled a list of recommended IGI Global readings. These references will provide additional information and guidance to further enrich your knowledge and assist you with your own research and future publications.

Ajiboye, O. E., & Yusuff, O. S. (2017). Foreign Land Acquisition: Food Security and Food Chains – The Nigerian Experience. In I. Management Association (Ed.), Natural Resources Management: Concepts, Methodologies, Tools, and Applications (pp. 1524-1545). Hershey, PA: IGI Global. https://doi.org/ doi:10.4018/978-1-5225-0803-8.ch072

Alapiki, H. E., & Amadi, L. A. (2018). Sustainable Food Consumption in the Neoliberal Order: Challenges and Policy Implications. In A. Obayelu (Ed.), *Food Systems Sustainability and Environmental Policies in Modern Economies* (pp. 90–123). Hershey, PA: IGI Global. doi:10.4018/978-1-5225-3631-4.ch005

Altaş, A. (2018). Geographical Information System Applications Utilized in Museums in Turkey Within the Scope of the Cultural Heritage Tourism: A Case Study of Mobile Application of Müze Asist. In S. Chaudhuri & N. Ray (Eds.), *GIS Applications in the Tourism and Hospitality Industry* (pp. 42–60). Hershey, PA: IGI Global. doi:10.4018/978-1-5225-5088-4.ch002

Andreea, I. R. (2018). Beyond Macroeconomics of Food and Nutrition Security. *International Journal of Sustainable Economies Management, 7*(1), 13–22. doi:10.4018/IJSEM.2018010102

Anwar, J. (2017). Reproductive and Mental Health during Natural Disaster: Implications and Issues for Women in Developing Nations – A Case Example. In I. Management Association (Ed.), Gaming and Technology Addiction: Breakthroughs in Research and Practice (pp. 446-472). Hershey, PA: IGI Global. https://doi.org/ doi:10.4018/978-1-5225-0778-9.ch021

Awadh, H., Aksissou, M., Benhardouze, W., Darasi, F., & Snaiki, J. (2018). Socioeconomic Status of Artisanal Fishers in the West Part of Moroccan Mediterranean. *International Journal of Social Ecology and Sustainable Development, 9*(1), 40–52. doi:10.4018/IJSESD.2018010104

Aye, G. C., & Haruna, R. F. (2018). Effect of Climate Change on Crop Productivity and Prices in Benue State, Nigeria: Implications for Food Security. In V. Erokhin (Ed.), *Establishing Food Security and Alternatives to International Trade in Emerging Economies* (pp. 244–268). Hershey, PA: IGI Global. doi:10.4018/978-1-5225-2733-6.ch012

Azizan, S. A., & Suki, N. M. (2017). Consumers' Intentions to Purchase Organic Food Products. In T. Esakki (Ed.), *Green Marketing and Environmental Responsibility in Modern Corporations* (pp. 68–82). Hershey, PA: IGI Global. doi:10.4018/978-1-5225-2331-4.ch005

Barakabitze, A. A., Fue, K. G., Kitindi, E. J., & Sanga, C. A. (2017). Developing a Framework for Next Generation Integrated Agro Food-Advisory Systems in Developing Countries. In I. Management Association (Ed.), Agri-Food Supply Chain Management: Breakthroughs in Research and Practice (pp. 47-67). Hershey, PA: IGI Global. doi:10.4018/978-1-5225-1629-3.ch004

Beachcroft-Shaw, H., & Ellis, D. (2018). Using Successful Cases to Promote Environmental Sustainability: A Social Marketing Approach. In I. Management Association (Ed.), Sustainable Development: Concepts, Methodologies, Tools, and Applications (pp. 936-953). Hershey, PA: IGI Global. doi:10.4018/978-1-5225-3817-2.ch042

Behnassi, M., Kahime, K., Boussaa, S., Boumezzough, A., & Messouli, M. (2017). Infectious Diseases and Climate Vulnerability in Morocco: Governance and Adaptation Options. In I. Management Association (Ed.), Public Health and Welfare: Concepts, Methodologies, Tools, and Applications (pp. 91-109). Hershey, PA: IGI Global. https://doi.org/ doi:10.4018/978-1-5225-1674-3.ch005

Bekele, F., & Bekele, I. (2017). Social and Environmental Impacts on Agricultural Development. In W. Ganpat, R. Dyer, & W. Isaac (Eds.), *Agricultural Development and Food Security in Developing Nations* (pp. 21–56). Hershey, PA: IGI Global. doi:10.4018/978-1-5225-0942-4.ch002

Benaouda, A., & García-Peñalvo, F. J. (2018). Towards an Intelligent System for the Territorial Planning: Agricultural Case. In F. García-Peñalvo (Ed.), *Global Implications of Emerging Technology Trends* (pp. 158–178). Hershey, PA: IGI Global. doi:10.4018/978-1-5225-4944-4.ch010

Bhaskar, A., Rao, G. B., & Vencatesan, J. (2017). Characterization and Management Concerns of Water Resources around Pallikaranai Marsh, South Chennai. In P. Rao & Y. Patil (Eds.), *Reconsidering the Impact of Climate Change on Global Water Supply, Use, and Management* (pp. 102–121). Hershey, PA: IGI Global. doi:10.4018/978-1-5225-1046-8.ch007

Bhyan, P., Shrivastava, B., & Kumar, N. (2022). Requisite Sustainable Development Contemplating Buildings: Economic and Environmental Sustainability. In A. Hussain, K. Tiwari, & A. Gupta (Eds.), *Addressing Environmental Challenges Through Spatial Planning* (pp. 269–288). IGI Global. https://doi.org/10.4018/978-1-7998-8331-9.ch014

Bogataj, D., & Drobne, D. (2017). Control of Perishable Goods in Cold Logistic Chains by Bionanosensors. In S. Joo (Ed.), *Applying Nanotechnology for Environmental Sustainability* (pp. 376–402). Hershey, PA: IGI Global. doi:10.4018/978-1-5225-0585-3.ch016

Bogataj, D., & Drobne, D. (2017). Control of Perishable Goods in Cold Logistic Chains by Bionanosensors. In I. Management Association (Ed.), Materials Science and Engineering: Concepts, Methodologies, Tools, and Applications (pp. 471-497). Hershey, PA: IGI Global. https://doi.org/ doi:10.4018/978-1-5225-1798-6.ch019

Bogueva, D., & Marinova, D. (2018). What Is More Important: Perception of Masculinity or Personal Health and the Environment? In D. Bogueva, D. Marinova, & T. Raphaely (Eds.), *Handbook of Research on Social Marketing and Its Influence on Animal Origin Food Product Consumption* (pp. 148–162). Hershey, PA: IGI Global. doi:10.4018/978-1-5225-4757-0.ch010

Bouzid, M. (2017). Waterborne Diseases and Climate Change: Impact and Implications. In M. Bouzid (Ed.), *Examining the Role of Environmental Change on Emerging Infectious Diseases and Pandemics* (pp. 89–108). Hershey, PA: IGI Global. doi:10.4018/978-1-5225-0553-2.ch004

Bowles, D. C. (2017). Climate Change-Associated Conflict and Infectious Disease. In M. Bouzid (Ed.), *Examining the Role of Environmental Change on Emerging Infectious Diseases and Pandemics* (pp. 68–88). Hershey, PA: IGI Global. doi:10.4018/978-1-5225-0553-2.ch003

Buck, J. J., & Lowry, R. K. (2017). Oceanographic Data Management: Quills and Free Text to the Digital Age and "Big Data". In P. Diviacco, A. Leadbetter, & H. Glaves (Eds.), *Oceanographic and Marine Cross-Domain Data Management for Sustainable Development* (pp. 1–22). Hershey, PA: IGI Global. doi:10.4018/978-1-5225-0700-0.ch001

Buse, C. G. (2017). Are Climate Change Adaptation Policies a Game Changer?: A Case Study of Perspectives from Public Health Officials in Ontario, Canada. In M. Bouzid (Ed.), *Examining the Role of Environmental Change on Emerging Infectious Diseases and Pandemics* (pp. 230–257). Hershey, PA: IGI Global. doi:10.4018/978-1-5225-0553-2.ch010

Calderon, F. A., Giolo, E. G., Frau, C. D., Rengel, M. G., Rodriguez, H., Tornello, M., ... Gallucci, R. (2018). Seismic Microzonation and Site Effects Detection Through Microtremors Measures: A Review. In N. Ceryan (Ed.), *Handbook of Research on Trends and Digital Advances in Engineering Geology* (pp. 326–349). Hershey, PA: IGI Global. doi:10.4018/978-1-5225-2709-1.ch009

Carfì, D., Donato, A., & Panuccio, D. (2018). A Game Theory Coopetitive Perspective for Sustainability of Global Feeding: Agreements Among Vegan and Non-Vegan Food Firms. In I. Management Association (Ed.), Game Theory: Breakthroughs in Research and Practice (pp. 71-104). Hershey, PA: IGI Global. https://doi.org/ doi:10.4018/978-1-5225-2594-3.ch004

Castagnolo, V. (2018). Analyzing, Classifying, Safeguarding: Drawing for the Borgo Murattiano Neighbourhood of Bari. In G. Carlone, N. Martinelli, & F. Rotondo (Eds.), *Designing Grid Cities for Optimized Urban Development and Planning* (pp. 93–108). Hershey, PA: IGI Global. doi:10.4018/978-1-5225-3613-0.ch006

Chekima, B. (2018). The Dilemma of Purchase Intention: A Conceptual Framework for Understanding Actual Consumption of Organic Food. *International Journal of Sustainable Economies Management, 7*(2), 1–13. doi:10.4018/IJSEM.2018040101

Chen, Y. (2017). Sustainable Supply Chains and International Soft Landings: A Case of Wetland Entrepreneurship. In B. Christiansen & F. Kasarcı (Eds.), *Corporate Espionage, Geopolitics, and Diplomacy Issues in International Business* (pp. 232–247). Hershey, PA: IGI Global. doi:10.4018/978-1-5225-1031-4.ch013

Çıtak, L., Akel, V., & Ersoy, E. (2018). Investors' Reactions to the Announcement of New Constituents of BIST Sustainability Index: An Analysis by Event Study and Mean-Median Tests. In M. Risso & S. Testarmata (Eds.), *Value Sharing for Sustainable and Inclusive Development* (pp. 270–289). Hershey, PA: IGI Global. doi:10.4018/978-1-5225-3147-0.ch012

D'Aleo, V., D'Aleo, F., & Bonanno, R. (2018). New Food Industries Toward a New Level of Sustainable Supply: Success Stories, Business Models, and Strategies. In V. Erokhin (Ed.), *Establishing Food Security and Alternatives to International Trade in Emerging Economies* (pp. 74–97). Hershey, PA: IGI Global. doi:10.4018/978-1-5225-2733-6.ch004

Dagevos, H., & Reinders, M. J. (2018). Flexitarianism and Social Marketing: Reflections on Eating Meat in Moderation. In D. Bogueva, D. Marinova, & T. Raphaely (Eds.), *Handbook of Research on Social Marketing and Its Influence on Animal Origin Food Product Consumption* (pp. 105–120). Hershey, PA: IGI Global. doi:10.4018/978-1-5225-4757-0.ch007

Danisman, G. O. (2022). What Drives Eco-Design Innovations in European SMEs? In U. Akkucuk (Ed.), *Disruptive Technologies and Eco-Innovation for Sustainable Development* (pp. 191–206). IGI Global. https://doi.org/10.4018/978-1-7998-8900-7.ch011

Deenapanray, P. N., & Ramma, I. (2017). Adaptations to Climate Change and Climate Variability in the Agriculture Sector in Mauritius: Lessons from a Technical Needs Assessment. In I. Management Association (Ed.), Natural Resources Management: Concepts, Methodologies, Tools, and Applications (pp. 655-680). Hershey, PA: IGI Global. https://doi.org/ doi:10.4018/978-1-5225-0803-8.ch030

Deenapanray, P. N., & Ramma, I. (2017). Adaptations to Climate Change and Climate Variability in the Agriculture Sector in Mauritius: Lessons from a Technical Needs Assessment. In I. Management Association (Ed.), Natural Resources Management: Concepts, Methodologies, Tools, and Applications (pp. 655-680). Hershey, PA: IGI Global. https://doi.org/ doi:10.4018/978-1-5225-0803-8.ch030

Deshpande, S., Basu, S. K., Li, X., & Chen, X. (2017). Smart, Innovative and Intelligent Technologies Used in Drug Designing. In I. Management Association (Ed.), Pharmaceutical Sciences: Breakthroughs in Research and Practice (pp. 1175-1191). Hershey, PA: IGI Global. https://doi.org/ doi:10.4018/978-1-5225-1762-7.ch045

Dlamini, P. N. (2017). Use of Information and Communication Technologies Tools to Capture, Store, and Disseminate Indigenous Knowledge: A Literature Review. In P. Ngulube (Ed.), *Handbook of Research on Theoretical Perspectives on Indigenous Knowledge Systems in Developing Countries* (pp. 225–247). Hershey, PA: IGI Global. doi:10.4018/978-1-5225-0833-5.ch010

Dolejsova, M., & Kera, D. (2017). The Fermentation GutHub Project and the Internet of Microbes. In S. Konomi & G. Roussos (Eds.), *Enriching Urban Spaces with Ambient Computing, the Internet of Things, and Smart City Design* (pp. 25–46). Hershey, PA: IGI Global. doi:10.4018/978-1-5225-0827-4.ch002

Dolunay, O. (2018). A Paradigm Shift: Empowering Farmers to Eliminate the Waste in the Form of Fresh Water and Energy Through the Implementation of 4R+T. In I. Management Association (Ed.), Sustainable Development: Concepts, Methodologies, Tools, and Applications (pp. 882-892). Hershey, PA: IGI Global. https://doi.org/ doi:10.4018/978-1-5225-3817-2.ch039

Dube, P., Heijman, W. J., Ihle, R., & Ochieng, J. (2018). The Potential of Traditional Leafy Vegetables for Improving Food Security in Africa. In V. Erokhin (Ed.), *Establishing Food Security and Alternatives to International Trade in Emerging Economies* (pp. 220–243). Hershey, PA: IGI Global. doi:10.4018/978-1-5225-2733-6.ch011

Duruji, M. M., & Urenma, D. F. (2017). The Environmentalism and Politics of Climate Change: A Study of the Process of Global Convergence through UNFCCC Conferences. In I. Management Association (Ed.), *Natural Resources Management: Concepts, Methodologies, Tools, and Applications* (pp. 77-108). Hershey, PA: IGI Global. 10.4018/978-1-5225-0803-8.ch004

Dutta, U. (2017). Agro-Geoinformatics, Potato Cultivation, and Climate Change. In S. Londhe (Ed.), *Sustainable Potato Production and the Impact of Climate Change* (pp. 247–271). Hershey, PA: IGI Global. doi:10.4018/978-1-5225-1715-3.ch012

Edirisinghe, R., Stranieri, A., & Wickramasinghe, N. (2017). A Taxonomy for mHealth. In N. Wickramasinghe (Ed.), *Handbook of Research on Healthcare Administration and Management* (pp. 596–615). Hershey, PA: IGI Global. doi:10.4018/978-1-5225-0920-2.ch036

Ekpeni, N. M., & Ayeni, A. O. (2018). Global Natural Hazard and Disaster Vulnerability Management. In A. Eneanya (Ed.), *Handbook of Research on Environmental Policies for Emergency Management and Public Safety* (pp. 83–104). Hershey, PA: IGI Global. doi:10.4018/978-1-5225-3194-4.ch005

Ene, C., Voica, M. C., & Panait, M. (2017). Green Investments and Food Security: Opportunities and Future Directions in the Context of Sustainable Development. In M. Mieila (Ed.), *Measuring Sustainable Development and Green Investments in Contemporary Economies* (pp. 163–200). Hershey, PA: IGI Global. doi:10.4018/978-1-5225-2081-8.ch007

Escamilla, I., Ruíz, M. T., Ibarra, M. M., Soto, V. L., Quintero, R., & Guzmán, G. (2018). Geocoding Tweets Based on Semantic Web and Ontologies. In M. Lytras, N. Aljohani, E. Damiani, & K. Chui (Eds.), *Innovations, Developments, and Applications of Semantic Web and Information Systems* (pp. 372–392). Hershey, PA: IGI Global. doi:10.4018/978-1-5225-5042-6.ch014

Escribano, A. J. (2018). Marketing Strategies for Trendy Animal Products: Sustainability as a Core. In F. Quoquab, R. Thurasamy, & J. Mohammad (Eds.), *Driving Green Consumerism Through Strategic Sustainability Marketing* (pp. 169–203). Hershey, PA: IGI Global. doi:10.4018/978-1-5225-2912-5.ch010

Eudoxie, G., & Roopnarine, R. (2017). Climate Change Adaptation and Disaster Risk Management in the Caribbean. In W. Ganpat & W. Isaac (Eds.), *Environmental Sustainability and Climate Change Adaptation Strategies* (pp. 97–125). Hershey, PA: IGI Global. doi:10.4018/978-1-5225-1607-1.ch004

Farmer, L. S. (2017). Data Analytics for Strategic Management: Getting the Right Data. In V. Wang (Ed.), *Encyclopedia of Strategic Leadership and Management* (pp. 810–822). Hershey, PA: IGI Global. doi:10.4018/978-1-5225-1049-9.ch056

Fattal, L. R. (2017). Catastrophe: An Uncanny Catalyst for Creativity. In R. Shin (Ed.), *Convergence of Contemporary Art, Visual Culture, and Global Civic Engagement* (pp. 244–262). Hershey, PA: IGI Global. doi:10.4018/978-1-5225-1665-1.ch014

Forti, I. (2017). A Cross Reading of Landscape through Digital Landscape Models: The Case of Southern Garda. In A. Ippolito (Ed.), *Handbook of Research on Emerging Technologies for Architectural and Archaeological Heritage* (pp. 532–561). Hershey, PA: IGI Global. doi:10.4018/978-1-5225-0675-1.ch018

Gharbi, A., De Runz, C., & Akdag, H. (2017). Urban Development Modelling: A Survey. In S. Faiz & K. Mahmoudi (Eds.), *Handbook of Research on Geographic Information Systems Applications and Advancements* (pp. 96–124). Hershey, PA: IGI Global. doi:10.4018/978-1-5225-0937-0.ch004

Ghosh, I., & Ghoshal, I. (2018). Implications of Trade Liberalization for Food Security Under the ASEAN-India Strategic Partnership: A Gravity Model Approach. In V. Erokhin (Ed.), *Establishing Food Security and Alternatives to International Trade in Emerging Economies* (pp. 98–118). Hershey, PA: IGI Global. doi:10.4018/978-1-5225-2733-6.ch005

Glaves, H. M. (2017). Developing a Common Global Framework for Marine Data Management. In P. Diviacco, A. Leadbetter, & H. Glaves (Eds.), *Oceanographic and Marine Cross-Domain Data Management for Sustainable Development* (pp. 47–68). Hershey, PA: IGI Global. doi:10.4018/978-1-5225-0700-0.ch003

Godulla, A., & Wolf, C. (2018). Future of Food: Transmedia Strategies of National Geographic. In R. Gambarato & G. Alzamora (Eds.), *Exploring Transmedia Journalism in the Digital Age* (pp. 162–182). Hershey, PA: IGI Global. doi:10.4018/978-1-5225-3781-6.ch010

Gomes, P. P. (2018). Food and Environment: A Review on the Sustainability of Six Different Dietary Patterns. In A. Obayelu (Ed.), *Food Systems Sustainability and Environmental Policies in Modern Economies* (pp. 15–31). Hershey, PA: IGI Global. doi:10.4018/978-1-5225-3631-4.ch002

Gonzalez-Feliu, J. (2018). Sustainability Evaluation of Green Urban Logistics Systems: Literature Overview and Proposed Framework. In A. Paul, D. Bhattacharyya, & S. Anand (Eds.), *Green Initiatives for Business Sustainability and Value Creation* (pp. 103–134). Hershey, PA: IGI Global. doi:10.4018/978-1-5225-2662-9.ch005

Goodland, R. (2017). A Fresh Look at Livestock Greenhouse Gas Emissions and Mitigation. In I. Management Association (Ed.), Natural Resources Management: Concepts, Methodologies, Tools, and Applications (pp. 124-139). Hershey, PA: IGI Global. doi:10.4018/978-1-5225-0803-8.ch006

Goundar, S., & Appana, S. (2018). Mainstreaming Development Policies for Climate Change in Fiji: A Policy Gap Analysis and the Role of ICTs. In I. Management Association (Ed.), Sustainable Development: Concepts, Methodologies, Tools, and Applications (pp. 402-432). Hershey, PA: IGI Global. doi:10.4018/978-1-5225-3817-2.ch020

Granell-Canut, C., & Aguilar-Moreno, E. (2018). Geospatial Influence in Science Mapping. In M. Khosrow-Pour, D.B.A. (Ed.), Encyclopedia of Information Science and Technology, Fourth Edition (pp. 3473-3483). Hershey, PA: IGI Global. doi:10.4018/978-1-5225-2255-3.ch302

Grigelis, A., Blažauskas, N., Gelumbauskaitė, L. Ž., Gulbinskas, S., Suzdalev, S., & Ferrarin, C. (2017). Marine Environment Data Management Related to the Human Activity in the South-Eastern Baltic Sea (The Lithuanian Sector). In P. Diviacco, A. Leadbetter, & H. Glaves (Eds.), *Oceanographic and Marine Cross-Domain Data Management for Sustainable Development* (pp. 282–302). Hershey, PA: IGI Global. doi:10.4018/978-1-5225-0700-0.ch012

Guma, I. P., Rwashana, A. S., & Oyo, B. (2018). Food Security Policy Analysis Using System Dynamics: The Case of Uganda. *International Journal of Information Technologies and Systems Approach, 11*(1), 72–90. doi:10.4018/IJITSA.2018010104

Guma, I. P., Rwashana, A. S., & Oyo, B. (2018). Food Security Indicators for Subsistence Farmers Sustainability: A System Dynamics Approach. *International Journal of System Dynamics Applications, 7*(1), 45–64. doi:10.4018/IJSDA.2018010103

Gupta, P., & Goyal, S. (2017). Wildlife Habitat Evaluation. In A. Santra & S. Mitra (Eds.), *Remote Sensing Techniques and GIS Applications in Earth and Environmental Studies* (pp. 258–264). Hershey, PA: IGI Global. doi:10.4018/978-1-5225-1814-3.ch013

Hanson, T., & Hildebrand, E. (2018). GPS Travel Diaries in Rural Transportation Research: A Focus on Older Drivers. In I. Management Association (Ed.), Intelligent Transportation and Planning: Breakthroughs in Research and Practice (pp. 609-625). Hershey, PA: IGI Global. doi:10.4018/978-1-5225-5210-9.ch027

Hartman, M. B. (2017). Research-Based Climate Change Public Education Programs. In I. Management Association (Ed.), Natural Resources Management: Concepts, Methodologies, Tools, and Applications (pp. 992-1003). Hershey, PA: IGI Global. doi:10.4018/978-1-5225-0803-8.ch046

Hashim, N. (2017). Zanzibari Seaweed: Global Climate Change and the Promise of Adaptation. In I. Management Association (Ed.), Natural Resources Management: Concepts, Methodologies, Tools, and Applications (pp. 365-391). Hershey, PA: IGI Global. https://doi.org/ doi:10.4018/978-1-5225-0803-8.ch019

Herrera, J. E., Argüello, L. V., Gonzalez-Feliu, J., & Jaimes, W. A. (2017). Decision Support System Design Requirements, Information Management, and Urban Logistics Efficiency: Case Study of Bogotá, Colombia. In G. Jamil, A. Soares, & C. Pessoa (Eds.), *Handbook of Research on Information Management for Effective Logistics and Supply Chains* (pp. 223–238). Hershey, PA: IGI Global. doi:10.4018/978-1-5225-0973-8.ch012

Huizinga, T., Ayanso, A., Smoor, M., & Wronski, T. (2017). Exploring Insurance and Natural Disaster Tweets Using Text Analytics. *International Journal of Business Analytics, 4*(1), 1–17. doi:10.4018/IJBAN.2017010101

Hung, K., Kalantari, M., & Rajabifard, A. (2017). An Integrated Method for Assessing the Text Content Quality of Volunteered Geographic Information in Disaster Management. *International Journal of Information Systems for Crisis Response and Management, 9*(2), 1–17. doi:10.4018/IJISCRAM.2017040101

Husnain, A., & Avdic, A. (2018). Identifying the Contemporary Status of E-Service Sustainability Research. In I. Management Association (Ed.), Sustainable Development: Concepts, Methodologies, Tools, and Applications (pp. 467-485). Hershey, PA: IGI Global. https://doi.org/ doi:10.4018/978-1-5225-3817-2.ch022

Iarossi, M. P., & Ferro, L. (2017). "The Past is Never Dead. It's Not Even Past": Virtual Archaeological Promenade. In A. Ippolito & M. Cigola (Eds.), *Handbook of Research on Emerging Technologies for Digital Preservation and Information Modeling* (pp. 228–255). Hershey, PA: IGI Global. doi:10.4018/978-1-5225-0680-5.ch010

Ignjatijević, S., & Cvijanović, D. (2018). Analysis of Serbian Honey Production and Exports. In *Exploring the Global Competitiveness of Agri-Food Sectors and Serbia's Dominant Presence: Emerging Research and Opportunities* (pp. 109–139). Hershey, PA: IGI Global. doi:10.4018/978-1-5225-2762-6.ch005

Jana, S. K., & Karmakar, A. K. (2017). Globalization, Governance, and Food Security: The Case of BRICS. In I. Management Association (Ed.), Natural Resources Management: Concepts, Methodologies, Tools, and Applications (pp. 692-712). Hershey, PA: IGI Global. https://doi.org/ doi:10.4018/978-1-5225-0803-8.ch032

Jana, S. K., & Karmakar, A. K. (2017). Food Security in Asia: Is There Convergence? In I. Management Association (Ed.), Natural Resources Management: Concepts, Methodologies, Tools, and Applications (pp. 109-123). Hershey, PA: IGI Global. https://doi.org/ doi:10.4018/978-1-5225-0803-8.ch005

John, J., & Kumar, S. (2018). A Locational Decision Making Framework for Shipbreaking Under Multiple Criteria. In I. Management Association (Ed.), Operations and Service Management: Concepts, Methodologies, Tools, and Applications (pp. 504-527). Hershey, PA: IGI Global. doi:10.4018/978-1-5225-3909-4.ch024

John, J., & Srivastava, R. K. (2018). Decision Insights for Shipbreaking using Environmental Impact Assessment: Review and Perspectives. *International Journal of Strategic Decision Sciences*, *9*(1), 45–62. doi:10.4018/IJSDS.2018010104

Joshi, Y., & Rahman, Z. (2018). Determinants of Sustainable Consumption Behaviour: Review and Conceptual Framework. In A. Paul, D. Bhattacharyya, & S. Anand (Eds.), *Green Initiatives for Business Sustainability and Value Creation* (pp. 239–262). Hershey, PA: IGI Global. doi:10.4018/978-1-5225-2662-9.ch011

Juma, D. W., Reuben, M., Wang, H., & Li, F. (2018). Adaptive Coevolution: Realigning the Water Governance Regime to the Changing Climate. In I. Management Association (Ed.), Hydrology and Water Resource Management: Breakthroughs in Research and Practice (pp. 346-357). Hershey, PA: IGI Global. https://doi.org/ doi:10.4018/978-1-5225-3427-3.ch014

K., S., & Tripathy, B. K. (2018). Neighborhood Rough-Sets-Based Spatial Data Analytics. In M. Khosrow-Pour, D.B.A. (Ed.), *Encyclopedia of Information Science and Technology, Fourth Edition* (pp. 1835-1844). Hershey, PA: IGI Global. https://doi.org/ doi:10.4018/978-1-5225-2255-3.ch160

Kabir, F. (2018). Towards a More Gender-Inclusive Climate Change Policy. In N. Mahtab, T. Haque, I. Khan, M. Islam, & I. Wahid (Eds.), *Handbook of Research on Women's Issues and Rights in the Developing World* (pp. 354–369). Hershey, PA: IGI Global. doi:10.4018/978-1-5225-3018-3.ch022

Kabir, F. (2018). Towards a More Gender-Inclusive Climate Change Policy. In I. Management Association (Ed.), Climate Change and Environmental Concerns: Breakthroughs in Research and Practice (pp. 525-540). Hershey, PA: IGI Global. https://doi.org/ doi:10.4018/978-1-5225-5487-5.ch027

Kanyamuka, J. S., Jumbe, C. B., & Ricker-Gilbert, J. (2018). Making Agricultural Input Subsidies More Effective and Profitable in Africa: The Role of Complementary Interventions. In A. Obayelu (Ed.), *Food Systems Sustainability and Environmental Policies in Modern Economies* (pp. 172–187). Hershey, PA: IGI Global. doi:10.4018/978-1-5225-3631-4.ch008

Karimi, H., & Gholamrezafahimi, F. (2017). Study of Integrated Coastal Zone Management and Its Environmental Effects: A Case of Iran. In R. Singh, A. Singh, & V. Srivastava (Eds.), *Environmental Issues Surrounding Human Overpopulation* (pp. 64–88). Hershey, PA: IGI Global. doi:10.4018/978-1-5225-1683-5.ch004

Kaya, I. R., Hutabarat, J., & Bambang, A. N. (2018). "Sasi": A New Path to Sustain Seaweed Farming From Up-Stream to Down-Stream in Kotania Bay, Molucass. *International Journal of Social Ecology and Sustainable Development*, *9*(2), 28–36. doi:10.4018/IJSESD.2018040103

Khader, V. (2018). Technologies for Food, Health, Livelihood, and Nutrition Security. In I. Management Association (Ed.), Food Science and Nutrition: Breakthroughs in Research and Practice (pp. 94-112). Hershey, PA: IGI Global. https://doi.org/ doi:10.4018/978-1-5225-5207-9.ch005

Kocadağlı, A. Y. (2017). The Temporal and Spatial Development of Organic Agriculture in Turkey. In W. Ganpat, R. Dyer, & W. Isaac (Eds.), *Agricultural Development and Food Security in Developing Nations* (pp. 130–156). Hershey, PA: IGI Global. doi:10.4018/978-1-5225-0942-4.ch006

Koundouri, P., Giannouli, A., & Souliotis, I. (2017). An Integrated Approach for Sustainable Environmental and Socio-Economic Development Using Offshore Infrastructure. In I. Management Association (Ed.), Renewable and Alternative Energy: Concepts, Methodologies, Tools, and Applications (pp. 1581-1601). Hershey, PA: IGI Global. doi:10.4018/978-1-5225-1671-2.ch056

Kumar, A., & Dash, M. K. (2017). Sustainability and Future Generation Infrastructure on Digital Platform: A Study of Generation Y. In N. Ray (Ed.), *Business Infrastructure for Sustainability in Developing Economies* (pp. 124–142). Hershey, PA: IGI Global. doi:10.4018/978-1-5225-2041-2.ch007

Kumar, A., Mukherjee, A. B., & Krishna, A. P. (2017). Application of Conventional Data Mining Techniques and Web Mining to Aid Disaster Management. In A. Kumar (Ed.), *Web Usage Mining Techniques and Applications Across Industries* (pp. 138–167). Hershey, PA: IGI Global. doi:10.4018/978-1-5225-0613-3.ch006

Kumar, C. P. (2017). Impact of Climate Change on Groundwater Resources. In I. Management Association (Ed.), Natural Resources Management: Concepts, Methodologies, Tools, and Applications (pp. 1094-1120). Hershey, PA: IGI Global. doi:10.4018/978-1-5225-0803-8.ch052

Kumari, S., & Patil, Y. (2017). Achieving Climate Smart Agriculture with a Sustainable Use of Water: A Conceptual Framework for Sustaining the Use of Water for Agriculture in the Era of Climate Change. In P. Rao & Y. Patil (Eds.), *Reconsidering the Impact of Climate Change on Global Water Supply, Use, and Management* (pp. 122–143). Hershey, PA: IGI Global. doi:10.4018/978-1-5225-1046-8.ch008

Kumari, S., & Patil, Y. (2018). Achieving Climate Smart Agriculture With a Sustainable Use of Water: A Conceptual Framework for Sustaining the Use of Water for Agriculture in the Era of Climate Change. In I. Management Association (Ed.), Climate Change and Environmental Concerns: Breakthroughs in Research and Practice (pp. 111-133). Hershey, PA: IGI Global. doi:10.4018/978-1-5225-5487-5.ch006

Kursah, M. B. (2017). Least-Cost Pipeline using Geographic Information System: The Limit to Technicalities. *International Journal of Applied Geospatial Research*, *8*(3), 1–15. doi:10.4018/ijagr.2017070101

Lahiri, S., Ghosh, D., & Bhakta, J. N. (2017). Role of Microbes in Eco-Remediation of Perturbed Aquatic Ecosystem. In J. Bhakta (Ed.), *Handbook of Research on Inventive Bioremediation Techniques* (pp. 70–107). Hershey, PA: IGI Global. doi:10.4018/978-1-5225-2325-3.ch004

Lallo, C. H., Smalling, S., Facey, A., & Hughes, M. (2018). The Impact of Climate Change on Small Ruminant Performance in Caribbean Communities. In I. Management Association (Ed.), Climate Change and Environmental Concerns: Breakthroughs in Research and Practice (pp. 193-218). Hershey, PA: IGI Global. doi:10.4018/978-1-5225-5487-5.ch010

Laurini, R. (2017). Nature of Geographic Knowledge Bases. In S. Faiz & K. Mahmoudi (Eds.), *Handbook of Research on Geographic Information Systems Applications and Advancements* (pp. 29–60). Hershey, PA: IGI Global. doi:10.4018/978-1-5225-0937-0.ch002

Lawrence, J., Simpson, L., & Piggott, A. (2017). Protected Agriculture: A Climate Change Adaptation for Food and Nutrition Security. In I. Management Association (Ed.), Natural Resources Management: Concepts, Methodologies, Tools, and Applications (pp. 140-158). Hershey, PA: IGI Global. doi:10.4018/978-1-5225-0803-8.ch007

Leadbetter, A., Cheatham, M., Shepherd, A., & Thomas, R. (2017). Linked Ocean Data 2.0. In P. Diviacco, A. Leadbetter, & H. Glaves (Eds.), *Oceanographic and Marine Cross-Domain Data Management for Sustainable Development* (pp. 69–99). Hershey, PA: IGI Global. doi:10.4018/978-1-5225-0700-0.ch004

Lucas, M. R., Rego, C., Vieira, C., & Vieira, I. (2017). Proximity and Cooperation for Innovative Regional Development: The Case of the Science and Technology Park of Alentejo. In L. Carvalho (Ed.), *Handbook of Research on Entrepreneurial Development and Innovation Within Smart Cities* (pp. 199–228). Hershey, PA: IGI Global. doi:10.4018/978-1-5225-1978-2.ch010

Ma, X., Beaulieu, S. E., Fu, L., Fox, P., Di Stefano, M., & West, P. (2017). Documenting Provenance for Reproducible Marine Ecosystem Assessment in Open Science. In P. Diviacco, A. Leadbetter, & H. Glaves (Eds.), *Oceanographic and Marine Cross-Domain Data Management for Sustainable Development* (pp. 100–126). Hershey, PA: IGI Global. doi:10.4018/978-1-5225-0700-0.ch005

Mabe, L. K., & Oladele, O. I. (2017). Application of Information Communication Technologies for Agricultural Development through Extension Services: A Review. In T. Tossy (Ed.), *Information Technology Integration for Socio-Economic Development* (pp. 52–101). Hershey, PA: IGI Global. doi:10.4018/978-1-5225-0539-6.ch003

Malomo, B. I. (2018). A Review of Psychological Resilience as a Response to Natural Hazards in Nigeria. In A. Eneanya (Ed.), *Handbook of Research on Environmental Policies for Emergency Management and Public Safety* (pp. 147–165). Hershey, PA: IGI Global. doi:10.4018/978-1-5225-3194-4.ch008

Manchiraju, S. (2018). Predicting Behavioral Intentions Toward Sustainable Fashion Consumption: A Comparison of Attitude-Behavior and Value-Behavior Consistency Models. In I. Management Association (Ed.), Fashion and Textiles: Breakthroughs in Research and Practice (pp. 1-21). Hershey, PA: IGI Global. doi:10.4018/978-1-5225-3432-7.ch001

Manzella, G. M., Bartolini, R., Bustaffa, F., D'Angelo, P., De Mattei, M., Frontini, F., ... Spada, A. (2017). Semantic Search Engine for Data Management and Sustainable Development: Marine Planning Service Platform. In P. Diviacco, A. Leadbetter, & H. Glaves (Eds.), *Oceanographic and Marine Cross-Domain Data Management for Sustainable Development* (pp. 127–154). Hershey, PA: IGI Global. doi:10.4018/978-1-5225-0700-0.ch006

Mbonigaba, J. (2018). Comparing the Effects of Unsustainable Production and Consumption of Food on Health and Policy Across Developed and Less Developed Countries. In A. Obayelu (Ed.), *Food Systems Sustainability and Environmental Policies in Modern Economies* (pp. 124–158). Hershey, PA: IGI Global. doi:10.4018/978-1-5225-3631-4.ch006

McKeown, A. E. (2017). Nurses, Healthcare, and Environmental Pollution and Solutions: Breaking the Cycle of Harm. In I. Management Association (Ed.), Natural Resources Management: Concepts, Methodologies, Tools, and Applications (pp. 392-415). Hershey, PA: IGI Global. https://doi.org/ doi:10.4018/978-1-5225-0803-8.ch020

Mili, B., Barua, A., & Katyaini, S. (2017). Climate Change and Adaptation through the Lens of Capability Approach: A Case Study from Darjeeling, Eastern Himalaya. In I. Management Association (Ed.), Natural Resources Management: Concepts, Methodologies, Tools, and Applications (pp. 1351-1365). Hershey, PA: IGI Global. https://doi.org/ doi:10.4018/978-1-5225-0803-8.ch064

Mir, S. A., Shah, M. A., Mir, M. M., & Iqbal, U. (2017). New Horizons of Nanotechnology in Agriculture and Food Processing Industry. In B. Nayak, A. Nanda, & M. Bhat (Eds.), *Integrating Biologically-Inspired Nanotechnology into Medical Practice* (pp. 230–258). Hershey, PA: IGI Global. doi:10.4018/978-1-5225-0610-2.ch009

Moallem, M., Sterrett, W. L., Gordon, C. R., Sukhera, S. M., Mahmood, A., & Bashir, A. (2021). An Investigation of the Effects of Integrating Computing and Project- or Problem-Based Learning in the Context of Environmental Sciences: A Case of Pakistani STEM Teachers. In S. Schroth & J. Daniels (Eds.), *Building STEM Skills Through Environmental Education* (pp. 49–89). IGI Global. https://doi.org/10.4018/978-1-7998-2711-5.ch003

Mujere, N., & Moyce, W. (2017). Climate Change Impacts on Surface Water Quality. In W. Ganpat & W. Isaac (Eds.), *Environmental Sustainability and Climate Change Adaptation Strategies* (pp. 322–340). Hershey, PA: IGI Global. doi:10.4018/978-1-5225-1607-1.ch012

Mukherjee, A. B., Krishna, A. P., & Patel, N. (2018). Geospatial Technology for Urban Sciences. In *Geospatial Technologies in Urban System Development: Emerging Research and Opportunities* (pp. 99–120). Hershey, PA: IGI Global. doi:10.4018/978-1-5225-3683-3.ch005

Nagarajan, S. K., & Sangaiah, A. K. (2017). Vegetation Index: Ideas, Methods, Influences, and Trends. In N. Kumar, A. Sangaiah, M. Arun, & S. Anand (Eds.), *Advanced Image Processing Techniques and Applications* (pp. 347–386). Hershey, PA: IGI Global. doi:10.4018/978-1-5225-2053-5.ch016

Naraine, L., & Meehan, K. (2017). Strengthening Food Security with Sustainable Practices by Smallholder Farmers in Lesser Developed Economies. In W. Ganpat, R. Dyer, & W. Isaac (Eds.), *Agricultural Development and Food Security in Developing Nations* (pp. 57–81). Hershey, PA: IGI Global. doi:10.4018/978-1-5225-0942-4.ch003

Naraine, L., & Meehan, K. (2017). Strengthening Food Security with Sustainable Practices by Smallholder Farmers in Lesser Developed Economies. In W. Ganpat, R. Dyer, & W. Isaac (Eds.), *Agricultural Development and Food Security in Developing Nations* (pp. 57–81). Hershey, PA: IGI Global. doi:10.4018/978-1-5225-0942-4.ch003

Naraine, L., & Meehan, K. (2017). Strengthening Food Security with Sustainable Practices by Smallholder Farmers in Lesser Developed Economies. In W. Ganpat, R. Dyer, & W. Isaac (Eds.), *Agricultural Development and Food Security in Developing Nations* (pp. 57–81). Hershey, PA: IGI Global. doi:10.4018/978-1-5225-0942-4.ch003

Nikolaou, K., Tsakiridou, E., Anastasiadis, F., & Mattas, K. (2017). Exploring Alternative Distribution Channels of Agricultural Products. *International Journal of Food and Beverage Manufacturing and Business Models*, *2*(2), 36–66. doi:10.4018/IJFBMBM.2017070103

Nishat, K. J., & Rahman, M. S. (2018). Disaster, Vulnerability, and Violence Against Women: Global Findings and a Research Agenda for Bangladesh. In N. Mahtab, T. Haque, I. Khan, M. Islam, & I. Wahid (Eds.), *Handbook of Research on Women's Issues and Rights in the Developing World* (pp. 235–250). Hershey, PA: IGI Global. doi:10.4018/978-1-5225-3018-3.ch014

O'Hara, S., Jones, D., & Trobman, H. B. (2018). Building an Urban Food System Through UDC Food Hubs. In A. Burtin, J. Fleming, & P. Hampton-Garland (Eds.), *Changing Urban Landscapes Through Public Higher Education* (pp. 116–143). Hershey, PA: IGI Global. doi:10.4018/978-1-5225-3454-9.ch006

Obayelu, A. E. (2018). Integrating Environment, Food Systems, and Sustainability in Feeding the Growing Population in Developing Countries. In A. Obayelu (Ed.), *Food Systems Sustainability and Environmental Policies in Modern Economies* (pp. 1–14). Hershey, PA: IGI Global. doi:10.4018/978-1-5225-3631-4.ch001

Othman, R., Nath, N., & Laswad, F. (2018). Environmental Reporting and Accounting: Sustainability Hybridisation. In G. Azevedo, J. da Silva Oliveira, R. Marques, & A. Ferreira (Eds.), *Handbook of Research on Modernization and Accountability in Public Sector Management* (pp. 130–158). Hershey, PA: IGI Global. doi:10.4018/978-1-5225-3731-1.ch007

Ouadi, A., & Zitouni, A. (2021). Phasor Measurement Improvement Using Digital Filter in a Smart Grid. In A. Recioui & H. Bentarzi (Eds.), *Optimizing and Measuring Smart Grid Operation and Control* (pp. 100–117). IGI Global. https://doi.org/10.4018/978-1-7998-4027-5.ch005

Padigala, B. S. (2018). Traditional Water Management System for Climate Change Adaptation in Mountain Ecosystems. In I. Management Association (Ed.), Climate Change and Environmental Concerns: Breakthroughs in Research and Practice (pp. 630-655). Hershey, PA: IGI Global. doi:10.4018/978-1-5225-5487-5.ch033

Panda, C. K. (2018). Mobile Phone Usage in Agricultural Extension in India: The Current and Future Perspective. In F. Mtenzi, G. Oreku, D. Lupiana, & J. Yonazi (Eds.), *Mobile Technologies and Socio-Economic Development in Emerging Nations* (pp. 1–21). Hershey, PA: IGI Global. doi:10.4018/978-1-5225-4029-8.ch001

Pandian, S. L., Yarrakula, K., & Chaudhury, P. (2018). GIS-Based Decision Support System for Village Level: A Case Study in Andhra Pradesh. In S. Chaudhuri & N. Ray (Eds.), *GIS Applications in the Tourism and Hospitality Industry* (pp. 275–295). Hershey, PA: IGI Global. doi:10.4018/978-1-5225-5088-4.ch012

Quaranta, G., & Salvia, R. (2017). Social-Based Product Innovation and Governance in The Milk Sector: The Case of Carciocacio and Innonatura. In T. Tarnanidis, M. Vlachopoulou, & J. Papathanasiou (Eds.), *Driving Agribusiness With Technology Innovations* (pp. 293–310). Hershey, PA: IGI Global. doi:10.4018/978-1-5225-2107-5.ch015

Rahman, M. K., Schmidlin, T. W., Munro-Stasiuk, M. J., & Curtis, A. (2017). Geospatial Analysis of Land Loss, Land Cover Change, and Landuse Patterns of Kutubdia Island, Bangladesh. *International Journal of Applied Geospatial Research*, *8*(2), 45–60. doi:10.4018/IJAGR.2017040104

Rajack-Talley, T. A. (2017). Agriculture, Trade Liberalization and Poverty in the ACP Countries. In W. Ganpat, R. Dyer, & W. Isaac (Eds.), *Agricultural Development and Food Security in Developing Nations* (pp. 1–20). Hershey, PA: IGI Global. doi:10.4018/978-1-5225-0942-4.ch001

Rajamanickam, S. (2018). Exploring Landscapes in Regional Convergence: Environment and Sustainable Development in South Asia. In I. Management Association (Ed.), Sustainable Development: Concepts, Methodologies, Tools, and Applications (pp. 1051-1087). Hershey, PA: IGI Global. https://doi.org/ doi:10.4018/978-1-5225-3817-2.ch047

Rizvi, S. M., & Dearden, A. (2018). KHETI: ICT Solution for Agriculture Extension and Its Replication in Open and Distance Learning. In U. Pandey & V. Indrakanti (Eds.), *Open and Distance Learning Initiatives for Sustainable Development* (pp. 163–174). Hershey, PA: IGI Global. doi:10.4018/978-1-5225-2621-6.ch008

Roşu, L., & Macarov, L. I. (2017). Management of Drought and Floods in the Dobrogea Region. In I. Management Association (Ed.), Agri-Food Supply Chain Management: Breakthroughs in Research and Practice (pp. 372-403). Hershey, PA: IGI Global. https://doi.org/ doi:10.4018/978-1-5225-1629-3.ch016

Rouzbehani, K., & Rouzbehani, S. (2018). Mapping Women's World: GIS and the Case of Breast Cancer in the US. *International Journal of Public Health Management and Ethics*, *3*(1), 14–25. doi:10.4018/IJPHME.2018010102

Roy, D. (2018). Success Factors of Adoption of Mobile Applications in Rural India: Effect of Service Characteristics on Conceptual Model. In M. Khosrow-Pour, D.B.A. (Ed.), Green Computing Strategies for Competitive Advantage and Business Sustainability (pp. 211-238). Hershey, PA: IGI Global. https://doi.org/ doi:10.4018/978-1-5225-5017-4.ch010

Sajeva, M., Lemon, M., & Sahota, P. S. (2017). Governance for Food Security: A Framework for Social Learning and Scenario Building. *International Journal of Food and Beverage Manufacturing and Business Models*, *2*(2), 67–84. doi:10.4018/IJFBMBM.2017070104

Sambhanthan, A., & Potdar, V. (2017). A Study of the Parameters Impacting Sustainability in Information Technology Organizations. *International Journal of Knowledge-Based Organizations*, *7*(3), 27–39. doi:10.4018/IJKBO.2017070103

Sanga, C., Kalungwizi, V. J., & Msuya, C. P. (2017). Bridging Gender Gaps in Provision of Agricultural Extension Service Using ICT: Experiences from Sokoine University of Agriculture (SUA) Farmer Voice Radio (FVR) Project in Tanzania. In I. Management Association (Ed.), Discrimination and Diversity: Concepts, Methodologies, Tools, and Applications (pp. 682-697). Hershey, PA: IGI Global. https://doi.org/ doi:10.4018/978-1-5225-1933-1.ch031

Santra, A., & Mitra, D. (2017). Role of Remote Sensing in Potential Fishing Zone Forecast. In A. Santra & S. Mitra (Eds.), *Remote Sensing Techniques and GIS Applications in Earth and Environmental Studies* (pp. 243–257). Hershey, PA: IGI Global. doi:10.4018/978-1-5225-1814-3.ch012

Schaap, D. (2017). SeaDataNet: Towards a Pan-European Infrastructure for Marine and Ocean Data Management. In P. Diviacco, A. Leadbetter, & H. Glaves (Eds.), *Oceanographic and Marine Cross-Domain Data Management for Sustainable Development* (pp. 155–177). Hershey, PA: IGI Global. doi:10.4018/978-1-5225-0700-0.ch007

Seckin-Celik, T. (2017). Sustainability Reporting and Sustainability in the Turkish Business Context. In U. Akkucuk (Ed.), *Ethics and Sustainability in Global Supply Chain Management* (pp. 115–132). Hershey, PA: IGI Global. doi:10.4018/978-1-5225-2036-8.ch006

Segbefia, A. Y., Barnes, V. R., Akpalu, L. A., & Mensah, M. (2018). Environmental Location Assessment for Seaweed Cultivation in Ghana: A Spatial Multi-Criteria Approach. *International Journal of Applied Geospatial Research*, *9*(1), 51–64. doi:10.4018/IJAGR.2018010104

Sen, Y. (2018). How to Manage Sustainability: A Framework for Corporate Sustainability Tools. In I. Management Association (Ed.), Sustainable Development: Concepts, Methodologies, Tools, and Applications (pp. 568-589). Hershey, PA: IGI Global. doi:10.4018/978-1-5225-3817-2.ch026

Shamshiry, E., Abdulai, A. M., Mokhtar, M. B., & Komoo, I. (2015). Regional Landfill Site Selection with GIS and Analytical Hierarchy Process Techniques: A Case Study of Langkawi Island, Malaysia. In P. Thomas, M. Srihari, & S. Kaur (Eds.), *Handbook of Research on Cultural and Economic Impacts of the Information Society* (pp. 248–282). Hershey, PA: IGI Global. doi:10.4018/978-1-4666-8598-7.ch011

Sharma, Y. K., Mangla, S. K., Patil, P. P., & Uniyal, S. (2018). Analyzing Sustainable Food Supply Chain Management Challenges in India. In M. Ram & J. Davim (Eds.), *Soft Computing Techniques and Applications in Mechanical Engineering* (pp. 162–180). Hershey, PA: IGI Global. doi:10.4018/978-1-5225-3035-0.ch008

Silvestrelli, P. (2018). The Impact of Events: To Which Extent Are They Sustainable for Tourist Destinations? Some Evidences From Expo Milano 2015. In M. Risso & S. Testarmata (Eds.), *Value Sharing for Sustainable and Inclusive Development* (pp. 185–204). Hershey, PA: IGI Global. doi:10.4018/978-1-5225-3147-0.ch008

Silvius, G. (2018). Sustainability Evaluation of IT/IS Projects. In I. Management Association (Ed.), Sustainable Development: Concepts, Methodologies, Tools, and Applications (pp. 26-40). Hershey, PA: IGI Global. https://doi.org/ doi:10.4018/978-1-5225-3817-2.ch002

Singh, R., Srivastava, P., Singh, P., Upadhyay, S., & Raghubanshi, A. S. (2017). Human Overpopulation and Food Security: Challenges for the Agriculture Sustainability. In R. Singh, A. Singh, & V. Srivastava (Eds.), *Environmental Issues Surrounding Human Overpopulation* (pp. 12–39). Hershey, PA: IGI Global. doi:10.4018/978-1-5225-1683-5.ch002

Singh, R., Srivastava, P., Singh, P., Upadhyay, S., & Raghubanshi, A. S. (2017). Human Overpopulation and Food Security: Challenges for the Agriculture Sustainability. In R. Singh, A. Singh, & V. Srivastava (Eds.), *Environmental Issues Surrounding Human Overpopulation* (pp. 12–39). Hershey, PA: IGI Global. doi:10.4018/978-1-5225-1683-5.ch002

Srivastava, N. (2017). Climate Change Mitigation: Collective Efforts and Responsibly. In I. Management Association (Ed.), Natural Resources Management: Concepts, Methodologies, Tools, and Applications (pp. 64-76). Hershey, PA: IGI Global. https://doi.org/ doi:10.4018/978-1-5225-0803-8.ch003

Stanganelli, M., & Gerundo, C. (2017). Understanding the Role of Urban Morphology and Green Areas Configuration During Heat Waves. *International Journal of Agricultural and Environmental Information Systems*, *8*(2), 50–64. doi:10.4018/IJAEIS.2017040104

Stewart, M. K., Hagood, D., & Ching, C. C. (2017). Virtual Games and Real-World Communities: Environments that Constrain and Enable Physical Activity in Games for Health. *International Journal of Game-Based Learning*, *7*(1), 1–19. doi:10.4018/IJGBL.2017010101

Stone, R. J. (2017). Modelling the Frequency of Tropical Cyclones in the Lower Caribbean Region. In W. Ganpat & W. Isaac (Eds.), *Environmental Sustainability and Climate Change Adaptation Strategies* (pp. 341–349). Hershey, PA: IGI Global. doi:10.4018/978-1-5225-1607-1.ch013

Syed, A., & Jabeen, U. A. (2018). Climate Change Impact on Agriculture and Food Security. In A. Eneanya (Ed.), *Handbook of Research on Environmental Policies for Emergency Management and Public Safety* (pp. 223–237). Hershey, PA: IGI Global. doi:10.4018/978-1-5225-3194-4.ch012

Tam, G. C. (2017). The Global View of Sustainability. In *Managerial Strategies and Green Solutions for Project Sustainability* (pp. 1–24). Hershey, PA: IGI Global. doi:10.4018/978-1-5225-2371-0.ch001

Tam, G. C. (2017). Understanding Project Sustainability. In *Managerial Strategies and Green Solutions for Project Sustainability* (pp. 110–139). Hershey, PA: IGI Global. doi:10.4018/978-1-5225-2371-0.ch005

Tam, G. C. (2017). Perspectives on Sustainability. In *Managerial Strategies and Green Solutions for Project Sustainability* (pp. 53–76). Hershey, PA: IGI Global. doi:10.4018/978-1-5225-2371-0.ch003

Tang, M., & Karunanithi, A. T. (2018). Visual Logic Maps (vLms). In *Advanced Concept Maps in STEM Education: Emerging Research and Opportunities* (pp. 108–149). Hershey, PA: IGI Global. doi:10.4018/978-1-5225-2184-6.ch005

Taşçıoğlu, M., & Yener, D. (2018). The Value and Scope of GIS in Marketing and Tourism Management. In S. Chaudhuri & N. Ray (Eds.), *GIS Applications in the Tourism and Hospitality Industry* (pp. 189–211). Hershey, PA: IGI Global. doi:10.4018/978-1-5225-5088-4.ch009

Thanh Tung, B. (2021). Pharmacology and Therapeutic Applications of Resveratrol. In A. Hussain & S. Behl (Eds.), *Treating Endocrine and Metabolic Disorders With Herbal Medicines* (pp. 321–333). IGI Global. https://doi.org/10.4018/978-1-7998-4808-0.ch014

Tianming, G. (2018). Food Security and Rural Development on Emerging Markets of Northeast Asia: Cases of Chinese North and Russian Far East. In V. Erokhin (Ed.), *Establishing Food Security and Alternatives to International Trade in Emerging Economies* (pp. 155–176). Hershey, PA: IGI Global. doi:10.4018/978-1-5225-2733-6.ch008

Tiftikçigil, B. Y., Yaşgül, Y. S., & Güriş, B. (2017). Sustainability of Foreign Trade Deficit in Energy: The Case of Turkey. In N. Ray (Ed.), *Business Infrastructure for Sustainability in Developing Economies* (pp. 94–109). Hershey, PA: IGI Global. doi:10.4018/978-1-5225-2041-2.ch005

Tiwari, S., Vaish, B., & Singh, P. (2018). Population and Global Food Security: Issues Related to Climate Change. In I. Management Association (Ed.), Climate Change and Environmental Concerns: Breakthroughs in Research and Practice (pp. 41-64). Hershey, PA: IGI Global. https://doi.org/ doi:10.4018/978-1-5225-5487-5.ch003

Toujani, A., & Achour, H. (2018). A Data Mining Framework for Forest Fire Mapping. In I. Management Association (Ed.), Information Retrieval and Management: Concepts, Methodologies, Tools, and Applications (pp. 771-794). Hershey, PA: IGI Global. doi:10.4018/978-1-5225-5191-1.ch033

Tripathy, B., & K., S. B. (2017). Rough Fuzzy Set Theory and Neighbourhood Approximation Based Modelling for Spatial Epidemiology. In I. Management Association (Ed.), *Public Health and Welfare: Concepts, Methodologies, Tools, and Applications* (pp. 1257-1268). Hershey, PA: IGI Global. https://doi.org/ doi:10.4018/978-1-5225-1674-3.ch058

Trukhachev, A. (2017). New Approaches to Regional Branding through Green Production and Utilization of Existing Natural Advantages. In I. Management Association (Ed.), Advertising and Branding: Concepts, Methodologies, Tools, and Applications (pp. 1758-1778). Hershey, PA: IGI Global. doi:10.4018/978-1-5225-1793-1.ch081

Tsobanoglou, G. O., & Vlachopoulou, E. I. (2017). Social-Ecological Systems in Local Fisheries Communities. In G. Korres, E. Kourliouros, & M. Michailidis (Eds.), *Handbook of Research on Policies and Practices for Sustainable Economic Growth and Regional Development* (pp. 306–316). Hershey, PA: IGI Global. doi:10.4018/978-1-5225-2458-8.ch026

Tuydes-Yaman, H., & Karatas, P. (2018). Evaluation of Walkability and Pedestrian Level of Service. In I. Management Association (Ed.), Intelligent Transportation and Planning: Breakthroughs in Research and Practice (pp. 264-291). Hershey, PA: IGI Global. doi:10.4018/978-1-5225-5210-9.ch012

Uddin, S., Chakravorty, S., Ray, A., & Sherpa, K. S. (2018). Optimal Location of Sub-Station Using Q-GIS and Multi-Criteria Decision Making Approach. *International Journal of Decision Support System Technology*, *10*(2), 65–79. doi:10.4018/IJDSST.2018040104

Uddin, S., Chakravorty, S., Sherpa, K. S., & Ray, A. (2018). Power Distribution System Planning Using Q-GIS. *International Journal of Energy Optimization and Engineering*, *7*(2), 61–75. doi:10.4018/IJEOE.2018040103

Uzun, F. V. (2018). Natural Resources Management. In A. Eneanya (Ed.), *Handbook of Research on Environmental Policies for Emergency Management and Public Safety* (pp. 1–21). Hershey, PA: IGI Global. doi:10.4018/978-1-5225-3194-4.ch001

V, M., Agrawal, R., Sharma, V., & T.N., K. (2018). Supply Chain Social Sustainability and Manufacturing. In I. Management Association (Ed.), *Technology Adoption and Social Issues: Concepts, Methodologies, Tools, and Applications* (pp. 226-252). Hershey, PA: IGI Global. https://doi.org/ doi:10.4018/978-1-5225-5201-7.ch011

van der Vliet-Bakker, J. M. (2017). Environmentally Forced Migration and Human Rights. In C. Akrivopoulou (Ed.), *Defending Human Rights and Democracy in the Era of Globalization* (pp. 146–180). Hershey, PA: IGI Global. doi:10.4018/978-1-5225-0723-9.ch007

Vargas-Hernández, J. G. (2022). A Comprehensive Entrepreneurship Model for the Internationalization of Green Innovation Businesses. In U. Akkucuk (Ed.), *Disruptive Technologies and Eco-Innovation for Sustainable Development* (pp. 131–149). IGI Global. https://doi.org/10.4018/978-1-7998-8900-7.ch008

Vaskov, A. G., Lin, Z. Y., Tyagunov, M. G., Shestopalova, T. A., & Deryugina, G. V. (2018). Design of Renewable Sources GIS for ASEAN Countries. In V. Kharchenko & P. Vasant (Eds.), *Handbook of Research on Renewable Energy and Electric Resources for Sustainable Rural Development* (pp. 1–25). Hershey, PA: IGI Global. doi:10.4018/978-1-5225-3867-7.ch001

Vázquez, D. G., & Gil, M. T. (2017). Sustainability in Smart Cities: The Case of Vitoria-Gasteiz (Spain) – A Commitment to a New Urban Paradigm. In L. Carvalho (Ed.), *Handbook of Research on Entrepreneurial Development and Innovation Within Smart Cities* (pp. 248–268). Hershey, PA: IGI Global. doi:10.4018/978-1-5225-1978-2.ch012

Wahab, I. N., & Soonthodu, S. (2018). Geographical Information System in Eco-Tourism. In S. Chaudhuri & N. Ray (Eds.), *GIS Applications in the Tourism and Hospitality Industry* (pp. 61–75). Hershey, PA: IGI Global. doi:10.4018/978-1-5225-5088-4.ch003

Weiss-Randall, D. (2018). Cultivating Environmental Justice. In *Utilizing Innovative Technologies to Address the Public Health Impact of Climate Change: Emerging Research and Opportunities* (pp. 110–143). Hershey, PA: IGI Global. doi:10.4018/978-1-5225-3414-3.ch004

Weiss-Randall, D. (2018). Cultivating Resilience. In *Utilizing Innovative Technologies to Address the Public Health Impact of Climate Change: Emerging Research and Opportunities* (pp. 204–235). Hershey, PA: IGI Global. doi:10.4018/978-1-5225-3414-3.ch007

Weiss-Randall, D. (2018). Climate Change Solutions: Where Do We Go From Here? In *Utilizing Innovative Technologies to Address the Public Health Impact of Climate Change: Emerging Research and Opportunities* (pp. 236–268). Hershey, PA: IGI Global. doi:10.4018/978-1-5225-3414-3.ch008

Whyte, K. P., List, M., Stone, J. V., Grooms, D., Gasteyer, S., Thompson, P. B., . . . Bouri, H. (2018). Uberveillance, Standards, and Anticipation: A Case Study on Nanobiosensors in U.S. Cattle. In I. Management Association (Ed.), Biomedical Engineering: Concepts, Methodologies, Tools, and Applications (pp. 577-596). Hershey, PA: IGI Global. https://doi.org/ doi:10.4018/978-1-5225-3158-6.ch025

Wulff, E. (2017). Data and Operational Oceanography: A Review in Support of Responsible Fisheries and Aquaculture. In P. Diviacco, A. Leadbetter, & H. Glaves (Eds.), *Oceanographic and Marine Cross-Domain Data Management for Sustainable Development* (pp. 303–324). Hershey, PA: IGI Global. doi:10.4018/978-1-5225-0700-0.ch013

Yener, D. (2017). Geographic Information Systems and Its Applications in Marketing Literature. In S. Faiz & K. Mahmoudi (Eds.), *Handbook of Research on Geographic Information Systems Applications and Advancements* (pp. 158–172). Hershey, PA: IGI Global. doi:10.4018/978-1-5225-0937-0.ch006

Yu, K., Liu, Y., & Sharma, A. (2021). Analyze the Effectiveness of the Algorithm for Agricultural Product Delivery Vehicle Routing Problem Based on Mathematical Model. *International Journal of Agricultural and Environmental Information Systems, 12*(3), 26–38. https://doi.org/10.4018/IJAEIS.2021070103

Zhou, X., Sharma, A., & Mohindru, V. (2021). Research on Linear Programming Algorithm for Mathematical Model of Agricultural Machinery Allocation. *International Journal of Agricultural and Environmental Information Systems, 12*(3), 1–12. https://doi.org/10.4018/IJAEIS.2021070101

About the Contributors

Jamal Mabrouki received his PhD in Process and Environmental Engineering at Mohammed V University in Rabat, specializing in artificial intelligence and smart automatic systems. He completed the Bachelor of Science in Physics and Chemistry with honors from Hassan II University in Casablanca, Morocco and the engineer in Environment and smart system. His research is on intelligent monitoring, control, and management systems and more particularly on sensing and supervising remote intoxication systems, smart self-supervised systems and recurrent neural networks. He has published several papers in conferences and indexed journals, most of them related to artificial intelligent systems, internet of things or big data and mining. Jamal will currently work in environment, energy and smart system professor at Mohammed V University in Rabat, Faculty of Science. Jamal is scientific committee member of numerous national and international conferences. He is also a reviewer of Modeling Earth Systems and Environment; International Journal of Environmental Analytical Chemistry; International Journal of Modeling, Simulation, and Scientific Computing; The Journal of Supercomputing, Energy & Environment and Big Data Mining and Analytics.

Mourade Azrour received his PhD from Faculty of sciences and Techniques, Moulay Ismail University of Meknes, Morocco. He has received his MS in computer and distributed systems from Faculty of Sciences, Ibn Zouhr University, Agadir, Morocco in 2014. Mourade currently works as computer sciences professor at the Department of Computer Science, Faculty of Sciences and Techniques, Moulay Ismail University of Meknès. His research interests include Authentication protocol, Computer Security, Internet of things, Smart systems, Machine learning and so ones. Mourade is member of the member of the scientific committee of numerous international conferences. He is also a reviewer of various scientific journals. He has published more than 60 scientific papers and book chapters. Mourade Has edited 2 scientific books “IoT and Smart Devices for Sustainable Environment” and “Advanced Technology for Smart Environment and Energy”. Finally, he has served as guest editor in journals “EAI Endorsed Transactions on Internet of Things”, “Tsinghua Science and Technology”, “Applied Sciences MDPI” and “Sustainability MDPI.”

A.V. Senthil Kumar has industrial experience for five years and teaching experience of 27 years. He has also received his Doctor of Science (D.Sc in Computer Science). He has to his credit 33 Book Chapters, 220 papers in International and National Journals 60 papers in International Conferences in International and National Conferences, and edited 12 books (IGI Global, USA). He is as Associate Editor of IEEE Access. He is an Editor-in-Chief for many journals and Key Member for India, Machine Intelligence Research Lab (MIR Labs). He is an Editorial Board Member and Reviewer for various International Journals. He is also a Committee member for various International Conferences. He is a Life member of International Association of Engineers (IAENG), Systems Society of India (SSI), member of The Indian Science Congress Association, member of Internet Society (ISOC), International Association of Computer Science and Information Technology (IACSIT), Indian Association for Research in Computing Science (IARCS), and committee member for various International Conferences.

Mounia Mezouara, a PhD student affiliated with the Laboratory of Natural Resources and Sustainable Development at the Faculty of Sciences, Ibn Tofaïl University—Kenitra, Morocco, brings a wealth of knowledge and expertise to her research endeavors. Holding a Master's degree in Human Neurocognition and Population Health, she demonstrates a keen specialization in diabetes within her thesis work. Her comprehensive understanding of the field of biology further enhances the depth of her contributions to scientific inquiry.

Dimitris Michalopoulos was born in Athens, in 1952. After graduating from the Italian School of Athens (1970) and from the History and Archaeological Department of the University of Athens (1974), he obtained his Ph.D. in 1978, in Paris, at the École des Hautes Études en Sciences Sociales (Social and Economic History). From 1980 to 1982 he was the Curator of the Archives of the President of the Hellenic Republic. He taught Diplomatic History and Balkan Issues at the Law School of the University of Salonica (first as a lecturer and afterwards as an assistant professor [1982-1994]). Further, he taught Naval and European History at the Naval War College of Greece (1989 – 1997), and Naval History at the Naval Academy of Greece (1994-1997). During the years 1990-2000 he was the director of the Museum of the City of Athens; and from 2004 to 2011 he was the director of the "Historical Institute for Studies on Eleutherios Veniselos and his Era". In May 1027, he was Visiting Professor at the University of Bucharest. Now he is Voluntary Professor of Greek History at the People's University, Athens, historical adviser to the "Institute of Hellenic Maritime History", Piraeus, and titular member of the Académie des arts et des sciences de la mer (France).

Mohamed Rachdi is a full professor at ENSAD, Hassan II University in Casablanca. With a doctoral degree in Information Retrieval from the same university, his research focuses on artificial intelligence, machine learning, deep learning, information retrieval, and prediction systems. Pr. Rachdi is dedicated to fostering the development of the next generation of computer scientists through teaching and mentorship.

Samrat Ray is current working at IIMS Pune as Dean (International Relations). He has completed his PHD from global top 100 university according to QS ranking in St Petersburg Russia in the field of economics and management from Peter the Great Saint Petersburg Polytechnic University, Russia working under Russian government in various advanced fields of Neuromarketing, sensory marketing, behavioral economics and big data innovations in top 10 Russian research framework laboratory in Moscow. His thesis is on improving innovation efficiency in start ups in emerging countries like Russia and India.He has completed three masters degrees with distinction from premier European and Indian Universities in advanced management sciences. He is a prolific author and researcher having more than 50 Scopus indexed publications in peer reviewed journals from reputed publishers like Sage, Elseiver, Frontiers, Hindawi belonging to Q1 segment . He has been author of 6 International and national patents on innovation and marketing management from USA, Australia, India and Canada. He has served as keynote speaker and conducted management development programs for various fortune 500 companies in Europe as well as Dubai .He has also been author of two books on machine learning and Artificial intelligence. He has international experience working with Dubai Education departments in starting new courses and training programs.He has also been involved with Russian Innovation network and Russian education forums. His research interests are Psychology in marketing, Poverty eradication, behavioral economics and innovation. He has also been involved with Incubations in IIM Calcutta for strategy and policy making for new start ups and provided consulting for new start up companies .Currently he was ranked in top 2 percent of global scientists and 47th in Russia in management sciences by ADscienceresearch index for global ranking 2023 for scientists .He has also been recipient of various honors and prizes globally for his contribution to management sciences.

Index

IGI Global
Publishing Tomorrow's Research Today
www.igi-global.com

Publishing Tomorrow's Research Today

Uncover Current Insights and Future Trends in

Business & Management

with IGI Global's Cutting-Edge Recommended Books

Print Only, E-Book Only, or Print + E-Book.

Order direct through IGI Global's Online Bookstore at **www.igi-global.com** or through your preferred provider.

ISBN: 9798369306444
© 2023; 436 pp.
List Price: US$ **230**

ISBN: 9798369300084
© 2023; 358 pp.
List Price: US$ **250**

ISBN: 9781668458594
© 2023; 366 pp.
List Price: US$ **240**

ISBN: 9781668486344
© 2023; 256 pp.
List Price: US$ **280**

ISBN: 9781668493243
© 2024; 318 pp.
List Price: US$ **250**

ISBN: 9798369304181
© 2023; 415 pp.
List Price: US$ **250**

Do you want to stay current on the latest research trends, product announcements, news, and special offers?

Join IGI Global's mailing list to receive customized recommendations, exclusive discounts, and more.

Sign up at: **www.igi-global.com/newsletters.**

Scan the QR Code here to view more related titles in Business & Management.

www.igi-global.com Sign up at www.igi-global.com/newsletters facebook.com/igiglobal twitter.com/igiglobal linkedin.com/igiglobal

Ensure Quality Research is Introduced to the Academic Community

Become a Reviewer for IGI Global Authored Book Projects

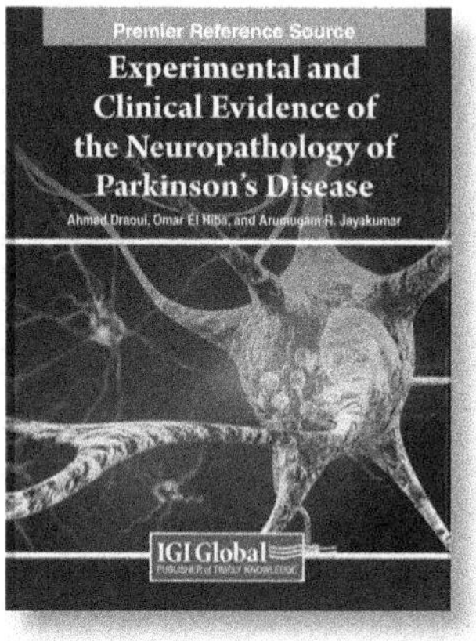

The overall success of an authored book project is dependent on quality and timely manuscript evaluations.

Applications and Inquiries may be sent to:
development@igi-global.com

Applicants must have a doctorate (or equivalent degree) as well as publishing, research, and reviewing experience. Authored Book Evaluators are appointed for one-year terms and are expected to complete at least three evaluations per term. Upon successful completion of this term, evaluators can be considered for an additional term.

If you have a colleague that may be interested in this opportunity, we encourage you to share this information with them.

IGI Global
Publishing Tomorrow's Research Today
www.igi-global.com

IGI Global Open Access Journal Program

Publishing Tomorrow's Research Today

IGI Global's Open Access Journal Program

Including Nearly 200 Peer-Reviewed, Gold (Full) Open Access Journals across IGI Global's Three Academic Subject Areas: Business & Management; Scientific, Technical, and Medical (STM); and Education

Consider Submitting Your Manuscript to One of These Nearly 200 Open Access Journals for to Increase Their Discoverability & Citation Impact

Web of Science Impact Factor 6.5

Web of Science Impact Factor 4.7

Web of Science Impact Factor 3.2

Web of Science Impact Factor 2.6

Choosing IGI Global's Open Access Journal Program Can Greatly Increase the Reach of Your Research

Higher Usage
Open access papers are 2-3 times more likely to be read than non-open access papers.

Higher Download Rates
Open access papers benefit from 89% higher download rates than non-open access papers.

Higher Citation Rates
Open access papers are 47% more likely to be cited than non-open access papers.

Submitting an article to a journal offers an invaluable opportunity for you to share your work with the broader academic community, fostering knowledge dissemination and constructive feedback.

Submit an Article and Browse the IGI Global Call for Papers Pages

We can work with you to find the journal most well-suited for your next research manuscript.
For open access publishing support, contact: journaleditor@igi-global.com

Publishing Tomorrow's Research Today

IGI Global

e-Book Collection

Including Essential Reference Books Within Three Fundamental Academic Areas

Business & Management
Scientific, Technical, & Medical (STM)
Education

- Acquisition options include Perpetual, Subscription, and Read & Publish
- No Additional Charge for Multi-User Licensing
- No Maintenance, Hosting, or Archiving Fees
- Continually Enhanced Accessibility Compliance Features (WCAG)

Over **150,000+** Chapters | Contributions From **200,000+** Scholars Worldwide | More Than **1,000,000+** Citations | Majority of e-Books Indexed in Web of Science & Scopus | Consists of Tomorrow's Research Available Today!

Recommended Titles from our e-Book Collection

Innovation Capabilities and Entrepreneurial Opportunities of Smart Working
ISBN: 9781799887973

Advanced Applications of Generative AI and Natural Language Processing Models
ISBN: 9798369305027

Using Influencer Marketing as a Digital Business Strategy
ISBN: 9798369305515

Human-Centered Approaches in Industry 5.0
ISBN: 9798369326473

Modeling and Monitoring Extreme Hydrometeorological Events
ISBN: 9781668487716

Data-Driven Intelligent Business Sustainability
ISBN: 9798369300497

Information Logistics for Organizational Empowerment and Effective Supply Chain Management
ISBN: 9798369301593

Data Envelopment Analysis (DEA) Methods for Maximizing Efficiency
ISBN: 9798369302552

Request More Information, or Recommend the IGI Global e-Book Collection to Your Institution's Librarian

For More Information or to Request a Free Trial, Contact IGI Global's e-Collections Team: eresources@igi-global.com | 1-866-342-6657 ext. 100 | 717-533-8845 ext. 100

Are You Ready to

Publish Your Research?

IGI Global
Publishing Tomorrow's Research Today

IGI Global offers book authorship and editorship opportunities across three major subject areas, including Business, STM, and Education.

Benefits of Publishing with IGI Global:

- Free one-on-one editorial and promotional support.
- Expedited publishing timelines that can take your book from start to finish in less than one (1) year.
- Choose from a variety of formats, including Edited and Authored References, Handbooks of Research, Encyclopedias, and Research Insights.
- Utilize IGI Global's eEditorial Discovery® submission system in support of conducting the submission and double-blind peer review process.
- IGI Global maintains a strict adherence to ethical practices due in part to our full membership with the Committee on Publication Ethics (COPE).
- Indexing potential in prestigious indices such as Scopus®, Web of Science™, PsycINFO®, and ERIC – Education Resources Information Center.
- Ability to connect your ORCID iD to your IGI Global publications.
- Earn honorariums and royalties on your full book publications as well as complimentary content and exclusive discounts.

Join Your Colleagues from Prestigious Institutions, Including:

Australian National University
Massachusetts Institute of Technology
Johns Hopkins University
Harvard University
Tsinghua University
Columbia University in the City of New York

Learn More at: www.igi-global.com/publish

or Contact IGI Global's Aquisitions Team at: acquisition@igi-global.com

www.ingramcontent.com/pod-product-compliance
Lightning Source LLC
LaVergne TN
LVHW080525300926
849162LV00007B/1535